Corrosion and Water Technology

Corrosion and Water Technology

for Petroleum Producers

Loyd W. Jones

"The forces of nature level the high mountains and seek to bring all things to a state of equality and eternal rest"—Ancient Philosophy

OGCI Publications
Oil & Gas Consultants International, Inc.
Tulsa

4554 South Harvard Avenue
Tulsa, Oklahoma 74135

Printed in the United States of America

Library of Congress Catalog Card Number: 88-060130
International Standard Book Number: 0-930972-09-0

Disclaimer

This book contains numerous recommendations and references to specific methods and materials which are advisory only. Mention of specific trade names, copyrighted or patented products and services are made as examples and should not be construed as recommendations for or against any company or their products and services. Neither Loyd W. Jones, CWT Consultants, Inc., nor Oil & Gas Consultants, Inc., its officers and employees accept any responsibility for the use of the methods and materials discussed herein. Use of the information is solely at the risk of the user.

Preface

About This Book

Much of the information presented herein was written in the form of class manuals for OGCI petroleum training courses on Corrosion Control and Oilfield Water Technology. These materials have been revised, updated and combined to produce this new edition.

Numerous explanations and practical guidelines based on first hand field and laboratory experience are included that are not readily available elsewhere. The result is primarily a digest of essential information for a special audience—oil and gas production personnel. The primary goal is to provide a basic level of understanding of corrosion and water problems as they exist in production operations and of the major control procedures that have evolved within the industry. An important side benefit should be improved ability of the reader to communicate with suppliers and consultants and to evaluate their recommendations. It is *not* an academic scientific treatise of corrosion and water chemistry.

These goals and objectives developed from extensive exposure to the special needs of field operations and engineering personnel. As a group, they generally have little time or inclination to study the voluminous and growing technical literature on corrosion and water. Yet, no one can argue the fact that more knowledge is needed by field production personnel. This is especially true since corrosion and water related problems together are at the top of the list of vital but generally non-profitable expenditures in petroleum production operations.

The two subjects, corrosion and water technology, are treated together for two reasons. First, essentially all corrosion of production equipment occurs in the presence of liquid water. Therefore, some knowledge of water properties and reactions is necessary for understanding how and where corrosion occurs. Although water disposal and injection systems have other difficulties in addition to corrosion, the same personnel that are responsible for field corrosion control programs must generally deal with water handling problems also. This book is dedicated to these hard working individuals.

The text is intentionally written in an easy to read style to encourage the reader to start at the beginning and to read through to the finish. This approach will provide answers to the questions most often asked by those new to the subjects and will help clarify points of confusion that often persist after some prior training. A comprehensive index is provided and cross references are placed in each section to show the inter-relationship of the various topics.

To those who may be approaching the study with some apprehension due to lack of scientific training, the author would simply offer you encouragement to proceed. The basic concepts are explained in easy to understand language and the practical insights you glean should prove helpful to you.

It must be pointed out, however, that the methods of studying and teaching corrosion processes have undergone drastic changes, particularly during the last twenty years. New techniques and instrumentation for electrochemical testing have become common place both in the field and laboratory. Technical articles appearing in so-called practical journals and books often contain data that are derived from test electrode polarization behavior, for example. While this book is not designed to train corrosion specialists, it does provide a working background in the electrochemical aspects of corrosion control procedures. Topics dealing with water and scale chemistry and microbiology are presented in the most practical terms possible while retaining a reasonable level of scientific accuracy.

About the Illustrations

Most of the illustrations were designed simply to augment the written discussion and to help the reader visualize the concept or the device being described. Details are purposely limited because ample technical information on such things as cathodic protection instruments, metals, coatings, inhibitors and water treating equipment are readily available from manufacturers, suppliers and service companies.

Several of the data plots, such as Figures 1.1-8 and 1.1-10, were created from a file of incidental data accumulated by the author during research studies for broader objectives. Additional point values computed from scattered literature sources were added and best fit curves were drawn to generate the printed figures. Although the results are believed to be valid and useful, they should be considered as practical illustrations rather than rigorously derived scientific data. Methods for approximate calculations of mineral scaling tendencies not otherwise referenced in Section 2 have similar origins.

Acknowledgments

A life time career in the technical fields of corrosion control and water technology can properly be equated to an equal time as a student. From this perspective, I am especially indebted to the members and staff of the National Association of Corrosion Engineers (NACE) for providing a constant flow of up-to-date knowledge.

Perhaps the single most difficult task in preparation of this text was deciding which items of information to include (or exclude) and how much each should be condensed for benefit of the intended readers. Guidance in this area was largely drawn from my former field associates who often called on Friday with a problem requiring an answer by Monday. Some of my individual co-workers who shared in that struggle (and from each of whom I learned a great deal) were Jack P. Barrett, John D. Alkire, W. D. Wakley, R. L. McGalliard, David Simms, Kenneth Schmidt, R. L. Robinson, Gene Donaldson, Lloyd Christensen, G. K. Gentis, S. M. Hockett, John M. Taylor, Jack S. Smart, K. C. Lunden, T. W. McSpadden, J. B. (Jim) Dobbs, H. Sue Carson, James Donham, Tom Elliott, and Dan H. Carpenter.

Special thanks are extended to T. O. (Tommy) Allen, Chief Executive Officer of OGCI for his wise guidance and trust, to Jean C. Allen-Ikeson for editing, to Verma Hughes for publication preparation, and to J. F. (Jim) Minter for helpful critique and suggestions.

Appreciation beyond words goes to my wife Judith A. (Jan) Jones, who translated my handwritten verbiage to the typewriter in proper form and provided the essential inspiration to keep going.

February, 1987 *Loyd W. Jones*

Contents

SECTION 1 Corrosion Control

1	**Corrosion Mechanisms and Influencing Factors**	**1–17**
	Introduction	3
	1.01 Corrosion Requires Energy	3
	1.02 High Energy State of Refined Metals	3
	1.03 Energy Transfer During Aqueous Corrosion	3
	1.04 Characteristics of the Basic Corrosion Cell	5
	1.05 Anodic Half-Cell Reactions	5
	1.06 Cathodic Half-Cell Reactions	5
	1.07 Oxidation—Reduction (Redox) Reactions	6
	1.08 Electrical Properties of Corrosion Cells	6
	1.09 Electrode Potentials and Direction of Current Flow	7
	1.10 Polarization of Electrodes	7
	1.11 Corrosion Cells Created by Differences in Electrodes	8
	1.12 Tabulations of Relative EMF's or Solution Potentials of Metals	8
	1.13 Dissimilar Metal Couples	9
	1.14 Physical Conditions of Metals That Promote Corrosion	10
	1.15 Surface Oxide Films and Hydrogen Overvoltage Effects	11
	1.16 Corrosion Cells Created by Differences in Electrode Environments	12
	1.17 Corrosive Agents in the Environment	13
	1.18 Relative Corrosiveness of O_2, CO_2 and H_2S	14
	1.19 The Contribution of Dissolved Salts	15
	1.20 Effects of Microorganisms	15
	1.21 Stray and Induced Electrical Currents	16
	1.22 Concluding Comments	17
2	**Fundamentals of Corrosion Prevention Methods**	**19–20**
	Introduction	19
	2.01 Surface Barriers	19
	2.02 Removal of Corrosive Agents and Temperature-Velocity Reduction	19
	2.03 Special Chemical Treatments and Solids Removal	20
	2.04 Cathodic Protection	20
	2.05 Special Alloys	20
	2.06 Non-Metallic Materials	20
	2.07 Dehydration	20
3	**Chemical Inhibitors**	**21–25**
	Introduction	21
	3.01 The Nature of Corrosion Inhibitors	21
	3.02 Inhibitor Classifications	21
	3.03 Effect of Ionic Character on Inhibiting Ability	22
	3.04 Characteristics of Oily Amine Inhibitor Films	22
	3.05 Water Soluble and Volatile Amine Inhibitors	23
	3.06 Formulations of Oil Field Inhibitors	23
	3.07 General Principles of Inhibitor Application	24
	3.08 Acid Inhibitors	25

4	**Cathodic Protection**	**27–38**
	4.01 Theory and Principles	27
	4.02 Current Requirements	27
	4.03 Current Sources and Concepts of Current Flow	28
	4.04 Sacrificial or Galvanic Anodes	29
	4.05 Impressed Current Systems	29
	4.06 Effect of Anode Bed Distance from Structure	31
	4.07 Electrical Potentials and Protection Criteria	31
	4.08 Measurement of Structure Potentials	32
	4.09 Interference—Mechanism and Correction	33
	4.10 Offshore Structure Potential Surveys	34
	4.11 Surface Potential Surveys	34
	4.12 IR Drop Method	35
	4.13 Effect of C.P. on Structure Potentials	35
	4.14 Soil Resistivity Measurements	35
	4.15 Field Testing for Current Requirement	36
	4.16 Sacrificial vs. Impressed Current Systems	36
	4.17 Protection of Tanks and Oil Treating Vessels	37
	References	38
5	**Protective Coatings and Plastics**	**39–46**
	Introduction	39
	5.01 Common Generic Types of Coatings and Plastics	39
	5.02 Characteristics of the Generic Types	40
	5.03 Coatings Systems and Nomenclature	43
	5.04 Methods of Steel Surface Preparation Prior to Coating	43
	5.05 Coating Application and Inspection	43
	5.06 Performance Testing and Laboratory Inspection	45
	5.07 Cathodic Disbondment and Soil Stress—External Pipe Coatings	46
	5.08 Suggested Types of Coatings for Oil Production Operations	46
	References	46
6	**Removal of Corrosive Gases—Mechanical and Chemical Methods**	**47–51**
	Introduction	47
	6.01 Hydrogen Sulfide	47
	6.02 Carbon Dioxide	47
	6.03 Oxygen	48
	6.04 Oxygen Exclusion	48
	6.05 Oxygen Removal	48
	6.06 Gas Stripping	49
	6.07 Vacuum Deaeration	50
	6.08 Corrosion Protection in Deaerators	50
	6.09 Removal of Oxygen with Chemicals	50
	References	51
7	**Metal Selection, Design and Handling Considerations**	**53–60**
	Introduction	53
	Metal Properties and Cracking Failures	
	7.01 Elastic Limit—Ductility—Yield and Tensile Strength	53
	7.02 Hardness	53
	7.03 Fatigue Strength and Corrosion Fatigue	54
	7.04 Heat Treatment of Steels	54
	7.05 Sulfide Stress Cracking (SSC)	54
	7.06 Hydrogen Blistering (Internal Fracturing)	55
	7.07 Chloride Stress Cracking (CSC)	56
	7.08 Stress Corrosion Cracking (SCC)	56

	Design and Handling Considerations	
	7.09 Metals Selection	56
	7.10 Designing with Dissimilar Metals	59
	7.11 Equipment Handling Precautions	59
	7.12 Velocity Effects	59
	7.13 Cavitation	59
	References	60
8	**Specific Corrosion Control Procedures in Oil Production Operations**	**61–78**
	Introduction	61
	8.01 Drilling Equipment Corrosion	61
	8.02 Drillpipe Corrosion Control Methods	62
	8.03 Drilling Corrosion Test Methods	63
	8.04 External Casing Corrosion	63
	8.05 Wellhead Insulators	64
	8.06 Casing Protection by Cement and Completion Fluids	64
	8.07 Cathodic Protection of Casing	64
	8.08 Measuring External Casing Corrosion	65
	8.09 Casing Potential Profiles	65
	8.10 Surface E Log I Tests for Estimating Casing Current Requirement	65
	8.11 Internal Subsurface Corrosion	67
	8.12 Inhibitor Treatment Methods for Wells	68
	8.13 Protection of Surface Production Equipment	72
	8.14 Protection of Water Injection Systems	72
	8.15 Corrosion Monitoring in Water Injection Systems	73
	8.16 Gas Processing Equipment	74
	8.17 Offshore Drilling—Production Operations	75
	References	78
9	**Corrosion Control in Enhanced Recovery (EOR) Projects**	**79–85**
	9.01 Carbon Dioxide (CO_2) EOR Systems	79
	9.02 Steam Flooding and In-Situ Combustion Projects	83
	9.03 Micellar (Surfactant)/Polymer Flooding	84
	References	85
10	**Detecting and Monitoring Corrosion Activity**	**87–98**
	10.01 Early Detection	87
	10.02 Devices for Monitoring Fluid Corrosiveness	87
	10.03 Characteristics and Limitations of Corrosion Rate Monitors	87
	10.04 Iron Content Analysis of Produced Water and the COPRA Correlation	93
	10.05 After-the-Fact Corrosion Detection	94
	10.06 Visual Inspection and Analysis	96
	10.07 Tests for Identifying Corrosion Products	97
	10.08 Record Keeping and Economic Analyses	98
	References	98

SECTION 2 WATER TECHNOLOGY

General Introduction		**101**
1	**Water Properties and Components**	**103–115**
	1.01 Properties of the Water Substance	103
	1.02 Water Constituents	103
	1.03 Behavior of Dissolved Substances	104
	1.04 Hydrolysis Reactions	106
	1.05 Water Constituents Found by Analysis	106
	1.06 Some Properties of the Metal Cations and Their Salts	107
	1.07 Water Properties Determined by Analysis	107
	1.08 Units of Concentration	109
	1.09 Examples of Water Analysis Calculations and Interpretations	112
	1.10 Graphical Plots of Water Analysis Data	114
2	**Mineral Scales and Water Compatibility**	**117–127**
	2.01 Introduction	117
	2.02 Mineral Scales and Problems of Predicting Their Occurrence	117
	2.03 Solubility, Supersaturation and Deposition	117
	2.04 Importance of Scale Prediction Calculations	118
	2.05 Prediction of Calcium Carbonate Scaling	119
	2.06 Sulfate Scales—General	120
	2.07 Calculation of Sulfate Scale Solubilities	122
	2.08 Practical Estimation of Calcium Sulfate Solubility Directly from Water Analyses	122
	2.09 Barium and Strontium Sulfate Scales	123
	2.10 Water Compatibility for Mixing	125
	2.11 Determination of Compatibility	125
	2.12 Incompatibility and Production Well Scaling	125
	2.13 Scale Removal	126
	2.14 Scale Inhibition	126
	References	127
3	**Practical Microbiology**	**129–134**
	3.01 Terminology and Classifications	129
	3.02 Nutritional Requirements and Energy Sources	129
	3.03 Bacteria Related Problems and Mechanisms	129
	3.04 Occurence and Growth Habits	131
	3.05 H_2S Generation in Producing Reservoirs	131
	3.06 Detection and Assessment of Bacterial Activity	132
	3.07 Interpretation of Bacteria Counts	133
	3.08 Direct Examination of Deposits	133
	3.09 Bacterial Control Chemicals	134
	3.10 Field Application of Biocides	134
	References	134
4	**Water Quality for Injection**	**135–139**
	4.01 Introduction	135
	4.02 The Fate of Solids Entering Rock Interstices	135
	4.03 Sources and Properties of Suspended Solids	136
	4.04 Water Quality Rating Methods	137
	4.05 An Empirical Water Quality Rating Guide	138
	4.06 Clay Swelling and Peptization	139
	References	139
5	**Solids and Oil Removal**	**141–153**
	5.01 Introduction	141
	5.02 Solids Removal from Surface Supply Waters	141
	5.03 Open System Clarification	141
	5.04 Filters for Surface Waters	142

5.05 Clarification of Produced Waters 143
5.06 Closed System Filtration 144
5.07 The Problem of Oil Carryover 144
5.08 Factors Influencing Oil Dispersion and Stabilization 146
5.09 Removal of Oil and Oily Solids by Coalescence 147
5.10 Short Path Separators 147
5.11 Separation by Flotation 148
5.12 Oil Removal Process Control 149
5.13 Oily Waste Disposal and Injection Well Cleaning 151

APPENDIX

Conversion Factors **155**

1A **Simplified Outline of Corrosion Theory** **159–166**
Introduction 159
1A.1 Basic Definitions 159
1A.2 Fundamentals of Electricity and Electrochemistry 160
1A.3 Sign Conventions of Electrochemical Reactions 160
1A.4 Measurement of Cell Potential 161
1A.5 Standard Half Cell Potentials 161
1A.6 Non-Standard Potentials 162
1A.7 Polarization 162
1A.8 Polarization Curves and Mixed Potential Theory 162
1A.9 Experimental Determination of I_{corr} 163
1A.10 Linear Polarization 164
1A.11 Passivity and Anodic Protection 164
1A.12 EMF-pH (Pourbaix) Diagrams 165
References 166
Other Suggested Reading 166

2A **Additional Information on Corrosion Inhibiting Chemicals** **169–171**
2A.1 Oxygen Corrosion Inhibitors 169
2A.2 Cationic and Mixed Charge Organic Inhibitors 170
2A.3 Reference List of Inhibitor Chemicals 170

3A **Scaling Tendency Calculation Examples** **175–176**
3A.1 Calcium Carbonate 175
3A.2 Calcium Sulfate 176

4A **Field Determination of Approximate Bacteria Populations in Water** **179–180**

5A **Test Method**
Methods for Determining Water Quality for Subsurface Injection Using Membrane Filters **183–193**

6A **Procedures for Determination of Oil in Water** **197**

Index **199–202**

Illustrations

1.1-1.	Energy cycle of iron.	4
1.1-2.	Representation of the basic corrosion cell.	5
1.1-3.	Schematic representation of a corrosion cell as an electrical circuit.	7
1.1-4.	Cathode to anode area effect in iron/copper couple exposed to aerated 3% NaCl solution.	9
1.1-5.	Selective galvanic corrosion of aluminum encased in steel.	11
1.1-6.	Rust tubercle causing a differential oxygen cell.	12
1.1-7.	Corrosion on a pipeline due to differential aeration.	13
1.1-8.	Comparative corrosiveness of three common gases in water solutions.	15
1.1-9.	Solubility of O_2 in water solutions of sodium chloride in equilibrium with air.	16
1.1-10.	Corrosion rates of carbon steel in aerated salt waters.	17
1.3-1.	Corrosion of carbon steel in a flowing sour gas stream with entrained water.	23
1.3-2.	Sucker rods after 3 months use in treated and uninhibited sour oil wells.	24
1.4-1.	Sacrificial anode on a buried pipeline	29
1.4-2.	Impressed current c.p. system on a pipeline.	30
1.4-3.	Circuitry of a rectifier.	31
1.4-4.	Measurement of pipe-to-soil potential on a pipeline.	32
1.4-5.	Potential profile along a pipeline.	33
1.4-6.	Interference bond on insulating flange at a cathodically protected well casing.	34
1.4-7.	IR drop or current flow method for detecting anodic areas.	35
1.4-8.	Four pin method for soil resistivity measurement.	36
1.4-9.	Soil resistivity profile of a pipeline.	37
1.5-1.	Holiday detector for thin film internal pipe coatings.	45
1.6-1.	Concentration of dissolved oxygen in pure water in equilibrium with air at atmospheric pressure.	48
1.6-2.	Counter-current gas stripping tower for removal of dissolved oxygen.	49
1.6-3.	Vapor pressure of water with 0.5 Mg/L of dissolved air.	50
1.7-1.	Environment vs. endurance limit of a low alloy steel.	54
1.7-2.	Cross section of a hydrogen blister in the wall of a sour oil storage tank.	55
1.7-3.	Stress corrosion cracking (SCC) in a stainless steel drill collar.	57
1.7-4.	Internal corrosion along the weld seam in ERW pipe.	58
1.8-1.	Drill string corrosion test ring.	63
1.8-2.	Casing potential profile test equipment and sample data plot.	66
1.8-3.	Gas well tubing corroded by CO_2.	69
1.8-4.	Two types of dual string completions for inhibitor treatment of deep hot wells.	71

1.8-5. Cross section and installation of hydril flatpak injection line. 72
1.8-6. Installed downhole capillary Chemical Transmission System. 73
1.8-7. Measuring D.O. in a flowing stream with a Chemet® Ampoule. 74
1.8-8. Recommended test points in an injection system. 75
1.9-1. Typical CO_2/water injection well. 81
1.10-1. Two types of corrosion test coupons and holders. 89
1.10-2. Illustration of test coupon data variance with time. 89
1.10-3. Flat coupons installed in water trap welded to bottom of a pipe line. 90
1.10-4. Corrosion test nipples. 90
1.10-5. Electrical resistance probe. 91
1.10-6. Illustration of two and three element polarization resistance probes. 91
1.10-7. Flush mounted polarization resistance probe. 92
1.10-8. Hydrogen pressure probe. 92
1.10-9. COPRA Correlation for a group of gas wells before and after initiation of a corrosion inhibition program. 94
1.10-10. Anatomy of a sucker rod break due to stress corrosion fatigue. 96
1.10-11. Cross section of a sucker rod that failed by corrosion fatigue. 97
1.10-11. Sulfide pitting in a water injection flowline. H_2S created by sulfate reducing bacteria. 97
2.1-1. Structure of the water molecule. 104
2.1-2. Scale of pH values and the three zones of alkalinity. 108
2.1-3. Flow diagram of a surface scatter turbidimeter. 110
2.1-4. Example of a water analysis pattern (Stiff Method). 114
2.1-5. Graphical plot of water analyses. 115
2.2-1. Values of K at various ionic strengths. 120
2.2-2. Chart for converting concentrations of CA^{++} and (HCO_3^- + $CO_3^=$) to pCa and pAlk. 121
2.2-3. Values of S_f at various temperatures and system pressures. 121
2.2-4. Approximate pH vs. R values. 122
2.2-5. Approximate solubilities of strontium sulfate and barium sulfate in oilfield brines. 124
2.3-1. Diatoms and algae filaments. 130
2.3-2. Iron bacteria filaments. 130
2.3-3. Anaerobic conditions created at a metal surface by shielding effect of aerobic organism slime layers. 132
2.4-1. Illustration of differences in rock face area in injection wells. 136
2.4-2. Effect of particle concentration on membrane filter test plots. 137
2.4-3. Effect of particle size distribution. 138
2.5-1. Three element manual unit. 143
2.5-2. Typical downflow pressure bed filter. 144
2.5-3. Closed loop emulsion treating facility. 145
2.5-4. Gravity separation of crude oils from produced brines—typical rates. 146
2.5-5. Representation of liquid/liquid (LL) oil coalescing filter cartridge. 148
2.5-6. Oil/water separator using oleophilic foam. 149
2.5-7. Elements of a Short Path Separator. 150
2.5-8. Induced gas flotation machine for water clarification. 151
2.5-9. Induced gas flotation cell mechanism. 152
2.5-10. Static induced gas flotation cell. 152
1A-1. Graphical representation of Ohm's Law. 160
1A-2. Impressed current or electrolytic corrosion cell. 161
1A-3. Polarization diagram of zinc corroding in acid. 163
1A-4. E Log I Curve. 164
1A-5. Example of an Anodic Polarization Plot for an Active-Passive Metal. 165
1A-6. Gross features of an EMF-pH diagram for Fe-H_2O system. 166

Tables

1.1-1	Relative Single Electrode Potential Values	8
1.1-2	Dissimilar Metal Corrosion Guide	10
1.4-1	Typical Current Density Requirements	28
1.4-2	Classification of Soil Corrosivity by Resistivity	36
1.4-3	Approximate Current Output of Magnesium Anodes	37
1.5-1	Types of Zinc-Rich Primers	42
1.5-2	Methods of Steel Surface Preparation	44
1.5-3	Laboratory Tests for Evaluation of Coatings Properties and Comparative Performance	45
1.5-4	Coatings for Specific Field Uses	46
1.6-1	Materials of Construction Recommended for Gas Stripping Columns	50
1.7-1	Some Prospective Alloys for Hot Well Tubulars	56
1.8-1	Calculation of Electrical Resistance of Casing	67
1.8-2	Elements of a Planned Corrosion Prevention Program for New Wells	67
1.8-3	Summary of Proven Inhibitor Application Methods for Producing Wells	68
1.10-1	Devices for Measuring Fluid Corrosivity	88
1.10-2	Corrosion Rate Expressions	88
1.10-3	Scheme for Field Inspection of Equipment Failures	96
1.10-4	Characteristic Features of Some Corrosion-Related Failures	96
1.10-5	Acid Solutions for Deposit Testing	97
2.1-1	Some Physical Properties of Water	104
2.1-2	Classifications and General Characteristics of Water	104
2.1-3	Common Dissolved Constituents of Water	105
2.1-4	Suspended and Organic Matter Found in Water	107
2.1-5	Concentration Units for Water Constituents	110
2.1-6	Examples of Water Analyses	112
2.2-1	Common Oilfield Scales	118
2.2-2	Factors for Converting Ion Concentrations to Individual Ionic Strengths	119
2.2-3		123
2.3-1	Microorganisms Found in Oil Field Waters	130
2.3-2	A Guide for General Interpretation of Bacterial Count Data	133
2.3-3	Some Bacterial Control Agents	134
1A-1	Definitions	159
2A-1	Inhibitor Chemicals	171
3A-1	Determination of Saturation pH by Hand Calculator	175

SECTION 1

Corrosion Control

1 Corrosion Mechanisms and Influencing Factors

Introduction

The most widely accepted definition of corrosion describes it as "the deterioration of a material, usually a metal, by reaction with its environment." This sounds simple enough. What is being described is an action, or more properly, a reaction between at least two things, for example, a piece of metal and whatever the metal is exposed to. In this study we will deal primarily with steel exposed to water. Liquid water is present in all corrosive situations normally encountered in petroleum production (see Section 2, Chapter 1 for information on water properties).

The end result of corrosion may be a hole in the wall of a flowline or tank, a rough pitted pump plunger, a cracked drill collar, etc. Many field people tend to think of corrosion as being the final condition of the failed item of equipment. The corrosion scientist or engineer thinks of corrosion as the reactions that cause the failure. If the reactions can be understood, ways and means to prevent future failures can be analyzed.

In this first chapter, we will explore the basic mechanisms and reactions that take place during the corrosion process. The discussion is designed to provide a foundation of understanding for all the corrosion control procedures covered in subsequent chapters. Technical terms are defined and explained as they are used.

1.01 Corrosion Requires Energy

It is obvious that a length of pipe is simply a piece of metal with a hole through it. If, after the pipe has been used awhile, it has two holes through it, then we have a corrosion situation. How and why did the unwanted hole get there? To drill a hole through the pipe wall, one would have had to apply energy from some source. Corrosion also involves energy transfer to create the unwanted hole. Where does the energy come from? The answer is suggested in the definition given previously: "reaction (of a metal) with its environment." Both the metal and the environment can supply energy to promote the reaction.

1.02 High Energy State of Refined Metals

Corrosion can be considered a natural result of the energy stored in the metal when it was refined and fabricated. This is illustrated by Figure 1.1-1.

Iron and similar metals exist in nature in the combined form such as oxides, carbonates or sulfides. The ores are quite stable and are practically insoluble in ordinary water. To change the ore form to purer and more useful refined metals requires energy addition. Therefore, refined metals are at a higher energy level or an unnatural state as compared to native ores. Thus, there is a normal tendency for refined metal to revert back to its combined form as it occurred in nature. This occurs through reaction with moisture, carbon dioxide, hydrogen sulfide, oxygen and other components commonly found in the atmosphere, soil, water, and process streams.

The various metals have inherently different tendencies to revert to their stable, combined forms. These differences are used both in predicting corrosion and in controlling it. Later we will take a closer look at the chemical reactivity of various metals and learn to organize the differences in a useful way.

1.03 Energy Transfer During Aqueous Corrosion

To understand how energy is transferred between a metal and its environment during the corrosion process, one needs to be able to visualize the form of the energy. For the present, the basic unit of energy can be considered as the electrical charge on the smallest particle of matter, which is called the *electron*.

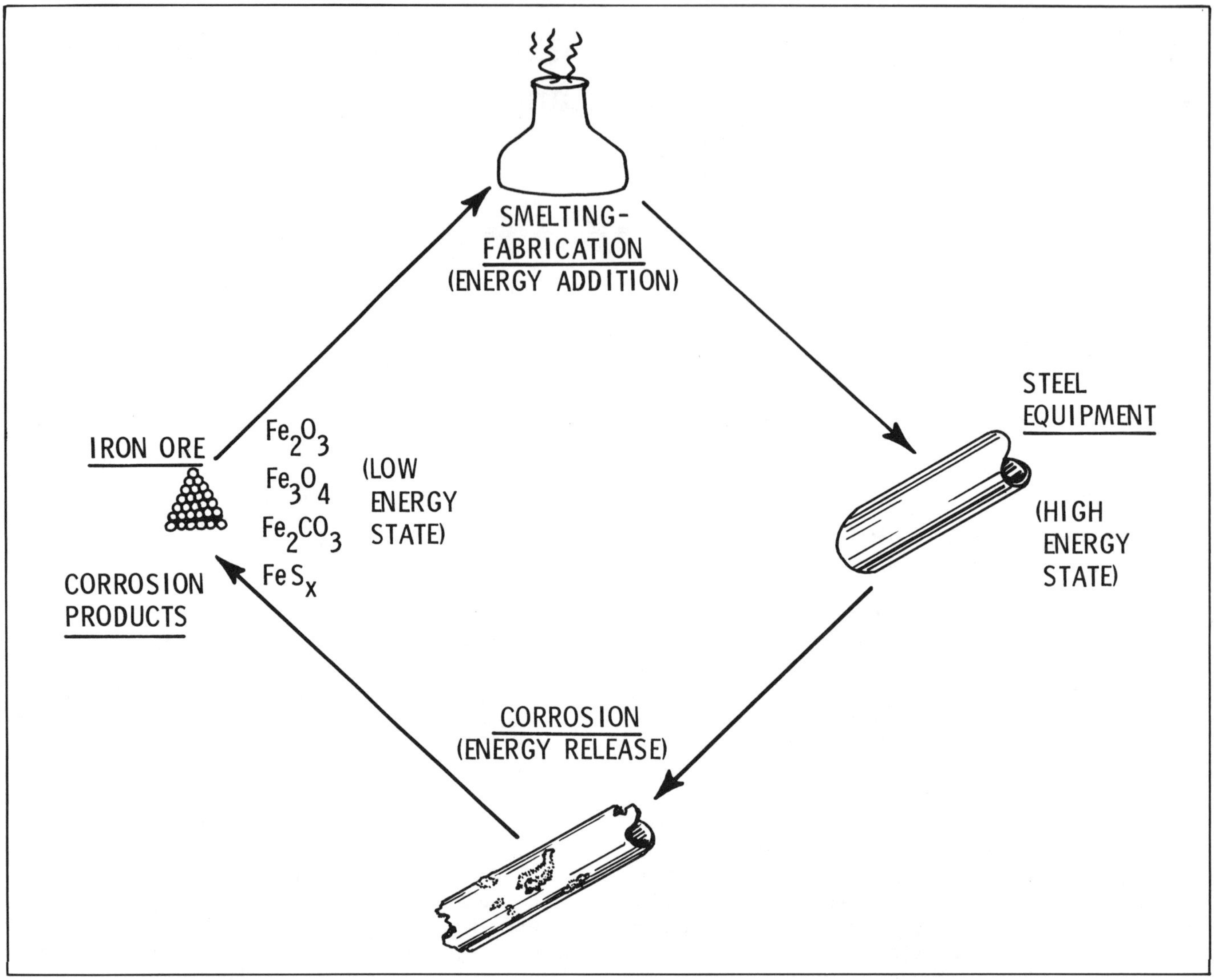

Figure 1.1-1. Energy cycle of iron.

Electrons are negatively-charged particles that form a part of all *atoms* that make up natural *elements*. Electrons generally do not exist alone and have essentially no mass. An electron is designated by the letter "e." Movement of electrons occurs quite easily through metals and this movement is called electricity.

Atoms are the smallest particles of each element and are made up of a dense positively-charged nucleus surrounded by electrons. When an atom of metal loses one or more electrons, the residue is called an *ion* and has a net positive charge (since some of its negative charges are missing).

$$Fe^{o} - 2e \rightarrow Fe^{++}$$
(iron atom) (ferrous ion)

$$Fe^{o} - 3e \rightarrow Fe^{+++}$$
(iron atom) (ferric ion)

Positively-charged ions are called *cations*.

Non-metal elements form *negatively*-charged ions (*anions*), usually by dissociation of *salts* or *acids* in solution. For example:

$$H_2S \rightarrow H^+ + HS^-$$
(hydrogen sulfide—an acid) (hydrogen ion—a cation) (sulfide ion—an anion)

$$CO_2 + H_2O \rightarrow H^+ + HCO_3^-$$
(carbon dioxide—an acid) (water) (bicarbonate ion)

$$H_2SO_4 \rightarrow 2H^+ + SO_4^=$$
(sulfuric acid) (sulfate ion)

$$H_2O \rightarrow H^+ + OH^-$$
(water) (hydroxyl ion)

$$NaCl \rightarrow Na^+ + Cl^-$$
(sodium chloride) (sodium ion) (chloride ion)

The existence of charged ions in solution in water allows the water to become electrically conductive. The solution is known as an *electrolyte*.

Corrosion in a water environment involves energy transfer by way of (1) electron flow within the metal and (2) an equal flow of electrical charges by movement of ions in the electrolyte. For this reason, corrosion is called an electrochemical process (see also Appendix 1A).

1.04 Characteristics of the Basic Corrosion Cell

The nature of electrochemical corrosion reactions can be more easily explained with a sketch of a typical corrosion "cell" such as that in Figure 1.1-2.

Electrochemists refer to the charge transfer between the metal surface and the electrolyte as *half-cell* reactions. Being electrical in nature, the half-cell reactions are governed by the fundamental laws of electricity. These will be treated in more detail later.

Figure 1.1-2 shows the surface of a piece of steel in contact with water (electrolyte) containing charged ions. Metal loss from the surface is occurring at a localized area called the *anode*. At the anode, electrically neutral atoms of iron, Fe^{o}, are going into solution and forming ferrous ions, Fe^{++}. This happens initially because the relatively pure iron is at a high energy level and has a measurable solution potential or tendency to dissolve.

1.05 Anodic Half-Cell Reactions

As each Fe^{++} ion is formed, two negatively-charged electrons, 2e, are left behind in the metal thus giving it a small negative charge. The half-cell reaction is:

$$\text{(a) } Fe^{o} \rightarrow Fe^{++} + 2e.$$

The iron atom that gives up its electrons is said to be *oxidized* even though oxygen may not be involved.

Continuation of the anodic oxidation (electron loss) reaction tends to cause accumulation of soluble Fe^{++} ions near the anode surface. If nothing happens to remove the Fe^{++}, the anode reaction starts to slow down. Further, the excess electrons must also be removed from the vicinity of the anode to keep the process going.

In oil producing operations it is common for certain substances to be present in the water that form insoluble precipitates with iron, thus removing it from solution. The most common of these are oxygen, hydrogen sulfide and carbon dioxide. In solution these gases react with iron, as indicated on the left side of Figure 1.1-2, to form hydrated iron oxides or rust ($Fe_2O_3 \cdot H_2O_x$), iron sulfides (FeS_x) or iron carbonate (Fe_2CO_3). Soluble iron ions are thus removed and the anodic reaction continues.

Release of excess electrons in the vicinity of the anode develops an *electrical pressure* or *potential*. This is called the anodic half-cell potential when it is measured relative to some other half-cell or reference electrode.

1.06 Cathodic Half-Cell Reactions

Due to the developed potential, the excess electrons flow away from the anode through the metal to a point where their charge can be neutralized or taken up. The consumption or gain of electrons in a chemical or electrochemical reaction is called *reduction* and the sub-

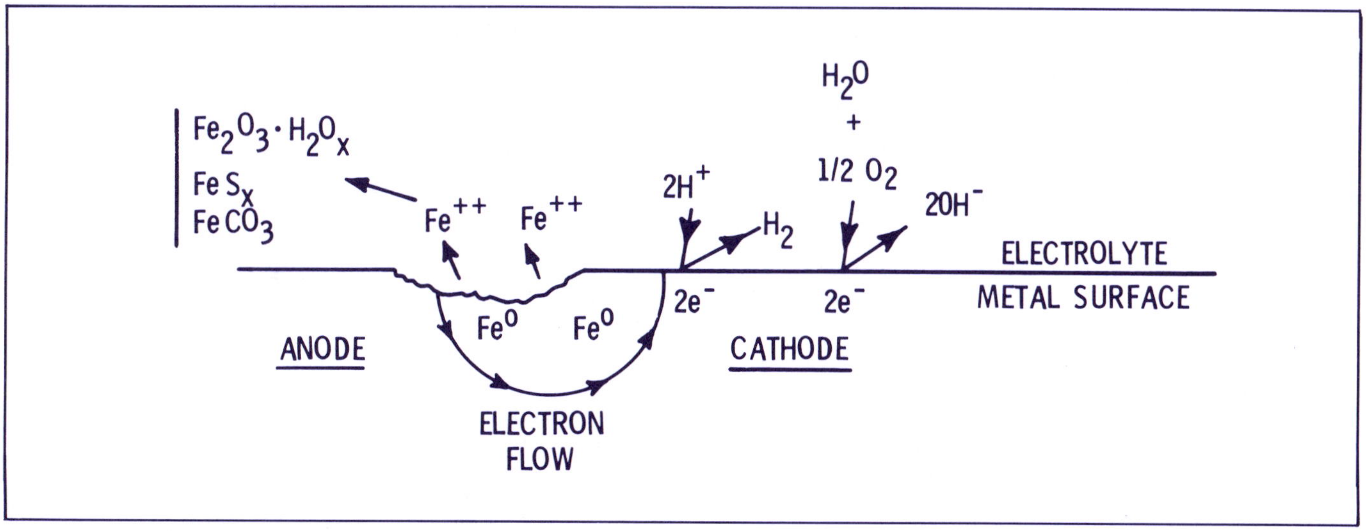

Figure 1.1-2. Representation of the basic corrosion cell.

stance accepting the electrons is said to be reduced. The site on a metal surface where reduction occurs is called the *cathode* and the process involved is the cathodic half-cell reaction.

Two common cathodic reduction (electron consuming) reactions are depicted in Figure 1.1-2. Dissolved oxygen is reduced at the cathode to form hydroxyl ions:

$$\text{(b)}\quad \underset{\text{(water)}}{H_2O} + \underset{\text{(oxygen)}}{{}^1/_2\,O_2} + \underset{\text{(electrons)}}{2e} \rightarrow \underset{\text{(hydroxyl ions)}}{2\,OH^-}\,.$$

If oxygen is *not* present but an acid such as H_2S or CO_2 is, the dominant cathode reaction is reduction of hydrogen ions to molecular hydrogen:

$$\text{(c)}\quad \underset{\text{(hydrogen ion)}}{2\,H^+} + 2e \rightarrow \underset{\text{(hydrogen gas)}}{H_2}.$$

Accumulation of molecular hydrogen gas at the cathode slows down the cathodic process.

1.07 Oxidation—Reduction (Redox) Reactions

It is extremely important that the nature of anodic oxidation and cathodic reduction reactions be understood. The half-cell reactions at the anode are called oxidation reactions because something is losing or giving up electrons. In reaction (a) of Item 1.05, iron (Fe^o) atoms lose 2 electrons to form ferrous ions (Fe^{++}). Therefore, the iron atoms are said to be oxidized. The Fe^{++} ions formed by oxidation of iron atoms can easily be oxidized still further. For example, if dissolved oxygen (O_2 from air) is present, the Fe^{++} ions are oxidized to Fe^{+++} ions by giving up electrons to the oxygen to form iron oxide or rust:

$$\text{(d)}\quad 2Fe^{++} + 1^1/_2\,O_2 \rightarrow \underset{\text{(insoluble rust)}}{Fe_2O_3}.$$

In Figure 1.1-2, the rust is represented as $Fe_2O_3 \cdot H_2O_x$ which means there is a variable (x) amount of H_2O combined with rust formed in the presence of water. In both equations (a) and (d) oxidation of iron is occurring because the iron is giving up electrons.

The rusting of iron is often referred to as oxidation or oxidation corrosion to distinguish it from other forms of corrosion that take place when elemental oxygen is *not* present. For example, in *air-free* water that contains H_2S, the overall corrosion reaction can be represented as:

$$\text{(e)}\quad Fe^o + H_2S \rightarrow FeS + H_2\,.$$

In this example, the iron is oxidized even though O_2 is absent. The oxidation step is that of iron going into solution as Fe^{++} ions, which then rapidly combine with H_2S to form insoluble iron (ferrous) sulfide, FeS. Notice that in this case, the second oxidation step of Fe^{++} converting to Fe^{+++} does not occur as it did when O_2 was present.

If the electrical oxidation potential (tendency to give up electrons) is sufficiently high, other oxidation (electron loss) reactions can occur at anode surfaces. For example, two oxygen atoms in water ($2H_2O$) may be oxidized at an anode to molecular O_2 gas by being forced to give up one electron each. Chloride (Cl^-) ions in salt water can be similarly oxidized to Cl_2 gas.

As explained previously, the electrons released by the anodic oxidation reactions migrate away through the metal conductor to other locations where they are taken up or consumed. In equation (b) of Item 1.06 the electrons are consumed by reaction with elemental oxygen in water to form hydroxyl ions. In equation (c), electrons are consumed by reaction with hydrogen ions to form hydrogen gas. Both of these are reduction reactions since electrons are consumed. Other reduction reactions can occur but these two are most prevalent in corrosion of oilfield equipment.

Given all this information, two important definitions can be established:

1. An *anode* is the site or electrode where *oxidation* (electron loss) reactions occur.

   If metal is the substance giving up electrons by dissolving and forming ions, corrosion is occurring.
2. A *cathode* is the site or electrode where *reduction* (electron consuming) reactions occur.

   Metal is not lost as long as the reduction reactions predominate.

1.08 Electrical Properties of Corrosion Cells

A corrosion cell consists of two half-cell *electrodes* (anode and cathode) immersed in an electrolyte and connected by a metallic path, as depicted schematically in Figure 1.1-3. The anode and cathode areas (electrodes) may be on a single piece of metal or they may be two separate metal items that are electrically connected (coupled) by a conductor.

Metal ions leave the anode by way of the electrolyte, and electrons travel from the anode to the cathode by way of the metallic path. *The anode suffers corrosion attack but the cathode is protected.* Very noble (least reactive) metals such as platinum and some non-metals such as carbon may function as anodes (sustain oxidation reactions) without corroding.

The electron current flows as a result of a potential difference (ΔE) between the anode and cathode. This potential difference (also called the voltage or electromotive force) results from (1) a difference in the electrodes, (2) a difference in the environments around the electrodes or (3) both.

The amount of *current* flow in a corrosion cell (and

thus the amount of corrosion at the anode) depends on the magnitude of the driving *voltage* and the total effective *resistance* or retardation of current flow in the cell. The relationship of these units in a simple electrical circuit composed of metal conductors is expressed by Ohm's Law:

$$I = \frac{E}{R} \text{ or } E = IR$$

in which:
- I = current in amperes
- E = potential or electromotive force (emf) in volts
- R = resistance in ohms.

Note in Figure 1.1-3 that four circuit resistances are represented. The boundary resistances R_3 and R_4 are actually apparent or effective resistances to current flow caused by both surface films and retardation of the electrochemical anode and cathode reactions previously discussed (see Appendix 1A, Item 1A.7). In practical corrosion control procedures, total cell resistance is taken as the sum of all resistances:

$$R = R_1 + R_2 + R_3 + R_4$$

where: R_1 = metallic path resistance (Note: This is usually low and is often neglected)
R_2 = electrolyte path resistance
R_3 = apparent or effective boundary resistance at the anode
R_4 = apparent or effective boundary resistance at the cathode

1.09 Electrode Potentials and Direction of Current Flow

The electrode potential difference, ΔE, between the anode and cathode is a measure of the driving voltage available to cause current flow and thus corrosion. Each of the electrodes exhibit half-cell potentials that are considered opposite in polarity with respect to each other; that is, one is positive (+) and the other is negative (−). By modern convention, the anode in a corrosion cell is considered to be negative (−) to the cathode (+).

Most practical corrosion literature discusses electrical current flow from anode to cathode *through the electrolyte*. This is depicted as "conventional current flow" in Figure 1.1-3 and the direction is opposite that of the electron flow in the metal. According to the conventional current flow concept, corrosion occurs where "current" leaves a metal structure. Where the "current" re-enters, corrosion is prevented. When reading books and articles about corrosion, conventional current flow direction (not electron flow) should be assumed unless otherwise indicated.

It is most important to remember, however, that the voltage signs (+ or −) of the anodes and cathodes are purely arbitrary and have meaning only in relation to some reference voltage. Potentials of both electrodes or half-cells may be negative (−) to a reference standard. The one with the lowest negative potential is considered to be positive in relation to the other. It is the difference (ΔE) between the electrode (half cell) potentials that is most important in corrosion work.

1.10 Polarization of Electrodes

The ΔE between anode and cathode is not a fixed value and can change easily. For example, referring again to Figure 1.1-3, the boundary resistances R_3 and R_4 can be increased by the chemical changes that are taking place at the surfaces of the anode and cathode electrodes while corrosion occurs. It was stated earlier (Items 1.05 and 1.06) that accumulation of Fe^{++} ions near the anode and H_2 gas molecules near the cathode tends to reduce corrosion rate. The reduction is related to changes in half-cell voltages whereby the anode becomes less negative, the cathode becomes less positive and the difference between the two is reduced. This *change* in half-cell voltage is called *polarization*. As corrosion progresses, both the anode and cathode tend to polarize, *but not equally*. Due to polarization effects, the voltage difference and the amount of current flow is diminished. How the polarization effect can be forced to reduce corrosion and is

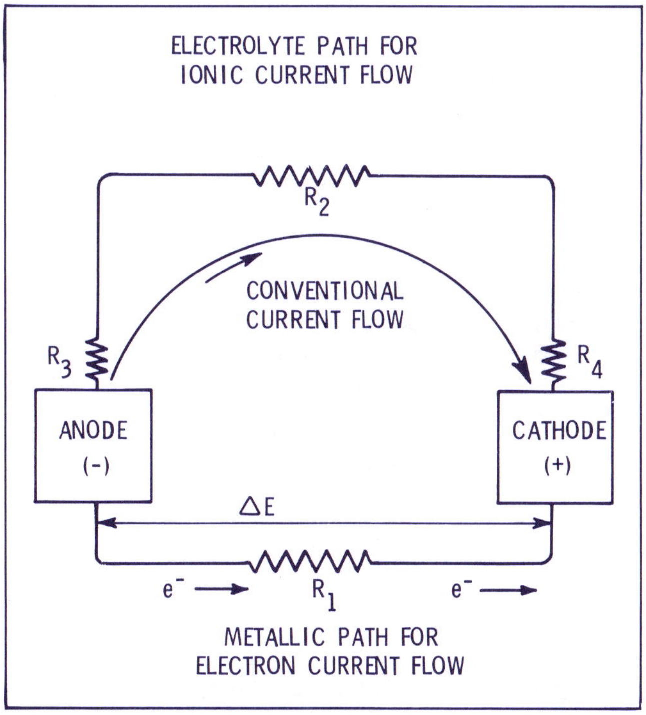

Figure 1.1-3. Schematic representation of a corrosion cell as an electrical circuit.

utilized in electronic instruments for measuring corrosion rates will be discussed later.

1.11 Corrosion Cells Created by Differences in Electrodes

In Item 1.08 it was pointed out that the driving force or voltage in a corrosion cell results from (1) differences in the anode and cathode electrodes, (2) differences in their environments, or (3) both conditions simultaneously. Condition (1), differences in electrodes, may be considered the fundamental or initiating cause of metal corrosion in almost all cases. This results from the fact that industrial metals are inherently heterogeneous and variable in composition and physical properties.

On a microscopic scale, ordinary steel is observed to consist of crystals or grains of relatively pure iron, Fe, intermixed with iron carbide, Fe_3C. Other metals and metal compounds are also present in various amounts depending on the grade of steel. Within the steel matrix there are a variety of small and relatively large inclusions of metal oxides and sulfides. The surfaces are usually coated with millscale (Fe_3O_4) or other corrosion products. These differences in composition create local anode/cathode cells.

The thermal (heat treatment) and mechanical history of a given piece or area of steel also causes it to be different than other pieces. For example, the body of a joint of tubing is usually different than that of the collars. High strength steels show differences in grain structure, chemical composition and physical properties in comparison with low strength steels. All these factors influence whether or not a particular piece of steel, or even a given spot on a piece of steel, becomes anodic and corrodes in a particular environment.

The underlying reason why a particular metal or some part of it starts to corrode is that its solution potential is higher than that of something else with which it is in electrical contact. A grain of pure Fe has a higher solution potential than that of an adjacent particle of Fe_3C. It becomes the anode in a tiny corrosion cell, if both are immersed in an electrolyte. Similarly, a joint of clean, new pipe placed in a string of older pipe covered with corrosion products becomes the anode in a much larger corrosion cell. A steel coupling placed in a copper line immediately becomes anodic and indeed will corrode at a higher than normal rate because it is coupled with the more cathodic copper.

All of the previous are examples of corrosion induced by electrode dissimilarities. The name "galvanic corrosion" is used to describe this type of attack.

1.12 Tabulations of Relative EMF's or Solution Potentials of Metals

In scientific literature and textbooks dealing with chemistry, there are tables or listings of various metals arranged in order of their chemical reactivity or tendency to corrode. These lists can be used beneficially to help explain the corrosion process and to help predict which metals will be more likely to corrode in a given environment. However, they can be quite confusing to the non-specialist and must be used carefully.

Table 1.1-1 was developed especially for this discussion to illustrate the principle of electrode potentials. The potential values (voltage numbers) are examples using four different reference electrodes.

The *standard* solution or oxidation potentials, E_o, in Table 1.1-1 were measured with individual metals in contact with a standard solution containing ions of the same metal and under precisely controlled conditions. These potential values change with the type, concentration, and temperature of the solutions to which the metals are exposed. For example, the potential of zinc can move down to below iron in the series when the zinc and its solution are heated to about 80°C (176°F). Chromium, in the passive state, exhibits a potential that is less negative or more noble than that of iron.

In general, the reactivity or tendency of metals to corrode decreases down the series. Metals at the top of the table go into solution relatively easily. Sodium metal, for example, will combine spontaneously and rapidly with

Table 1.1-1
Relative Single Electrode Potential Values

		Approximate Potential in Volts vs. Various Reference Electrodes			
Metal	*Ion*	E_o (1)	*SCE* (2)	$Cu/CuSO_4$ (3)	*Ag/AgCl* (4)
Sodium	Na^+	−2.71			
Magnesium	Mg^{++}	−2.37	−1.62	−1.6	−1.1
Magnesium Alloy	(6% Al)			−1.6	−1.2
Aluminum	Al^{+++}	−1.66	−0.9	−0.8	−0.7
Aluminum Alloy	(5% Zn)		−1.1	−1.0	−0.9
Zinc	Zn^{++}	−0.76	−1.0	−1.0	−0.9
Chromium	Cr^{+++}	−0.74			
Iron	Fe^{++}	−0.44	−0.7	−0.6	−0.5
Mild Steel (rusted)				−0.2 to −0.4	0 to −0.3
Nickel	Ni^{++}	−0.25	−0.3		
Tin	Sn^{++}	−0.14			
Lead	Pb^{++}	−0.13	−0.2		
Hydrogen	H^+	0.0			
Carbon (non-metal)			+0.3	+0.3	
Copper	Cu^{++}	+0.34		+0.5	
Silver	Ag^+	+0.80			
Platinum	Pt^{++}	+1.2	+0.2		
Gold	Au^{++}	+1.5			

(1) Standard Oxidation or Solution Potentials vs. Standard Hydrogen Electrode (SHE)
(2) Saturated Calomel Electrode in Sea Water
(3) Copper/Copper Sulfate Electrode
(4) Silver/Silver Chloride Electrode

water. In other words, sodium metal corrodes at an extremely rapid rate compared with iron and other metals that are lower in the list.

At the bottom of the list, we find platinum and gold. These metals will not dissolve (that is, will not corrode) even in most strong acids. The metals at the bottom of the list are the so-called "noble metals" and are found in their pure metallic state in nature.

When two different types of metal serve as electrodes in a corrosion cell, the metal higher in the electromotive series will generally be anodic and will corrode. The other metal, the cathode, usually is not harmed. Also, the farther the metals are apart in the series, the greater will be the voltage of the cell and the greater will be the corrosion tendency when the two metals are electrically coupled.

A metal high in the series can displace one with a lower E_o from solution. For example, a piece of iron or nickel placed in a water solution of copper sulfate will go into the solution in place of the copper, and the copper will precipitate onto the nickel or iron as a metallic coating. Zinc placed in even a dilute water solution of a mineral acid will rapidly displace hydrogen (as H_2 molecules) from the solution. Note that metals above hydrogen in the series have negative E_o values while those below have positive E_o values. The standard hydrogen electrode (SHE) is the reference (see Appendix 1A).

As previously mentioned, the numerical value and polarity (+ or − sign) of the electrode potentials given in Table 1.1-1 only have significance in relation to some reference standard. For example, in a corrosion cell consisting of an iron electrode and a copper electrode immersed in an electrolyte and connected by a metallic conductor, the iron is negative (anodic) to the copper and the initial potential difference (ΔE) is about 0.78 volts, calculated as follows:

$$\underset{\text{(Iron } E_o)}{-0.44} - \underset{\text{(Copper } E_o)}{(+0.34)} = \underset{(\Delta E)}{0.78 \text{ volts}}$$

As corrosion progresses the potential difference, ΔE, diminishes due to polarization effects.

1.13 Dissimilar Metal Couples

If any two of the metals in the series are electrically coupled in the same solution, the metal higher in the series will usually corrode at a higher rate than if it had been uncoupled. At the same time, the more noble (least reactive) metal corrodes less rapidly and sometimes not at all. This is called the galvanic or dissimilar metal effect. The greater the difference in solution potentials of the two metals, the more pronounced are these effects. For this reason metals widely spaced in the series should not be connected unless it is intended to purposely sacrifice one of the active metals such as aluminum, zinc or magnesium to protect important pieces of equipment. This sacrificing process is called cathodic protection. It is the reason for galvanizing steel and bonding magnesium to buried pipe. These practices are discussed in more detail in Chapter 4 of Section 1.

When dealing with combinations of different metals (dissimilar metal couples), it is important to know the effect of electrode size. An iron bolt in a sheet of copper will corrode very rapidly since all current is concentrated on the relatively small anode; viz., the *current density* is high. When the electrode sizes are reversed (large iron, small copper), the galvanic effect is spread over the large anode and the damage is less concentrated.

The data of Figure 1.1-4 illustrates the cathode to anode area effect in an iron/copper couple exposed to aerated 3% NaCl solution. Iron plate exposed alone to the aerated salt water corroded at about 500 milligrams/square decimeter/day (mdd). Coupled to a sheet of copper 20 times larger in size, the iron plate corroded at 6000 mdd. (See Chapter 10, Table 1.10-2, for comparison of corrosion rate expressions.)

Table 1.1-2 was specially prepared for this text as a practical guide to the corrosion consequences of coupling equipment made of different metals. When considering combinations of metals, the combinations listed as NG in Table 1.1-2 should generally not be made. Those listed as "questionable" can be made where one or the other metal parts is expendable, viz., can be replaced at low cost and without hazardous consequences. Combinations listed as "OK" can generally be used satisfactorily but the OK listings should not be interpreted as infallible. The data of Table 1.1-2 are presented only as a guide and the user must assume all risk of use.

Some of the most common places to find combinations of dissimilar metals in oilfield equipment is in downhole pumps, wellheads and brine pumps (see Section 1, Chapter 7, Item 7.10). In the latter, aluminum bronze or Ni-Resist pump heads are often mounted on carbon steel pump bodies. The combination is OK (galvanically) according to Table 1.1-2 if the relative area of carbon steel (anode) is large. When galvanic (dissim-

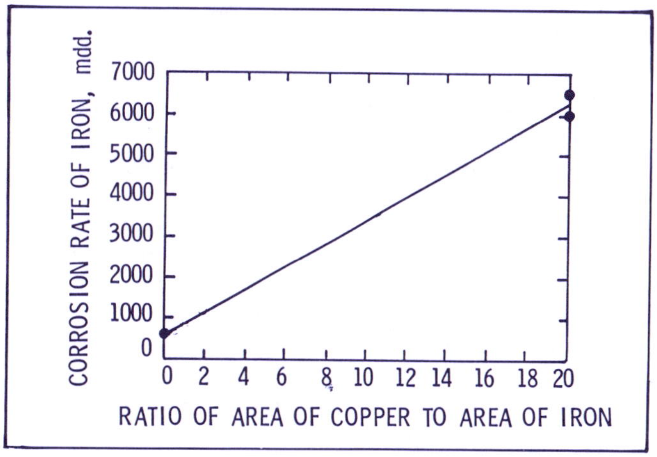

Figure 1.1-4. Cathode to anode area effect in iron/copper couple exposed to aerated 3% NaCl solution.

Table 1.1-2
Dissimilar Metal Corrosion Guide (Probable effect of coupling two metals in seawater at 25°C)

		Secondary Metal											
Primary Metal	*Rel. Area*[(1)]	*Al Alloys*	*Cadmium*	*Ni-Resist 1 & 2*	*Brass or Bronze*	*Copper Cu*	*90/10 Cu-Ni*	*70/30 Cu-Ni*	*410 SS*	*304 SS*	*316 SS*	*Monel*	*Carbon Steel*
Al Alloy	S	—	NG	NG	NG	NG	NG	NG	NG	NG	NG	NG	NG
	E	—	NG	NG	NG	NG	NG	NG	NG	NG	NG	NG	NG
	L	—	?	?	NG	NG	NG	NG	NG	?	?	NG	?
Carbon Steel	S	?	OK	NG	NG	NG	NG	NG	NG	NG	NG	NG	—
	E	NG	NG	?	NG	NG	NG	NG	NG	NG	NG	NG	—
	L	NG	OK	OK	OK	OK	OK	OK	OK	OK	OK	OK	—
90/10 Cu-Ni	S	NG	NG	OK	?	?	—	NG	NG	NG	NG	NG	OK
	E	NG	NG	NG	?	?	—	OK	NG	NG	?	?	NG
	L	NG	NG	NG	NG	NG	—	OK	NG	NG	OK	OK	NG
70/30 Cu-Ni	S	NG	NG	OK	?	?	OK	—	NG	NG	NG	NG	OK
	E	NG	NG	NG	?	?	OK	—	NG	NG	?	?	NG
	L	NG	NG	NG	NG	NG	NG	—	NG	NG	OK	OK	NG
Ni-Resist 1 and 2	S	?	OK	—	NG	NG	NG	NG	NG	NG	NG	NG	OK
	E	NG	NG	—	NG	NG	NG	NG	NG	?	OK	OK	?
	L	NG	NG	—	OK	OK	OK	OK	NG	OK	OK	OK	NG
Copper	S	NG	NG	OK	?	—	NG	NG	NG	NG	NG	NG	OK
	E	NG	NG	NG	?	—	?	?	NG	NG	NG	?	NG
	L	NG	NG	NG	NG	—	?	?	NG	NG	NG	OK	NG
Brass or Bronze	S	NG	NG	OK	—	NG	NG	NG	NG	NG	NG	NG	OK
	E	NG	NG	NG	—	?	?	?	NG	NG	NG	NG	NG
	L	NG	NG	NG	—	?	?	?	NG	NG	NG	OK	NG
304 SS	S	?	NG	OK	NG	NG	NG	NG	?	—	?	NG	OK
	E	NG	NG	NG	NG	NG	NG	NG	NG	—	NG	?	NG
	L	NG	NG	NG	NG	NG	NG	NG	NG	—	NG	?	NG
316 SS	S	?	NG	OK	?	NG	OK	OK	NG	?	—	NG	OK
	E	NG	NG	OK	?	NG	?	?	NG	NG	—	?	NG
	L	NG	NG	NG	NG	NG	NG	NG	NG	NG	—	?	NG
Monel	S	NG	NG	OK	OK	OK	OK	OK	NG	?	OK	—	OK
	E	NG	NG	OK	NG	?	?	?	NG	NG	?	—	NG
	L	NG	NG	NG	NG	NG	NG	NG	NG	NG	?	—	NG

[1]Relative Area of Primary Metal to Secondary Metal
S = Small E = About Equal L = Large
Symbol Legend: NG = One metal in couple will corrode
? = Questionable
OK = Neither metal is likely to corrode because of coupling. However, failure may occur due to other causes.

ilar metal) couples exist and only one part of the equipment is coated, it should be the cathodic metal. *Never coat the anode* since current density will be very high at coating pin holes and concentrated pitting will occur.

Figure 1.1-5 shows severe galvanic corrosion of the aluminum core of a lightweight free piston used to unload salt water in a low pressure gas producing well. The sharp edge and body of the outer steel casing is not corroded because the steel was the protected cathode in the dissimilar metal couple.

1.14 Physical Conditions of Metals That Promote Corrosion

Table 1.1-1 illustrates the large voltage differences that can exist between two chemically different types of metals. However, two pieces of steel with the same *chemical* composition may form a galvanic corrosion cell if the *physical* condition of one is different from the other. Following are some examples:

1. **Metals with different microstructure**

   Welding causes metal microstructure (grain size) changes near the weld bead. Unless properly post-weld heat treated, the heat affected zone (HAZ) alongside the weld will become anodic either to the weld itself or to the metal that remained cool. In some cases, the weld bead corrodes preferentially because its microstructure differs from the surrounding steel. Forging and other fabrication or shaping processes can change the metal grain size. The transition zone between large and small grain steel usually corrodes at a higher than normal rate. For example, the upset ends of J-55 well tubing often corrodes where the upset joins the body. This is prevented by full length normalizing (heat treatment) after upsetting. More on heat treatment in Chapter 7 of Section 1.

2. **Scratches and abrasions**

   Scratches or rough abraded areas are anodic to smooth surfaces. This effect is often seen on sucker rods and drill pipe. Special handling techniques are used to minimize this problem.

3. **Differential strain**

   A strained area such as a hammer dent or an indention left by a pipe wrench will corrode because it is anodic to surrounding unstrained areas. Again, special care should be taken when running tubing or sucker rods.

1.15 Surface Oxide Films and Hydrogen Overvoltage Effects

Before the study of metal characteristics that affect corrosion is completed, two other properties that have great influence should be considered. Fortunately, both tend to reduce corrosion under proper conditions.

First is passivating surface films. Under oxidizing conditions, some metals generate surface films that make them "passive" or less reactive in the particular environment. For example, it can be seen by examining Table 1.1-1 that aluminum has a high solution potential and indeed will corrode sacrifically when coupled with a metal lower in the electromotive series. However, an uncoupled piece of aluminum placed in neutral aerated water or soil generally withstands corrosion quite well. Aluminum is so reactive, in fact, that it quickly forms a tenacious protective layer of aluminum oxide on its surface. Some salts such as $CaCl_2$ in seawater tend to destroy the film but special aluminum alloys are available for such service.

The corrosion resistance of steel tubes in boilers is enhanced by dense iron oxide films formed at high temperature by reaction of iron with the salt-free water. Similar resistance to oxidizing acids is achieved by forced anodic polarization (see Appendix 1A).

Iron alloyed with 13% or more chromium forms the relatively *passive* stainless steels. Ternary (three component) alloys of iron, chromium (18%) and nickel (8%) are even more corrosion resistant in oxidizing environments. When air or other oxidizing agents are *absent*, especially where salts are present, stainless steel may become *active* and corrode at a high rate due to loss of protective films. For example:

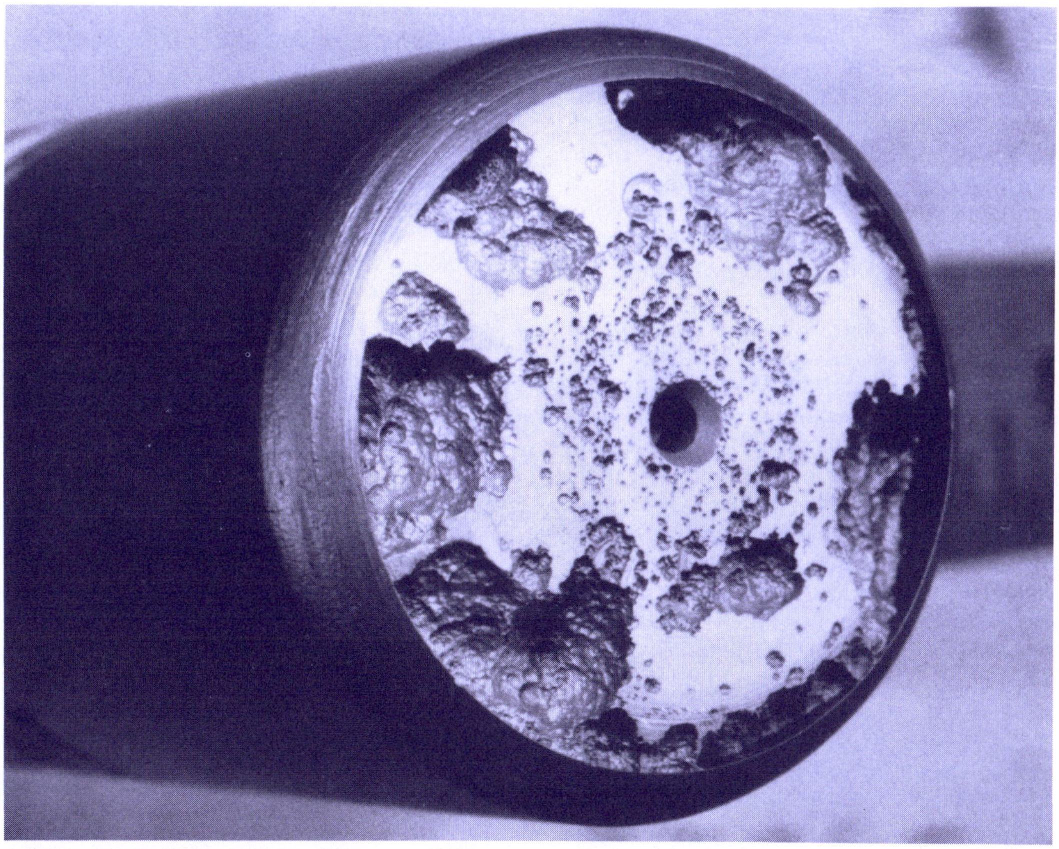

Figure 1.1-5. Selective galvanic corrosion of aluminum encased in steel.

Corrosion Rates of 18-8 Stainless steel in 5% H_2SO_4 Acid at 30°C

Air Saturated	*Air Free*
2 mdd	300 mdd

The solution potential of 18-8 stainless steel vs. a saturated calomel reference electrode (SCE) in open flowing seawater is near −0.05 volts. Active stainless steel in poorly aerated water, such as under sludge deposits, exhibits much more anodic potentials in the order of −0.5 volts and will pit.

Second, it was mentioned previously that metals above hydrogen in the electromotive series will corrode and displace hydrogen from solution. For example:

$$Fe^{++} + H_2S \rightarrow FeS + H_2$$

$$Zn^{++} + 2HCl \rightarrow ZnCl_2 + H_2$$

When a metal corrodes with evolution of hydrogen gas on the cathodic surfaces, there is a definite minimum voltage that must develop in order to plate out the hydrogen. If there is not sufficient voltage to plate out the hydrogen, the corrosion reaction is greatly reduced. Some metal surfaces require an additional voltage to make the hydrogen deposit on them. This additional voltage has been termed *overvoltage*. The overvoltage of hydrogen is different for each metal. On platinum it is very low, 0.09 volt; on mercury, it is high, 1.04 volt; overvoltage on iron is 0.40 volt. These voltage values are measured at very low current densities (amperes/unit area).

The fact that iron corrodes at a very slow rate in neutral water with no dissolved oxygen, is attributed to the overvoltage of hydrogen. The effective voltage of the local cells is so low that hydrogen is generated at only a slow rate and corrosion is minimal. Rapid corrosion and pitting of iron in air-free water can result from other factors such as growth of sulfide producing bacteria (see Section 2, Chapter 3).

Overvoltage increases with current density. The overvoltage of hydrogen on iron changes from 0.40 volt at a current of 1/1000 amp/cm^2 to 0.77 volt at 1 amp/cm^2. Therefore, if the cathodic area is small compared to the anodic area, current density on it will be relatively large, causing a high overvoltage of the hydrogen on the cathodic area. The result may be that the cell will not have enough voltage to force the hydrogen to plate out on the cathode. When this is the case, the corrosion reaction at the anode is also stifled. This effect is part of the reason why relative size is important in dissimilar metal couples as indicated in Table 1.1-2. For example, a small piece of copper (cathode) coupled to a large piece of iron (anode) will not accelerate corrosion of the iron very much.

In the case of corrosion of iron or steel in strong acid solution, overvoltage is of little help. The driving force (voltage) of the local anode/cathode couples in an acid environment is high enough that even the hydrogen overvoltage will not hold back the current. Hydrogen is evolved steadily and corrosion at the anodic areas proceeds without interruption. If, however, sodium arsenite corrosion inhibitor is added to the acid, a layer of metallic arsenic is plated out on the steel surface. Arsenic has a very high hydrogen overvoltage. Consequently, the thin film of arsenic metal on the steel surface inhibits evolution of hydrogen and thus the corrosion. Arsenic compounds are no longer widely used to inhibit acids used for oil well stimulation since arsenic poisons the platinum catalysts used in refineries and is toxic.

1.16 Corrosion Cells Created by Differences in Electrode Environments

Under Item 1.11 some of the more important aspects of corrosion initiated by differences in metals were discussed. The nature and concentration of substances dissolved or suspended in the electrolyte are equally important in understanding and controlling corrosion processes.

While metal differences often determine the location of anodes, components of the electrolyte generally determine the rate or intensity of the corrosion attack. This is especially true when a single type of metal such as steel is exposed to water. Even if the steel is uniform and homogeneous, differences in the electrolyte surrounding various parts of the steel can cause potential gradients and thus areas of localized corrosion. Following are three common cases:

1. **Differential Oxygen Concentration**

   When water containing dissolved oxygen flows through a steel pipe or the aerated water is held in a steel tank, oxygen concentration at all parts of the steel surface is *not* the same. Any kind of deposit such as sludge or rust can shield part of the metal and reduce availability of oxygen to that area. The portion of steel exposed to water with a *low* oxygen concentration becomes anodic and corrodes at a faster rate than connecting portions that have more oxygen available to them. Corrosion under a rust layer (tubercle) as depicted in Figure 1.1-6, serves to illustrate the effect.

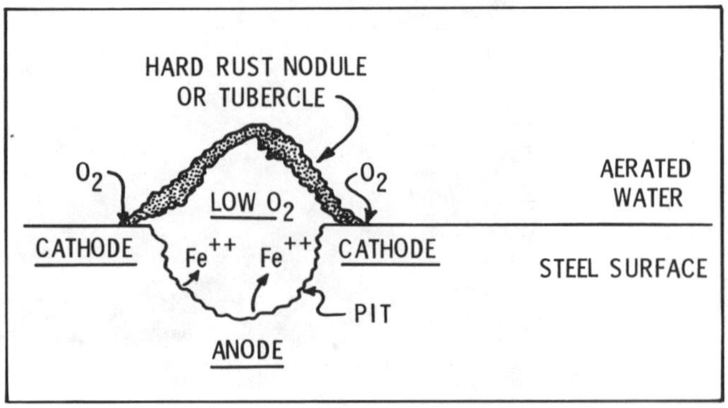

Figure 1.1-6. Rust tubercle causing a differential oxygen cell.

Initially a spot on the steel surface starts to corrode with iron ions going into solution. The Fe^{++} ions move out into the aerated water and are precipitated as rust. As the rust deposit accumulates, the area of steel under it is shielded and becomes oxygen deficient or anaerobic. Corrosion of the spot continues and is accelerated as a result of the electrical potential gradient between the steel where the pit is developing and the surrounding unshielded area. The pitting steel is anodic while the aerated steel around it is cathodic. Availability of ample oxygen favors, by mass action, the *cathodic* reaction:

$$H_2O + {}^1/_2O_2 + 2e \rightarrow 2OH^- \text{ (reduction)}$$

while at the *anode*:

$$Fe^o \rightarrow Fe^{++} + 2e \text{ (electrochemical oxidation)}$$

and

$$Fe^{++} + 2OH^- \rightarrow Fe(OH)_2$$

$$2Fe(OH)_2 + {}^1/_2O_2 \rightarrow Fe_2O_3 \cdot (H_2O)_2 \textit{ rust} \text{ (chemical oxidation)}$$

(Note that hydroxide, OH^-, formed at the cathode can assist in precipitating rust at adjacent anodes).

The latter reactions (precipitation of rust) occur a short distance away from the metal surface where Fe^{++} ions contact OH^- ions and oxygen. Migration of the iron ions away from the vicinity of the pit minimizes anodic polarization and thus promotes continued pit growth. Other types of deposits such as dirt, sand, bacterial slime or mineral scales can set up similar differential oxygen cells.

A much larger but similar differential oxygen or aeration cell is illustrated in Figure 1.1-7. The portion of the pipe exposed to dense clay with little or no oxygen becomes anodic to other pipe surfaces exposed to loose, aerated soil. In this case, the anodic area is relatively small and the cathodic areas are large. The small anodic area is called an electrical "hot spot" (see Section 1, Chapter 4, Item 4.08).

2. **Differential Concentration of Electrolytes**
When two portions of metal are immersed in salt solutions of different strength, the one exposed to the more *dilute* solution will generally be the anode. This effect can be reversed if some other more powerful corrosion promoting effect such as differential aeration occurs simultaneously.

3. **Temperature Differential**
A region of higher temperature is anodic to a cooler region if a more powerful influence is not present.

1.17 Corrosive Agents in the Environment

In oil producing operations, the environmental factors or agents that promote most of the corrosion are the three gases:

1. Oxygen O_2 ;
2. Hydrogen Sulfide H_2S ;
3. Carbon Dioxide CO_2 .

These substances are so important that their mode of action will be discussed individually. The discussion will be confined to the case of carbon steel exposed to salt water that is in equilibrium with gas containing one or more of the three corrosive gases.

1. **Oxygen**
Corrosion by oxygen generally occurs in surface equipment such as water injection systems handling aerated water or where air contamination of produced water has been permitted. The latter condition is prevalent and can cause rapid equipment loss if the produced water is also sour (contains H_2S). For this reason, air should be totally excluded from water sys-

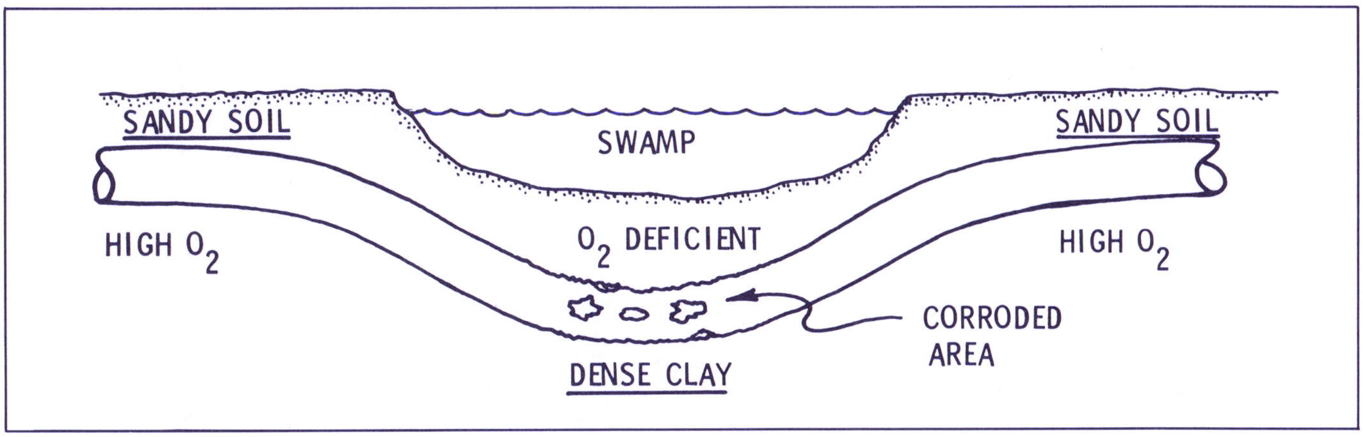

Figure 1.1-7. Corrosion on a pipeline due to differential aeration.

tems handling sour water (see Chapter 6 of Section 1).

Oxygen promotes corrosion in two ways. First, it is a powerful cathode depolarizer, viz., retards cathodic polarization. When an acid substance such as H_2S or CO_2 is present *but O_2 is not*, hydrogen gas tends to form on the cathode, viz.:

Ionization of H_2S

$$\text{(a) } H_2S \rightarrow H^+ + SH^-$$

Cathodic reduction

$$\text{(b) } 2H^+ + 2e \rightarrow \underset{\substack{\textit{Step 1} \\ \text{(atomic} \\ \text{hydrogen)}}}{2H^o} \rightarrow \underset{\substack{\textit{Step 2} \\ \text{(molecular} \\ \text{hydrogen)}}}{H_2}$$

Accumulation of gaseous H_2 tends to film over and polarize the cathode. However, if ample O_2 is present, the second step of equation (b) does not occur and instead:

$$\text{(c) } 2H^o + {}^1\!/_2O_2 \rightarrow H_2O.$$

Thus, the cathodic reaction and attendant corrosion is promoted because polarization by H_2 is minimized when oxygen is present.

Secondly, the oxygen removes iron by precipitation of iron oxides at the anode and thus prevents anodic polarization by Fe^{++} ions.

2. **Hydrogen Sulfide**

H_2S is a weak acid but is abundant in oil producing operations. The acidity, or ability of H_2S to generate H^+ ions, is indicated by its first ionization constant:

$$K_1 = \frac{[H^+][SH^-]}{[H_2S]} = 5.7 \times 10^{-8}.$$

Under standard conditions, each mol of H_2S in solution produces only 5.7×10^{-8} mols of H^+ ion. However, as the H^+ is removed by the cathodic reduction reaction (see Item 1.06), more is formed and hydrogen gas readily appears on steel exposed to *air-free* water containing H_2S in equation (b) previously given.

The anion, SH^- dissociates further to $S^=$ and H^+. The $S^=$ ion reacts with iron to form the black FeS corrosion product so commonly found.

3. **Carbon Dioxide**

CO_2 is a stronger acid than H_2S. It combines with water to form carbonic acid:

$$CO_2 + H_2O \rightarrow H_2CO_3$$

which ionizes first to:

$$H_2CO_3 \rightarrow H^+ + HCO_3^-$$

the ionization constant for which is:

$$K_1 = 4.31 \times 10^{-7}.$$

The major contribution of CO_2 in corrosion of production equipment is the increase in acidity by H^+ ion generation. Its corrosion product, $FeCO_3$ (siderite), is often found on the tubing of gas wells. (See also Chapter 9, Item 9.01).

Photos of corrosion pitting caused by H_2S and CO_2 appear in Section 1, Item 8.11 and Item 10.06.

1.18 Relative Corrosiveness of O_2, CO_2 and H_2S

Figure 1.1-8 shows a comparison of O_2, CO_2 and H_2S with respect to relative corrosiveness under specific conditions. Corrosiveness is defined here as a measure of corrosion damage at specific concentrations of the corrosive agent in water. The data of Figure 1.1-8 are based on corrosion rates measured by exposing carbon steel test specimens to water solutions containing various concentrations (parts per million) of each separate gas. Note that 1 ppm of O_2 is as corrosive as 100 ppm of H_2S or 50 ppm of CO_2 under the test conditions. At 8 ppm, O_2 is roughly 5 times more corrosive than 800 ppm of H_2S. Note also that corrosion rates are directly proportional to O_2 concentration, viz., corrosion rate increases or decreases in proportion to O_2 content of the water when O_2 is the only corrosive agent present.

This information is given to provide a general idea of the comparative corrosiveness of the three important gases on clean steel surfaces at low temperature. The data for O_2 correlates reasonably well with *overall* corrosion rates measured in the field on steel exposed at low temperatures to neutral aerated salt waters. Corrosion rates at *pits* (see Figure 1.1-6) can be several times greater. For H_2S, the laboratory to field data correlation is also fairly good. The corrosiveness of CO_2 at low temperature is quite variable and can fall above or below that of H_2S, depending to a large extent on the bicarbonate (HCO_3) alkalinity of the water (see Section 2, Chapter 1). The data for CO_2 shown here is for waters having low (less than 50 mg/liter) of bicarbonate alkalinity. When alkalinity is a few hundred mg/liter, the corrosivity of CO_2 at low temperatures is usually lower than that shown here. In hot high pressure wells producing water and gas rich in CO_2, corrosion can be intense. This will be discussed in greater detail in Chapters 8 and 9.

It should also be pointed out that combinations of O_2 with either H_2S or CO_2 can cause rapid corrosion, even at low temperatures. Even a trace amount of O_2 (less than 0.1 ppm) acts as a cathodic depolarizer as discussed previously and keeps the corrosion reactions going at a high rate.

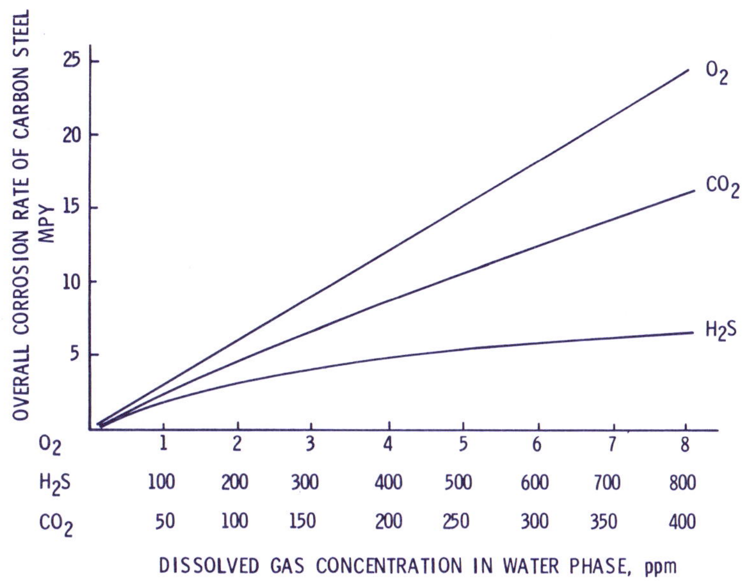

Figure 1.1-8. Comparative corrosiveness of three common gases in water solutions (25°C, 5–7 day exposure, 2–5 g/liter NaCl, HCO_3 alkalinity <50 mg/L—computed from several data sources).

1.19 The Contribution of Dissolved Salts

Salt water is usually considered to be more corrosive than low salinity or fresh water. However, neutral dissolved salts such as sodium chloride (NaCl) or calcium chloride ($CaCl_2$) are *not* the primary corroding agents.

They do promote or accelerate corrosion if one or more of the three *corrosive gases (O_2, H_2S or CO_2) are present*. The effect of the salts is primarily that of reducing the electrical resistance of the electrolyte (R_2 of Figure 1.1-3). By lowering total cell resistance, more current flows at a given potential gradient between anode and cathode, viz.:

$$I = \frac{E}{R}$$

If E is constant, then I is inversely proportional to R and I (salt water) > I (fresh water).

In low resistivity soils or water, not only is current flow greater but more distant anodic and cathodic areas can affect each other and total corrosion is accelerated.

Dissolved salts also have other effects that can be either good or bad. While they increase the electrical conductivity of the electrolyte, the salts reduce solubility of dissolved gases such as O_2 as indicated in Figure 1.1-9.

Figure 1.1-10 shows corrosion rates observed on low-alloy carbon steel test specimens exposed 16 days to oxygen saturated water solutions of neutral (non-acidic) salts. Maximum corrosion occurred at about three to four weight percent of salts, which is typically the range of dissolved salts in seawater.

Chloride salts tend to destroy the passivity of some stainless steels and can cause them to fail by rapid cracking under tensile stress at temperatures higher than about 80°C (176°F). This effect is called *chloride stress cracking* (CSC) and will be discussed in more detail in Chapter 7 of this Section.

Chromium steels with less than 22% Cr generally become less resistant to corrosion by H_2S and CO_2 at increasing sodium chloride concentrations.

1.20 Effects of Microorganisms

Accumulations of bacterial slimes, fungi and molds contribute to corrosion of steel equipment. In oil producing operations, the major bacteria-related corrosion problem is attack by hydrogen sulfide that is generated by metabolic processes of certain organisms. The most commonly found sulfide producers are the *Desulfovibrio* species such as *Desulfovibrio* desulfuricans (see Chapter 3, Section 2). These are referred to as sulfate reducers because they utilize $SO_4^{=}$ ions from salty waters and form H_2S. To a limited extent, sulfide corrosion results from the H_2S generated by species of the genus *Clostridia* from organic sulfur compounds. These are theromophilic (favor elevated temperature) and can be found in the bottom sludge of heater treaters operating at up to about 75°C (167°F).

More indirectly, bacterial accumulations or heavy slime deposits can promote corrosion by creating differential oxygen cells on steel in aerated waters (see Item 1.16).

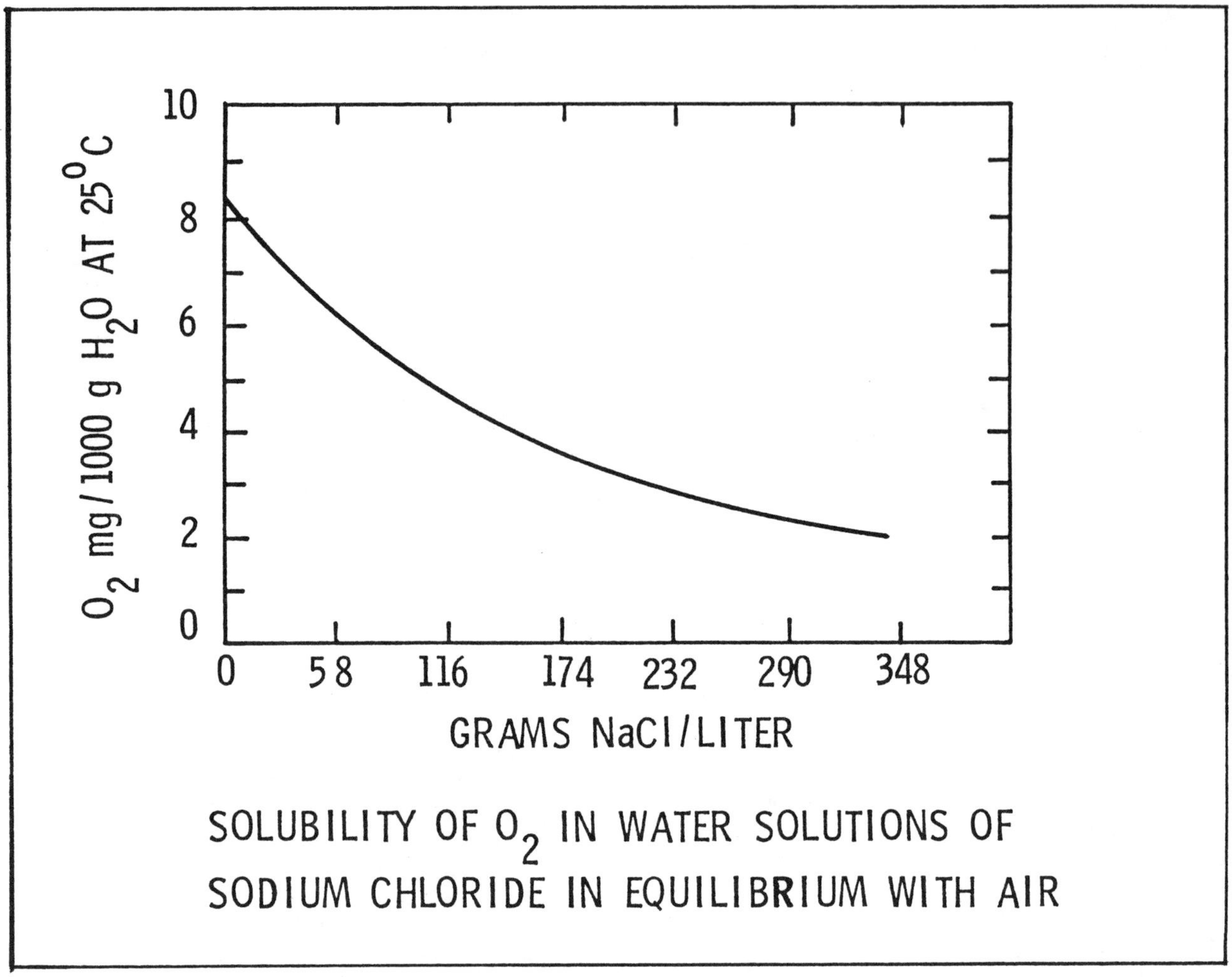

Figure 1.1-9. Solubility of O_2 in water solutions of sodium chloride in equilibrium with air.

1.21 Stray and Induced Electrical Currents

In the galvanic or spontaneous corrosion cells that have been discussed previously, electrical energy flows between anodes and cathodes through both the metal (by way of electrons) and the electrolyte (by way of soluble ions). The electrical current is generated in place when the corroding metal atoms undergo a chemical change from electrically neutral atoms to electron deficient ions (see Item 1.05). Current flow within the cell is proportional to the rate of metal loss. The opposite can also be stated; that is, *corrosion rate is proportional to current flow*.

Corrosion of metal equipment can also be caused by electrical currents originating from an outside source. Three common sources of extraneous electrical currents are:

1. Grounding of electrical equipment such as the motor of a pumping unit that is grounded to the well casing.
2. Current induced on a pipeline where it passes close to or under high voltage power transmission lines.
3. Current flow induced between an unprotected structure such as a pipeline or casing and nearby equipment that is under cathodic protection. This effect is called cathodic protection *interference* and will be discussed in more detail later (see Chapter 4, Section 1).

When the electrical current originates from an outside source the corrosion is called *electrolytic* attack. Corrosion rate or rate of metal loss is again dependent on the amount of current flowing. However, a number of other chemical reactions take place at the anode in addition to the metal going into solution. This is especially true where current density (amperes per unit of anode area) is very high.

The anodic electrochemical reactions (oxidation) that take place during electrolytic corrosion are quite complex but electron flow, away from the anode and back

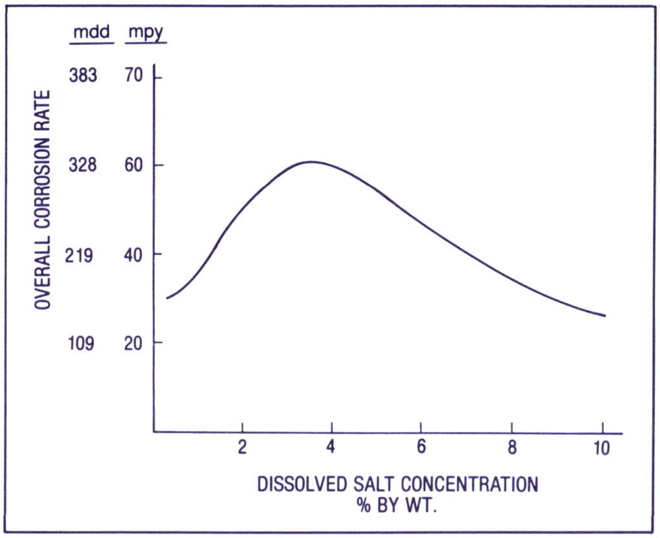

Figure 1.1-10. Corrosion rates of carbon steel in aerated salt waters. (NaCl 80%, $CaCl_2$ 20%, pH 6.5–7.0, 22–25°C).

through the metal, still occurs just as in spontaneous corrosion cells. Conventional electrical current is considered to be flowing in a direction opposite to that of electron flow as explained previously (see Figure 1.1-3 and Item 1.08). Therefore, the mechanism of electrolytic corrosion can be more easily grasped if the anodic chemical reactions are neglected and the corrosion is explained by the conventional current flow concept. Simply stated, corrosion occurs where conventional electrical current leaves the metal and enters the earth or an electrolyte such as salt water.

The so-called stray currents may be entering the metal from a grounding bond or may be induced currents as mentioned previously. Corrosion by stray currents take place only where the conventional current leaves the structure to enter the earth or electrolyte. However, local cell action may be taking place anywhere along the metal surface and this damage is in addition to the stray current effect.

One ampere of direct (dc) current flowing for one year from either a large or small area can cause loss of up to 9.1 kilograms (20.1 lb) of steel.

Metal loss due to alternating (ac) current discharge is generally only about 1% of this amount. However, the damage can be serious if the area of ac current discharge is small.

Stray current corrosion is usually evidenced by deep isolated pits with smooth sides resembling dimples in the metal. The pits are often aligned in a row along the underside of storage tanks or other vessels resting on the ground. Similar row patterns are occasionally seen along one side of the exterior of well casing where current flowing down the well leaves to reenter the earth. Additional information on casing corrosion will be given in Section 1, Chapter 8.

1.22 Concluding Comments

The preceding portions of this chapter provide the basic knowledge necessary to understand both the electrical and chemical nature of corrosion. From here on, the text will concentrate on practical control methods and will demonstrate how each is based on the fundamental principles just covered.

It should be understood that a large body of scientific information has been omitted and that which is given is purposely simplified for clarity and brevity. The science of corrosion is relatively young and growing rapidly. Modern theory recognizes that corroding systems are not at equilibrium and relatively simple instrumentation has been developed to monitor half-cell potential changes (polarization) during corrosion under variable conditions.

A condensed outline of some of the more important elements of corrosion theory is presented in Appendix 1A. Reading of this section is recommended for those desiring further insight on corrosion mechanisms and especially for persons attempting to keep abreast of current technical literature.

2 Fundamentals of Corrosion Prevention Methods

Introduction

In this section the general nature of the most popular techniques applied to slow down the corrosion process will be examined. The term "slow down" is used here because in practice it is almost impossible and usually not necessary to totally stop corrosion attack. The principal objective of corrosion engineering is simply to protect equipment during its useful life and to do this in the most economical way possible.

Following is a list of the most practical and time-proven methods used to combat corrosion in the oil industry:

Corrosion Inhibitors
Plastic and Cement Liners
External Pipe Coatings
Cathodic Protection
All Plastic Tubulars and Tanks
Special Alloys
Special Chemical Treatments and Solids Removal
Removal of Corrosive Gases
Reduction in Temperature and Velocity
Dehydration

Each of these methods will be dealt with in detail as the discussion progresses. At this point, the functional nature and how control techniques relate to corrosion processes will briefly be considered.

2.01 Surface Barriers

The metal surface reactions that occur during corrosion can be diminished if the metal is coated with something that: (a) increases the electrical resistance at the metal-electrolyte interface, (b) increases either anodic or cathodic polarization, or both (c) increases hydrogen overvoltage or (d) reduces diffusion of ions from the electrolyte to the metal.

Two basic types of metal surface barriers are used. These are:

1. Chemical inhibitors that coat the surface by adsorption from solution or dispersion in a liquid carrier.
2. Mechanically applied protective coatings.

The barriers on the metal can be visualized as follows:

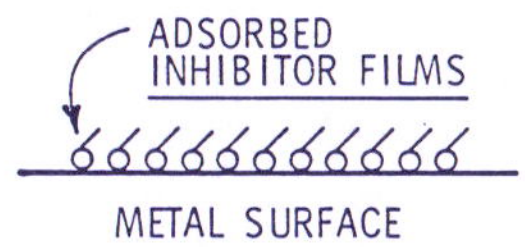

While there are similarities in the ways by which protective coatings and inhibitor films interfere with and slow corrosion, there are many differences in them. The coatings are typically organic polymers or metal films and are much thicker than the inhibitor films. Both generally separate the metal from the electrolyte and reduce electrical current flow. In general, they primarily affect the metal surface and do little or nothing to alter or change the water environment contacting the metal. Water altering chemicals are discussed under Item 2.03.

2.02 Removal of Corrosive Agents and Temperature-Velocity Reduction

It has been pointed out that most of the corrosion in the oil industry can be traced to the presence of O_2, H_2S or CO_2. If steps are taken to remove these gases, the corrosion they cause can be prevented. Methods for removing dissolved gases are covered in Section 1, Chapter 6.

The corrosivity of these gases in water solution is due to their chemical and electrochemical reactions with the metal. Like essentially all chemical reactions, the rate of reaction is increased by heat. Temperature reduction is, therefore, a viable means to slow corrosion.

The rate at which the corrosive gases in solution arrive at the metal surface also governs their ultimate rate of reaction. Thus, the velocity of the electrolyte relative to the metal is an important factor and can sometimes be employed for corrosion control. However, the effect is not simple since velocity at the surface where the reactions occur is not the same as the mass velocity.

In addition to affecting the rate of transport of corrosive agents to metal, velocity becomes important where solid particles are present. Impingement of solids in a moving stream scours the metal surface making it more reactive and may literally abrade away particles of metal.

2.03 Special Chemical Treatments and Solids Removal

Under this category are treatments to prevent bacterial growth, use of neutralizers to reduce acidity, scale inhibitors and processes to prevent or remove deposits of suspended solids. All of these are environment altering techniques.

Biocides retard growth and accumulation of microbial colonies that promote metal attack by differential aeration cells and generation of H_2S.

Sedimentation and filtration of water are beneficial in corrosion control programs to prevent deposits that promote local cell action.

Chemicals that inhibit deposition of scale solids can also help to keep surfaces clean and less susceptable to localized pitting.

Pigging or scraping of pipelines is a mechanical means to clean the metal and is often practiced in combination with the previously listed methods for control of bacteria, scale and solids deposits.

In some situations, neutralizing chemicals such as ammonia or soda ash can be employed to prevent corrosion by CO_2 (carbonic acid) or by spent acids from well stimulation operations.

2.04 Cathodic Protection

Cathodic protection is, as the name implies, a method by which metal equipment is protected by converting all of it into a cathode. In effect, cathodic protection makes steel behave more like a noble metal. This is accomplished by reversing or canceling the corrosion current by imposing a more powerful current from an outside source. More detail is given on this in Chapter 4 of Section 1.

2.05 Special Alloys

Steel is alloyed with other metals such as chromium and nickel to make it more noble or less susceptable to attack in specific environments. A protective film, probably a complex oxide layer, spontaneously forms on these alloys when oxygen is present.

Electrochemical measurements made under controlled laboratory conditions demonstrate the electrical properties (such as half-cell potentials vs. reference electrodes) of corrosion-resistant alloys exposed to acids or other agressive media. These polarization diagrams resemble in many respects the potential shift of ordinary steel under influence of anodic protection (see Appendix 1A). The comparison is made here simply to convey the concept that in both cases something is being done to alter the electrochemical properties of metal so that its tendency to corrode is reduced.

2.06 Non-Metallic Materials

Low pressure flowlines and tanks are often constructed entirely of plastic materials. Fiberglass reinforced polyesters are a popular type. PVC (polyvinyl chloride) plastic with clay or glass filler is widely used for low temperature water service as is ABS (acrylonitrile-butadiene-styrene). While these materials do not corrode electrochemically as steel does, they do deteriorate. The polyesters and epoxy plastics are degraded by ultraviolet light and should be protected from long exposure to sunlight for best service. Heat (temperatures above approximately 80°C or 170°F) is generally deleterious but more heat resistant plastics and laminates are being developed.

Ceramics and ceramic or glass coatings on steel are used in specific applications. Ceramic plungers are employed in pumps handling corrosive waters. More use of ceramics is anticipated in very corrosive and high-temperature environments.

2.07 Dehydration

Gas transported by pipelines is often dehydrated to prevent condensation of liquid water onto the metal surfaces. When no liquid water phase is present, dry H_2S, CO_2 or O_2 do not corrode steel at temperatures below about 200°C (390°F). Dehydration is the major defense against corrosion of gas transmission lines in CO_2 injection operations for enhanced oil recovery (EOR). (See Section 1, Item 9.01.)

3 Chemical Inhibitors

Introduction

Chemical inhibitors have been called the first line of defense against corrosion in the oil industry. Anyone working with corrosion control in oil producing operations should have a good understanding of the chemicals, how they are applied and what can be expected of them. Specific treating procedures for oilfield inhibitors are given in Section 1, Item 8.11. Additional chemical information and a reference list of several inhibitor chemicals are given in Appendix 2A.

The following discussion is provided to familiarize the reader with inhibitors in general, with emphasis on the types most used in production operations.

3.01 The Nature of Corrosion Inhibitors

A great deal of scientific study has been devoted to inhibitors. However, most of what is known has grown from trial and error experiments, both in the laboratories and field. No set of rules, equations, or theories has been developed to completely guide inhibitor development or use. The technology of corrosion inhibition is, therefore, largely an art.

Only a few generalizations can be made about the materials and how they work. However, these general concepts which follow, are helpful in dispelling some of their mystery and to aid an understanding of them.

1. Inhibitors are chemicals that, directly or indirectly, coat or film a metal surface to protect it from its environment.
2. Most inhibitors are adsorbed by the metal surface from a solution or dispersion but some are applied directly as a coating.
3. Inhibitors are generally either:
   a. reactive (chemically react with the metal or environment), or
   b. adsorptive film formers.
4. The protective films slow the corrosion process by:
   a. increasing either anodic or cathodic polarization, or both;
   b. reducing diffusion of ions to the metal surface;
   c. increasing electrical resistance at the metal-electrolyte interface;
   d. increasing hydrogen overvoltage (see Section 1, Item 1.15).

In theoretical discussions of inhibitor behavior, the compounds are often classified as "anodic" inhibitors or "cathodic" inhibitors. This distinction is based on laboratory tests that measure anode and cathode polarization effects under carefully controlled conditions (see Appendix, Item 1A.8).

The anodic and cathodic inhibitor designations will not be used here. Instead, they will be discussed in terms of their composition and behavior as this is more meaningful in field operations.

3.02 Inhibitor Classifications

To learn about the nature and functional behavior of inhibitor chemicals, it is helpful to group them as:

1. Inorganic,
2. Organic,
3. Anionic,
4. Cationic.

Inorganic compounds are derived from mineral sources and usually do not contain carbon in their structures. Organic inhibitors contain carbon to carbon (−C−C−) groups in their molecules and are usually derived from vegetable or animal sources.

1. **Inorganic** inhibitors are usually crystalline salts such as sodium chromate, sodium silicate, sodium phosphate or sodium molybdate. In water solution, these salts separate to form the positively charged sodium cations (Na^+) and negatively charged anions, viz., chromate ($CrO_4^=$), silicate ($SiO_3^=$), phosphate ($PO_4^=$), and molybdate

($MoO_4^{=}$). Only the negative anions of these compounds are involved in reducing metal corrosion. Therefore, these products are classed as inorganic anionic inhibitors.

When zinc is used instead of sodium, the zinc cation (Zn^{++}) is beneficial and the compounds may be called inorganic mixed-charge inhibitors. The primary application for inorganic anionic and mixed-charge inhibitors is in cooling waters.

2. **Organic anionic** inhibitors such as sodium sulfonates and sodium MBT are used in cooling water and antifreeze solutions. Sodium phosphonates fall into this class and are useful in cooling water inhibitor preparations to aid in both corrosion control and prevention of mineral scale precipitation.

3. **Organic cationic** inhibitor compounds in their concentrated forms, are typically oily liquids or wax-like solids. Their active portions are generally large carbon-hydrogen (hydrocarbon) chains or rings with positively charged amine nitrogen groups on one end. The amine structures can be represented as follows:

$R \cdot NH_2$	$R_2 \cdot NH$	$R_3 \cdot N$	$[R_4N]^+$
Primary	Secondary	Tertiary	Quaternary

where R represents the hydrocarbon portions of the molecules. In water or acids the terminal primary, secondary and tertiary amine groups take on an additional hydrogen (H^+) that gives them a net positive or cationic charge. Quite commonly there are two amine nitrogens present, one primary and one secondary, located at the end of the long R Group:

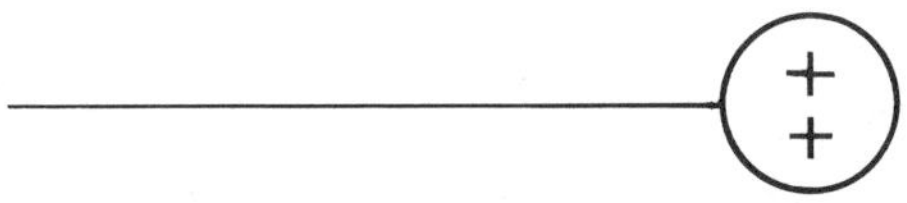

The molecules are unbalanced or dissymetrical with one portion being oil soluble or hydrophobic and the other being polar or hydrophilic (water loving). Since only part of the inhibitor molecules are hydrophilic they are said to be "semi-polar".

The polar amine groups adsorb to metal and the hydrocarbon portion forms an oily water repellant surface film. Because of their molecular shape (dissymetry), these materials act as surfactants and can stabilize emulsions of oil and water.

The long chain amines just described are the active constituent in most oil field inhibitor compositions (that will be discussed later in this chapter). "Inhibitors" are the multi-component mixtures supplied by the chemical service companies in ready to use form.

3.03 Effect of Ionic Character on Inhibiting Ability

With the information just given on the chemical nature of inhibitors, some additional generalizations may be drawn:

1. Most inhibitors used for control of corrosion caused by dissolved oxygen (O_2) are anionic (negatively charged) ions, either organic or inorganic. They are relatively ineffective against H_2S and CO_2.
2. Most products that are effective for controlling corrosion by hydrogen sulfide (H_2S) and carbon dioxide (CO_2) are cationic (positively charged) organic substances, typically long chain amines.

The amine groups adsorb strongly onto iron sulfide coated metal surfaces and amine inhibitors actually work best where H_2S is present and O_2 is absent. Amines, especially the volatile types, chemically react with H_2S and the complex that is formed helps to build the protective film.

3.04 Characteristics of Oily Amine Inhibitor Films

Most of the non-volatile amine-base oilfield inhibitors are preferentially oil soluble, viz., they partition between oil and water phases, typically about 9 to 1. This is true even if they are water dispersible. Their oily portions are solvated by oil molecules and some of the oil stays with the amines even after adsorption onto steel. Thus, the films they form have an oily, water-repellent character.

Classic adsorption theory dictates that the amines adsorb, or blanket, metal surfaces initially as a film that is only one molecule thick. This is only about 1×10^{-7} mm. However, the films with occluded oil build to such thicknesses that they become visible and easily detected.

The inhibitor films are so tough and oily that they lubricate rubbing metal parts such as sucker rods sliding on tubing walls.

Loss or dissipation of the adsorbed film is primarily a function of time, temperature and the concentration of inhibitor in fluids contacting the metal surface. When concentration in the produced fluids declines between intermittent applications, the film tends to desorb and go back into solution. Time required for film loss can be as short as a few hours at 160°C (320°F) or higher to as long as one or two weeks at 50°C (120°F) in producing oil wells. Film life in gas wells can be six months or longer where liquid production is low.

A general misconception exists in the industry as to the effects of fluid velocity on film life of oily amine inhibitors. It is generally known that *untreated corrosion rates*, without inhibitor, tend to increase with increasing mass flow velocity through tubulars. Consequently it is commonly assumed that when some "critical" velocity is exceeded, the inhibitor film will be swept away. However, research has shown that in flowing gas streams with entrained water, inhibitor films are not disrupted at mass velocities up to 100 ft (30 m)/sec. Curve "A" of Figure 1.3-1 illustrates the increase in *untreated* corrosion rates at high velocity in a wet sour gas. No loss of inhibitor film occurred in these tests at flow velocities up to 60 ft (18.3 m)/sec.

It should be noted, however, that abrasion by high velocity solid particles can tear away the films. This is often evidenced by corrosion-erosion on tube turns or elbows where solid particles impinge (Section 1, Item 7.12).

Concerning temperature effects, it has been demonstrated that the imidazoline compounds are generally more thermally stable than the mixed charge amine-acid salts and perform better at high (>125°C or 260°F) temperatures (see Appendix 2A).

3.05 Water Soluble and Volatile Amine Inhibitors

Low molecular weight alkyl (straight chain) and some cyclic amines are used as inhibitors in the absence of oxygen. Their volatility makes them useful for treatment of gas streams and vapor spaces in well casings. The films they form are not so tenacious as that of the oily, high molecular weight amines discussed previously. Consequently they must be replenished more often or even continuously.

In general, the lower alkyl amines such as diethylamine, $(C_2H_5)_2NH$, are more effective than the volatile cyclic amines for control of H_2S corrosion. The alkyl amines are alkaline, but due to complexing with H_2S and film formation, the amount of amine required for inhibition is less than that required to neutralize all the acid gases (H_2S and CO_2) present.

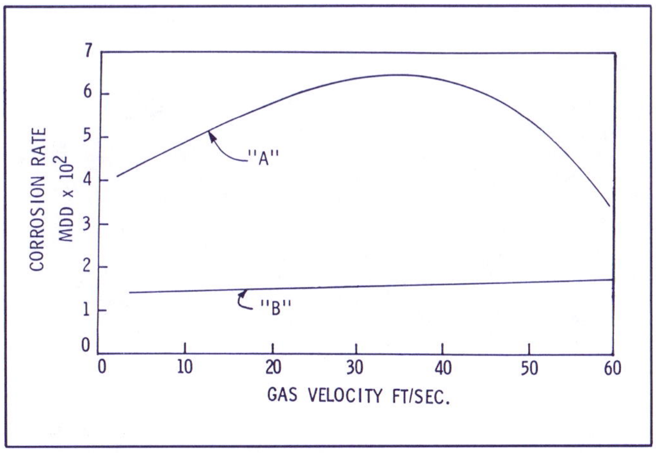

Figure 1.3-1. Corrosion of carbon steel in a flowing sour gas (10 vol % H_2S/CO_2) stream with entrained water. Temperature: 107°C, pressure: 1500 psig. Curve A: No inhibitor. Curve B: Metal filmed prior to each 24 test with 10% oil solution of imidazoline inhibitor.

Volatile cyclic amines such as morpholine and cyclohexylamine are used to inhibit CO_2 corrosion in steam condensate systems. They perform primarily as acid neutralizers. High molecular weight filming amines are also used for this service.

Certain non-volatile water soluble amines are true film formers (by adsorption). Most of these are the quaternary ammonium chlorides:

$$[R_4N]^+ \cdot [Cl]^-.$$

Due to their water solubility, these chemicals generally have short film life. They are often used as biocides. Some high molecular weight water soluble quaternaries are applied by continuous treatment for corrosion control in high temperature wells (Section 1, Item 8.11d).

3.06 Formulations of Oil Field Inhibitors

As stated, most of the inhibitor compositions prepared for the oil producing industry are based on the high molecular weight amines, typically the imidazolines or amine/acid salts (Appendix 2A).

To prepare field strength formulations, the amine base is commonly diluted with an aromatic solvent and alcohol is added to the lower pour point. The active amine base content is usually 20 to 40% by weight. The final mixture is usually oil soluble.

By adding emulsifiers, the oil soluble mixture is also given the ability to disperse temporarily into water. This helps to distribute the inhibitor into both the oil and water phases of producing wells.

In addition to emulsifiers added for water dispersibility, demulsifier compounds are used to help prevent stable emulsion formation. The additives for preventing creation of stable curdy emulsions are important for inhibitor effectiveness and also to avoid production upsets.

Many inhibitor blenders go a step further and prepare three phase treating compositions by blending (1) the oil-soluble amine base, (2) emulsifiers for water dispersibility and (3) a volatile low molecular weight amine for gas phase treatment (these are called dual phase inhibitors).

Water-soluble inhibitors such as the quaternary ammonium chlorides are generally supplied in an alcohol-water solution. These are often sold as combination corrosion inhibitors and biocides. The water-soluble quaternaries generally do not stablize emulsions and do not require demulsifier additives.

Combinations of corrosion and mineral scale inhibitors have been formulated for oil production use where both problems coexist. These are generally not highly effective and it is usually best to apply separate treatments where corrosion and scaling problems are severe.

3.07 General Principles of Inhibitor Application

The following general concepts and guidelines are helpful in planning all types of inhibitor treatments. (See Section 1, Item 8.11 for details on oilfield inhibitor applications.)

1. Determine the chemical nature and solubility characteristics of the inhibitors being considered. Do not think of the products only in terms of their trade names or numbers. Each inhibitor supplier should be required to provide data showing that his product is suitable for the systems to be treated.
2. The inhibitor must be able to contact the metal to be protected to be effective. This means the product must be soluble or dispersible in the fluid contacting the metal and must be applied in such a way that it can reach all the surfaces.
3. The concentration must be correct for the use conditions. This is determined most accurately by field trials.
4. Always initiate treatment with a high starting concentration that conditions the metal surfaces and then reduce to the required continuous treating level or frequency of batch application. This general practice is recommended for all inhibitors, whether water or oil soluble.

Intermittent or batch-type treatment with filming amine inhibitors is made possible by the persistency of the adsorbed films. An important point here is that the long lasting films must be laid down initially from a solution or dispersion that contains a high concentration of inhibitor. Some literature sources indicate that about 1000 ppm (or mg/L) of inhibitor is adequate for good film deposition. However, 3000 to 5000 ppm (0.3 to 0.5 volume percent) is recommended as the minimum batch treatment concentration. No increase in film life is obtained by using batch concentrations greater than about 10,000 ppm (1 vol %).

Once formed, the adsorbed film remains protective for a period of time depending on conditions, as discussed under Item 3.04. The film can be kept intact and protective if the fluids passing over it contains a small continuous concentration of inhibitor. Typically, 5 to 30 ppm continuous inhibitor concentration is all that is required for film maintenance. However, the initial film formation will not be achieved if only the low (5 to 30 ppm) concentrations are used without benefit of the initial slug or batch treatment. Several frequent batch treatments initially can provide even better establishment of the film and this should be done whenever economics permit.

Figure 1.3-2 illustrates the effectiveness of consistent intermittant batch treatments in a rod-pumped sour oil well. The upper rod was used for only three months without treatment. The lower rod was used for the same

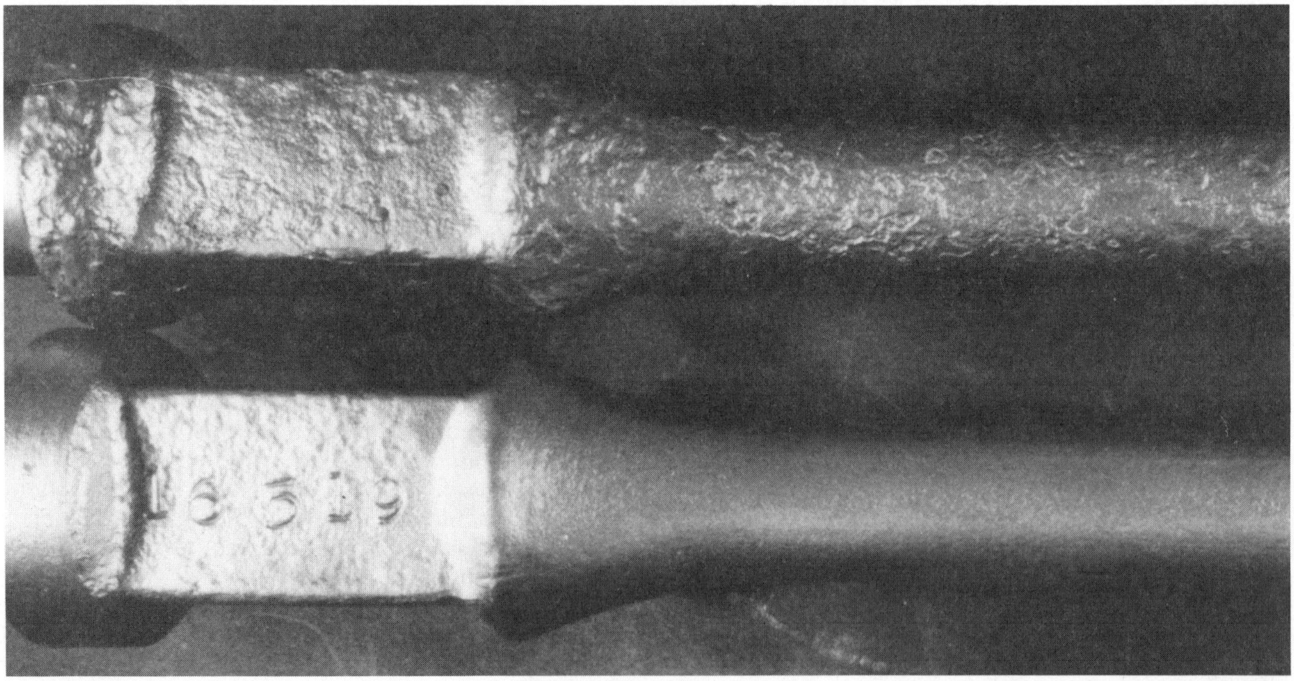

Figure 1.3-2. Sucker rods after 3 months use in treated and uninhibited sour oil wells. Upper: untreated. Lower: treated once each week with filming amine inhibitor.

period of time in a nearby well being treated once each week with 7.5 liters (2 gal) of a water dispersible amine base inhibitor.

3.08 Acid Inhibitors

The products used by service companies to inhibit stimulation and well cleanout acids are complex mixtures. Typically they contain high boiling pyridines and other amines together with acetylenic alcohols such as propargyl alcohol and hexynol. The latter are characterized by presence of the acetylenic linkage, $-C{\equiv}C-$. The acetylenic alcohol constituents apparently polymerize on metal surfaces to form a temporary protective film.

At well temperatures up to about 80°C (175°F) the acid inhibitors generally provide adequate protection for exposure times of 12 to 24 hours. At higher temperatures the maximum allowable exposure time to avoid serious damage drops rapidly and may be as short as four hours at 150°C (300°F).

Acid treatments should be carefully planned in cooperation with the service company supplying the inhibited acid. The acid should be thoroughly mixed just prior to injection to insure that the inhibitor is uniformly blended and has not separated. Some producers take samples for visual comparison at the start, middle and end of the batch injection.

Service companies generally recommend the amount of inhibitor to use (typically 0.5 to 3 volume percent) and the length of time the acid may be safely left in the well. Every reasonable effort should be made to unload the acid within the specified time and to produce the well long enough to flush out the spent acid. So-called spent acid is extremely corrosive. Serious corrosion of downhole equipment results when the acid-treated well is shut in before the pH of the produced water returns to near normal (pH 5 or higher).

4 Cathodic Protection

4.01 Theory and Principles

The fundamental and simple basis of cathodic protection can be readily explained by referring back to Figure 1.1-3. The metal *receiving* electrons is the *non-corroding cathode* in a corrosion cell. By providing an outside source of electrical energy that constantly supplies an adequate flow of electrons, a whole metal structure can be made into a cathode and is thus cathodically protected.

The quantity of electron flow or the amount of electric current required to maintain a structure (such as a buried pipeline or immersed portions of a marine platform) in the protected cathodic condition depends primarily on its environment. The current demand is a function of the rate of cathodic depolarization that, in turn, is governed by the kinetics of the electrochemical reactions occurring at the cathode-electrolyte interface (see item 1.06, Cathodic Reactions, Section 1).

In oil-producing operations, some common environmental conditions where corrosion is controlled by cathodic protection (abbreviated c.p.) are:

1. external exposure to soils with varying amounts of moisture, salts and dissolved oxygen,
2. external exposure to flowing, aerated seawater, and
3. interior of tanks exposed to water containing dissolved salts and O_2, H_2S and CO_2.

In all these situations, hydrogen ions (H^+) are available for the cathodic reduction reaction. Where oxygen is available to react with and remove the atomic hydrogen (H^o) formed at the cathode surface, the current demand is proportional to the oxygen supply. If the hydrogen ion concentration is high due to acid conditions, the electron flow required to maintain cathodic polarization will be high.

Ideally, cathodic protection eliminates current flow between parts of a structure by eliminating the potential differences that previously existed between the parts.

4.02 Current Requirements

The unit of measure of the current demand is the current density or amperes (A) per unit area of metal, i.e., amperes/square foot or per square meter of surface. Total current requirement is roughly equivalent to the sum of all the anode to cathode currents originally flowing in the metal.

Since corrosion currents are generally relatively low, the c.p. current density is usually expressed as milliamperes (mA)/ft^2 or m^2. Typical current density requirements for bare steel in various environments are listed in Table 1.4-1.

Variations in current density requirements between soft and hard waters and the influence of water velocity as indicated in Table 1.4-1 are effects that have to do with metal surface films. In low-velocity hard waters containing soluble calcium, c.p. causes a scale film to form on the protected metal. This is due primarily to precipitation of calcium by the hydroxide formed by the cathodic reaction:

$$\text{(a) } H_2O + {}^1/_2O_2 + 2e \rightarrow 2OH^-,$$

followed by:

$$\text{(b) } Ca^{++} + HCO_3^- + OH^- \rightarrow CaCO_3 + H_2O.$$

These reactions are important in the practical application of cathodic protection since the calcareous (calcium containing) scale coats over and effectively reduces the area of bare metal and thus the current required. When water velocity is high, the protective scale is not easily formed.

Another effect of high water velocity is rapid transport of dissolved oxygen to the metal. In gently flowing sea-

Table 1.4-1
Typical Current Density Requirements

Environment	mA/ft^2	mA/m^2
Moist soil	2–3	21–33
Soft Fresh Water(1)		
@ < 1 m/sec velocity	4–6	43–65
Hard Fresh Water(2)		
@ < 1 m/sec velocity	2–4	22–43
Hard Fresh Water		
@ > 5 m/sec velocity	10–12	107–129
Seawater or Brines		
@ < 1 m/sec velocity	8–10	86–107
Seawater or Brines		
@ > 5 m/sec velocity	15–80	161–860

(1)< 40 mg/liter Ca^{++}
(2)> 40 mg/liter Ca^{++}

water in locations such as the Gulf of Mexico, offshore structures can generally be protected with 8 to 12 mA/ft^2 (86 to 129 mA/m^2). In the swifter currents of the North Sea, current densities of about 15 mA/ft^2 (160 mA/m^2) or more are necessary for complete protection.[1]

These values are based on experience with actual systems and represent the overall current densities that have been applied. On complex structures such as drilling-production platforms in deep seawater, minimum current required for protection can vary at different locations on the structure and with time. Welded brace junctures near the water surface actually need more current than straight sections of bracing at lower depths. This results from the higher oxygen content of the upper water and the greater tendency of the welded joints to corrode.

The time factor mentioned is largely a function of the rate at which the calcareous scale accumulates. If initial current density is too low, the scale may never form and the structure may not "polarize". Application of very high currents, viz., 100 mA/ft^2 (1100 mA/m^2) for a few days causes rapid scale buildup, which may be maintained by 1/3 the current density that is traditionally applied to bare steel.

These variables emphasize the complexity of practical application of the basically simple scheme of cathodic protection. Others will become apparent as we progress through the technology.

4.03 Current Sources and Concepts of Current Flow

Referring again to Figure 1.1-3, note that in a natural or spontaneous corrosion cell, electrons flow from anode to cathode through the metal circuit. Conventional electric current is designated as flowing in the opposite direction from anode to cathode through the connecting electrolyte or soil. Only the conventional electric current flow path is considered in practical c.p. work, and the surface chemical reactions involving electrons are neglected.

In a natural corrosion cell, electrons are released as the anode metal dissolves and forms positively-charged (electron deficient) ions. The metal atoms that go into solution and lose electrons are said to be electrochemically oxidized. Therefore, the anode is designated as the site where electrochemical oxidation (electron-loss) reactions predominate. At the cathode surfaces exposed to the electrolyte, electrochemical reduction (electron-consuming) reactions occur. Equation (a) (Item 4.02) is a common cathodic reaction where dissolved oxygen is electrochemically reduced to form hydroxyl ($-OH$) ions. Generation of molecular hydrogen, H_2 from hydrogen ions, H^+, is another cathodic reduction reaction.

The same concepts of electron and electric current flow are valid for cathodic protection systems whereby a whole structure is converted to a large cathode by virtue of being connected to one or more separate anodes through a metal conductor. Regardless of the source of electrical energy used to apply c.p., electrochemical oxidation (electron-releasing) reactions occur at the anode surfaces and electrochemical reduction (electron-consuming) reactions occur at the cathode surfaces. Again, electric current is considered to be flowing from anode to cathode through the electrolyte (or soil) path. Therefore, *a continuous electrolyte path must always connect between anode and cathode for c.p. to work*. This means that c.p. applied to the outside of a pipe or tank will not protect the inside surfaces, and vice versa.

Two sources of c.p. current are used:

1. Sacrificial or galvanic anodes made of zinc, magnesium or aluminum, and
2. Electric generators—usually alternating (a.c.) power converted to direct (d.c.) current by means of rectifiers.

When sacrificial anodes are connected through a conductor wire to steel immersed in a connecting electrolyte, the steel spontaneously becomes the cathode in the dissimilar metal couple (Section 1, Item 1.11). The commonly used galvanic anode metals (zinc, magnesium or aluminum) naturally have higher solution potentials than iron or steel (see Table 1.1-1). For example, the standard oxidation potential of magnesium in the electromotive series is -2.37 volts vs. -0.44 volts for iron. When first coupled, there is an effective driving force, ΔE, between magnesium and steel of about 1.5 volts vs. a copper/copper sulfate reference to move current (depending on anode composition, temperature, etc.).

Figure 1.4-1 illustrates a sacrificial anode installed on a pipeline. By the conventional concept of current flow, current leaves the anode and travels through the soil electrolyte to the pipe, thus protecting it.

Another concept is that the point of current discharge to the electrolyte (soil) is transferred from the pipe surface to the anode surface. Current is thought of as being drained from the pipe through the connecting wire to the anode where it is discharged to the earth. Thus, c.p. an-

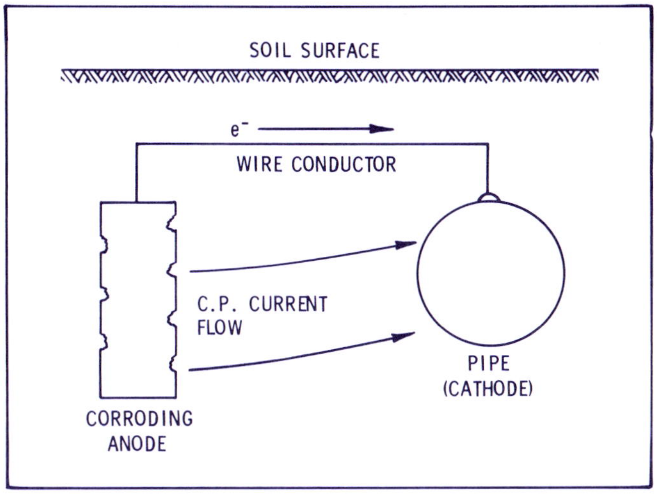

Figure 1.4-1. Sacrificial anode on a buried pipeline.

ode installations are sometimes referred to as *current drainage systems*.

When electrical generators or rectifiers are used, the c.p. installation is called an impressed current system. In this case, anodic oxidation reactions that supply electrons are forced to take place at the anode surfaces. For example, certain ions in the electrolyte that contain oxygen in their structure undergo electron loss (oxidation) and molecular oxygen (O_2) is generated at the anode surfaces. Several other ionic species can be oxidized depending on the water (electrolyte) composition and the operating voltage. However, in contrast to sacrificial anodes, the material of the impressed current anodes may not be corroded. In fact, the anodes in impressed current systems are specially selected for long life.

Common impressed current anode materials for use in soil are graphite (carbon) or cast iron with a high silicon content. Metal anodes that are externally coated with a thin layer of platinum are used in either soil or salt water. All these are capable of discharging large amounts of electrical current with little loss of anode material itself. As explained previously, the primary source of electrons is from electrochemical oxidation of ions in the surrounding electrolyte.

4.04 Sacrificial or Galvanic Anodes

The amount of c.p. current that can be supplied to (or drained from) a structure by sacrificial anodes is limited by:

1. low driving voltages,
2. electrical resistance of soil or electrolyte, and
3. current wasted by local cell corrosion on the anodes due to impurities.

Due to local cell corrosion, magnesium anodes are only about 50% efficient. That is, about 50% of the current generated by solution of the magnesium goes to support corrosion on the anode itself.

Special alloys and anode-casting techniques are used to increase efficiency of commercial magnesium and aluminum anodes. Iron and copper impurities are kept as low as possible. Zinc and mercury or indium (In) are alloyed with aluminum for increased efficiency. Aluminum anodes do not function well in soils and their primary use is in flowing seawater.

Typical current output capabilities follow for anode metals under loadings of 200-700 mA/ft^2 (based on anode area).

	Effective Ampere/Hours	
Anode Metal Composition	*per lb*	*per kg*
Al-Zn-Hg alloy	1270	2800
Mg(H-1 alloy)	500	1100
Zinc	370	820

Aluminum (Zn−Hg or Zn−In alloy) anodes are commonly used on offshore marine structures due to high current ouput per unit weight. The Al−Zn−In alloy is growing in popularity.

Proper anode distribution is critical so that all portions of the structure receive adequate current. The number and sizes (total weight) of anodes are selected for a given life span (10-20 years) based on design current density and exposed metal area. (See Section 1, Item 8.17).

Magnesium anodes are most often used for pipeline protection due to their higher driving potential (1.55 volts vs. Cu/$CuSO_4$ compared to 1.1 volts for zinc and aluminum). Installations are usually designed for a five to ten year life on coated pipe. Anode sizes vary from 8 to 23 kg (17 to 50 lb) each. Single or multiple anodes are placed at anodic areas located by line potential surveys and in low resistivity soils. Special mixtures of gypsum and clay are used as backfill around galvanic anodes to reduce resistance to current discharge. Prepackaged single anodes in sacks of backfill material are available.

Full length protection on well coated pipelines is achieved by groups of magnesium anodes spaced along the line for proper current distribution. Alternately, uniform current distribution is obtained with a magnesium ribbon anode that is laid parallel with the pipe and connected at intervals. The ribbon type anode is suitable only for near perfectly coated pipelines where total current requirement is low.

Zinc is generally effective only in salt waters and soils of no more than about 1000 ohm-centimeter resistivity (see Item 4.14, this chapter). Zinc bracelet anodes are used for long life on submerged marine lines that are coated with cement-asphalt mastics or other thick coatings.

4.05 Impressed Current Systems

Figure 1.4-2 illustrates a rectifier-driven anode bed supplying c.p. current to a section of a buried pipeline.

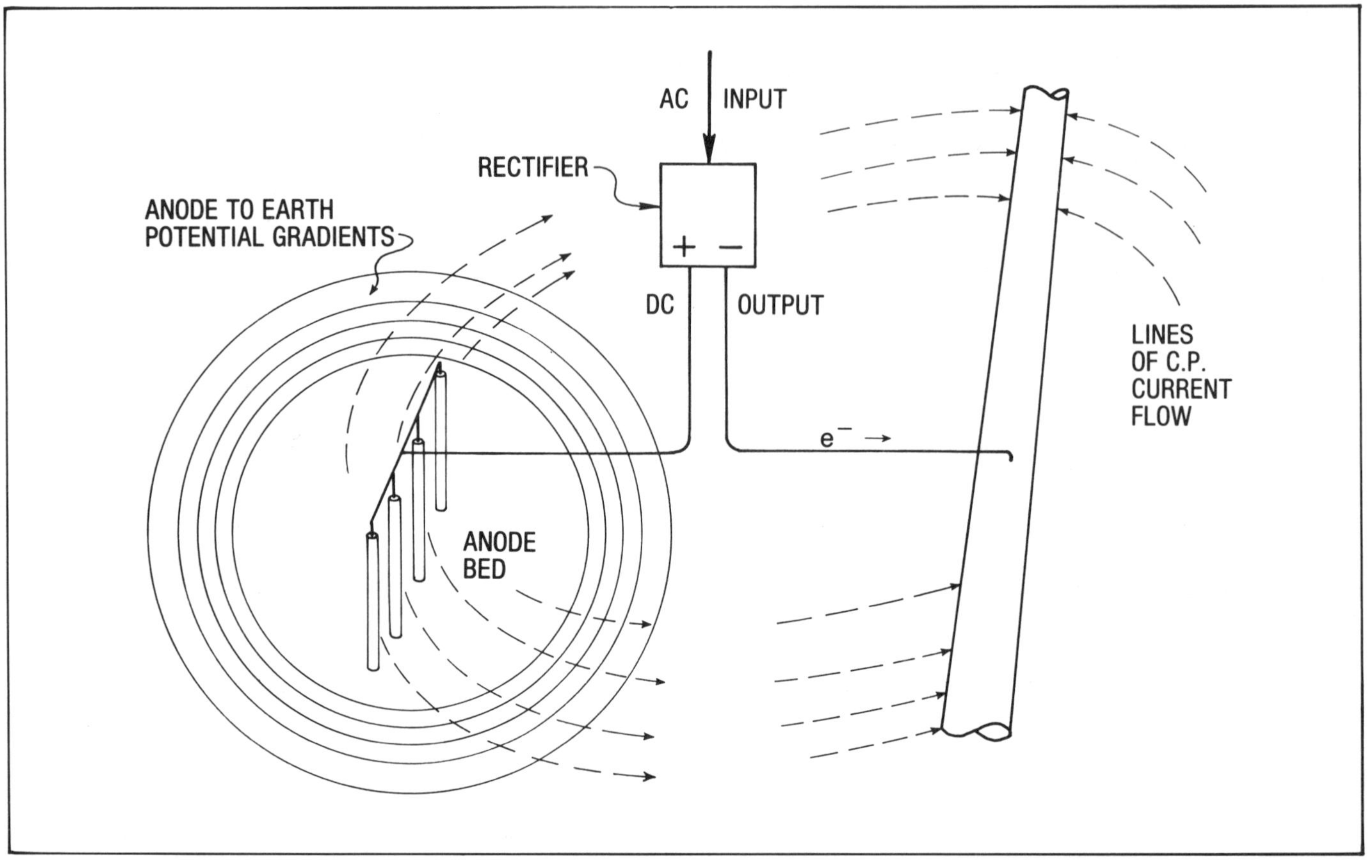

Figure 1.4-2. Impressed current c.p. system on a pipeline.

The individual anodes for soil service are typically cylindrical rods made of graphite or high silicon cast iron and are usually mounted vertically in groups (preferably in a straight row) alongside the pipeline. The bore holes for vertical rods are three to four times larger than the rod diameter and the annular space is filled with uniformly packed coke breeze or graphite powder to minimize anode to soil resistance. Cast iron anodes can be used without backfill in low resistivity soils.

If underlying rock prevents vertical burial, the rods can be laid horizontally at the depth of the pipe. In congested areas, single impressed current anode rods may be distributed along the structure to be protected instead of in bed clusters. Single deep vertical bore holes accomodating one or more impressed current anodes may also be used.[2]

A wide selection of rectifiers is available, including air-cooled and oil-immersion types. The range of voltage and current (amperage) output is nearly unlimited for various size installations. Automatic rectifiers are available that vary the current output as required to maintain a preset potential on the protected structure.

The circuitry of a single phase, full-wave bridge rectifier is shown schematically in Figure 1.4-3. Note that the positive (+) terminal of the d.c. output connects to the anodes and *the negative (−) terminal connects to the structure to be protected*. The + and − designations on the output terminals are for electrical reference only and should not be confused with positive and negative signs for electrochemical half-cell reactions. Special care must be taken to insure that the electrical polarity is correct (as illustrated in Fig. 1.4-2), otherwise corrosion of the structure will be accelerated.

In addition to rectified utility power, other direct current power sources may be used in remote locations. These include:

1. engine driven generators—either d.c. or a.c. with a rectifier,
2. thermoelectric generators using fuel-heated thermocouples to create electric current,
3. solar panel electric generators, and
4. wind generators with storage batteries.

The thermoelectric type is particularly convenient and effective for long gas transmission lines where the fuel is taken from the line being protected. A 10 volt, 10 ampere thermogenerator feeding to a ground bed with

1.0 ohm resistance will use 3000 to 3500 m^3 (106,000 to 123,600 ft^3) of natural gas/year.

Solar or photovoltaic power is being used for cathodic protection of remote pipelines and also for well casings.[3] Automatic types that track the sun are available.

Impressed current c.p. systems are used on some offshore drilling/production platforms, especially where current requirements are very high and the total weight of galvanic (aluminum) anodes is detrimental. Direct current is supplied by electric generators. Anodes are usually mounted on the platform legs and cross members at positions designed for good current distribution, especially to brace junctions where stress is highest. Platinum-coated titanium and lead with 6% silver are widely used anode materials for marine service.

4.06 Effect of Anode Bed Distance from Structure

Buried impressed current anode beds on pipelines or other large structures are usually located at an *electrically remote* distance away from the structure to facilitate current distribution. In Figure 1.4-2, the circles around the anode bed represent voltage or potential gradients in the soil. The anode-to-soil potential is highest close to the bed and diminishes with each increment of distance away. The potential change is inversely proportional to the volume of earth through which the current discharge must flow. For example, if the anode-to-soil potential is measured with a voltmeter and reference cell at a series of short intervals moving away from the bed, it will be found that the difference in potential readings between each interval becomes increasingly smaller with each measurement. Ultimately, a point is reached where the potential readings remain practically constant and show no further significant change. The protected structure also has a zone of potential gradients around it out to a constant potential point. The anode bed is considered electrically remote if it is far enough away that the zones of potential gradients around it and around the structure do not overlap.

The radius of the potential gradients is determined by the amount of current flowing and the resistivity of the soil. Single anodes discharge less current than groups of anodes and can therefore be placed closer to the structure and still remain electrically remote. Close positioning of distributed impressed current anodes is sometimes used in congested areas to avoid the problem of interference on adjacent unprotected structures (to be discussed later).

Close anodes protect less total structure area per anode because the current does not have sufficient distance through which to spread and provide coverage. Conversely, close-in anodes provide higher current density (amperes per unit area) with lower driving voltage since the electrolyte path is shorter and offers lower resistance.

On coated pipelines, close impressed current anode beds can cause excessively high potentials and coating disbondment on portions of the pipe that are nearest the anodes (see Section 1, Item 5.08).

Development of excessively high potentials does not occur when close galvanic anodes are used to protect bare uncoated steel. On offshore platforms, galvanic anodes (generally aluminum alloy) are fastened directly to the legs and cross braces with usually no more than 30 cm (12 in.) separation. The anodes are strategically positioned around the structure for uniform current distribution.

4.07 Electrical Potentials and Protection Criteria

Engineering design of c.p. installations requires knowledge of the current density required to achieve the desired state of protection. As previously indicated, the protected state is reached when potential differences (and thus current flow) between original anodic and cathodic

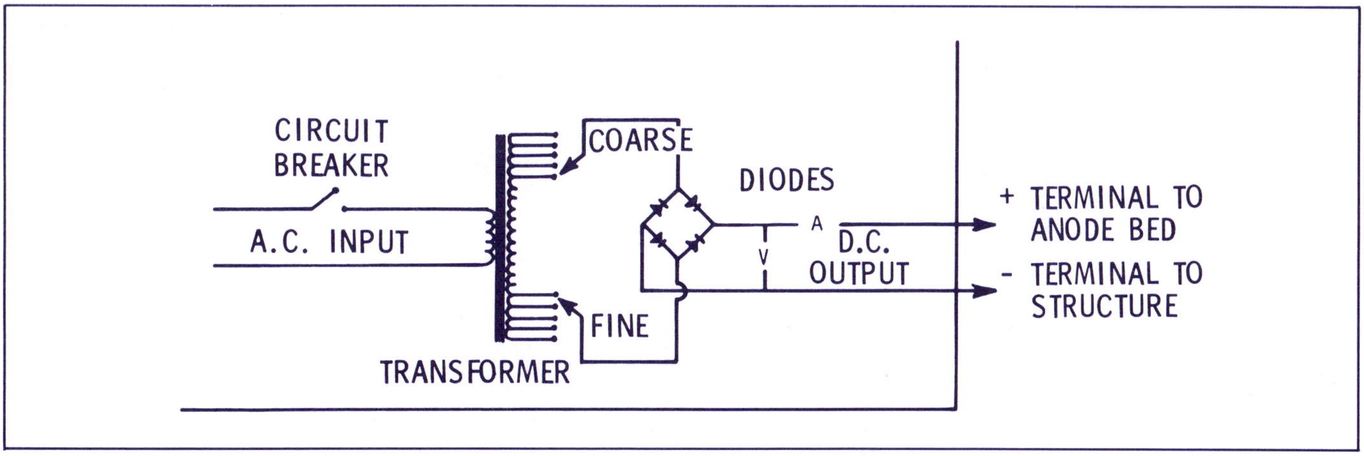

Figure 1.4-3. Circuitry of a rectifier.

sites are eliminated. Therefore, structure potential is normally used as the preferred standard or criterion of protection rather than relying on the calculated current density being applied. The problem has been and still is a matter of determining just what level of potential is acceptable for most environments. The basis for the three most widely used potential criteria is provided in the following discussion.

Anodic corroding areas on steel buried in moist soil or immersed in aerated water exhibit measured potentials of about −0.80 volts vs. a copper/copper sulfate (Cu/$CuSO_4$) reference electrode and about −0.73 volts vs. silver/silver chloride (Ag/AgCl). The cathodic areas have less negative (more positive) potentials. The ΔE between anodes and cathodes is most often controlled by the rate of oxygen arrival at the cathodic areas, viz., rate of cathodic depolarization (see Section 1, Items 1.06 and 1.16).

When a buried or immersed steel structure is connected to an external anode with a higher negative potential than that of the structure anodes or to a source of negative current (such as the negative pole of a rectifier), the cathodic areas on the steel tend to polarize to the potential of the anode sites. This requires sufficient current, supplied from the external source, to accomodate the cathodic surface reactions (see Section 1, Item 1.06) and no additional current is drawn from local anodes. The previously existing anode/cathode potential differences are now erased and the structure is protected. The whole structure is now at a relatively uniform potential but this potential is always less negative (or more positive) than that of the external anode supplying the current.

In practice, the protected structure potential is taken as −0.85 volts vs. Cu/$CuSO_4$ or −0.80 volts vs. Ag/AgCl to allow for error in the readings. These are the minimum measured potentials to assure adequate total protection and may be higher at some points. Truly uniform potentials are not usually obtained but the previous stated potentials are the most widely used standard criteria for determining protection.[4]

Alternate criteria for cathodic protection (in lieu of the previous) are acceptable.[4,5] One is a minimum negative (cathodic) voltage shift of 0.3 volts (300 millivolts) at all points on the structure after application of current. Another is measured by temporarily interrupting the protective current (rectifier turned off) and testing the structure potential over the next few minutes. When c.p. current is stopped the structure potential at the test point will immediately drop (become less negative) and then slowly drift down toward the initial unprotected potential. If the *off* potential (measured after the first rapid drop) is 100 mV more negative than the original unprotected potential at each test point, the structure is considered to be protected with the current on. This does not require that all potentials be above −0.85 vs. Cu/$CuSO_4$ with the current on. Recent studies[6] indicate the 100 mV polarization (off potential) criterion is probably the most generally valid indicator of protection in a variety of soils as commonly encountered by long pipelines. Programable automatic current interrupters are available for conducting "off potential" or "instant off" tests.

4.08 Measurement of Structure Potentials

Relative potential values around and along buried or immersed structures must be measured to determine:

1. if protection is needed,
2. where the anodes should be placed,
3. how much current must be supplied, and
4. whether or not protection has been achieved.

Potentials at various points are measured in comparison to a reference electrode which is simply a stable half-cell. The best reference electrodes for field use are physically sturdy and resist polarization (changes in potential) and are thus suitable as standards.

The copper/copper sulfate (Cu/$CuSO_4$) electrode or half-cell is most popular in c.p. work in soils, but the silver/silver chloride (Ag/AgCl) is most often used in seawater. Measured potential values must always state the reference such as X volts vs. Cu/$CuSO_4$, Ag/AgCl, saturated calomel, etc.

A Cu/$CuSO_4$ half-cell is simply a bright copper rod or tube immersed in a saturated solution of copper sulfate. The metal and solution are usually contained in a plastic housing that has a wire lead connection on top and a porous plug (for ion conduction) on the bottom.

Figure 1.4-4 depicts a set up for measuring potential at a point on a pipeline with a Cu/$CuSO_4$ reference electrode pressed into the soil directly above the line. Such values are referred to as *pipe-to-soil* (P/S) potentials. Differences in the potential readings at various points along the line are referred to as IR drops. This is acceptable since in Ohms Law:

$$\underset{\text{(potential, volts)}}{E} = \underset{\text{(current, amperes)}}{I} \times \underset{\text{(resistance, ohms)}}{R}$$

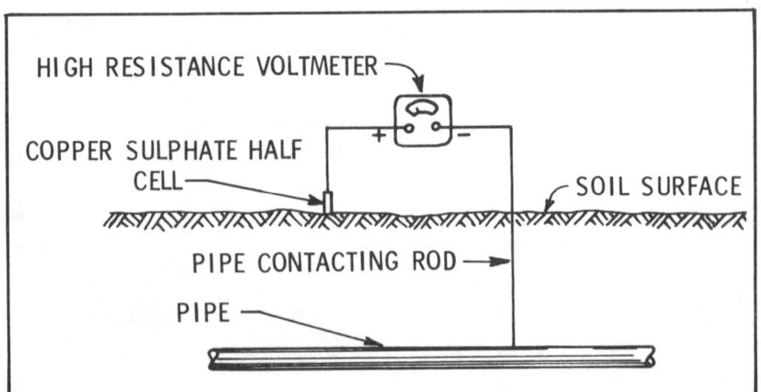

Figure 1.4-4. Measurement of pipe-to-soil potential on a pipeline.

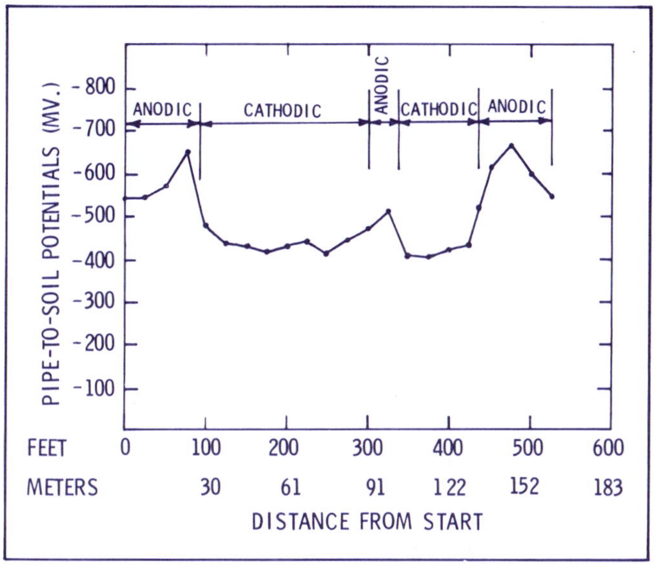

Figure 1.4-5. Potential profile along a pipeline.

The reference electrode is *always* connected to the positive (+) terminal of a millivoltmeter (50k ohm or greater input resistance) or a potentiometer. The measured pipe potential (mv) is recorded as a negative (−) value.

A plot of several P/S values taken over a distance along an *unprotected* bare pipeline may look similar to Figure 1.4-5. The line connecting the data points is called a potential profile. Note that the areas of most negative potential are the anodes or "hot spots". Usually, these occur in zones of lowest soil resistivity such as poorly aerated wet clays. For better resolution, repeat readings closer than the 30 m (100 ft) spacings shown in Figure 1.4-5 should be made through the areas indicated to be hot spots.

As a general rule, corrosion is occurring at the anodic hot spots indicated by the points of highest negative potential. An *exception* to this general rule can occur when an unprotected line or structure is being corroded (generally in a limited area) by:

1. interference from an adjacent structure which is under cathodic protection, or
2. discharge of induced (stray) current from high voltage power lines or from a strong electrical grounding system.

When interference, stray or grounding currents are present, a valley of unusally low (less negative) potentials will generally show on the profile plot and indicate current discharge (corrosion) at that point. Any unusual dips in the profile should be investigated to determine cause and corrective action required. Fluctuating or unsteady P/S readings should also be investigated since these indicate presence of variable induced stray currents.

When steady but unusually low potentials are found over a short distance of pipe, additional tests should be run by moving the Cu/$CuSO_4$ reference half-cell laterally (perpendicular) away from the pipe at that location. If the lateral readings taken about 8 to 10 m (or 25 to 30 ft) away are even less negative (more positive) than those taken directly over the pipe, an anodic hot spot exists and is probably due to interference effects.

4.09 Interference—Mechanism and Correction

At this point it is advisable to discuss interference in more detail since it is a growing problem in oil and gas production and pipeline operations.

A practical explanation of the interference effect between pipelines can be visualized in Figure 1.4-2. A pipeline being cathodically protected by a rectifier and anode groundbed installation is shown. Assume that a second pipeline in the vicinity passes diagonally near the groundbed at one point (through the zone of high anode-to-earth potentials) and near the protected pipeline some distance away. The second pipeline is not electrically connected to either the groundbed or the protected line. It is electrically isolated. However, it is intersecting the lines of c.p. current flow from the anodes and, being a better conductor than the soil, some of the anode groundbed current is picked up on it. The current is carried by the second pipeline to the point where it passes near the protected line. Here the current jumps from the second pipeline and travels the short distance to the protected line and back to the rectifier circuit. Corrosion occurs on the second pipeline where the current leaves it. Pipe-to-soil potential tests on both lines at their closest points may show a potential difference of a few tenths to greater than one volt. For example, the protected line potential may be −0.85 vs. Cu/$CuSO_4$ while the unprotected line (where it crosses or passes near the other) may be −0.3 volts vs. Cu/$CuSO_4$. As a general rule, the measured potential of the unprotected line at the point of interference will be 0.2 volts or more lower (less negative) than the potential readings taken on the same line 8 to 10 m (25 to 30 ft) on either side. This difference appears and disappears when the rectifier on the protected line is turned on and off.

In cases of interference such as just described, it is common practice to place a conductor wire between the two lines at the point of interference. Instead of flowing through the soil, the current then passes harmlessly through the conductor bond and the potential difference between the two lines at that point is eliminated. The amount of current passing through the bond can be adjusted by a resistor to elevate the spot potential of the unprotected line to a value just slightly higher (more negative) than the unprotected potentials measured on the same line 20 to 30 m (65 to 100 ft) away. This is usually acceptable and avoids excessive current drain from the rectifier especially where the second line is the largest. Interference bonding between lines belonging to separate companies must be approved by both parties and any legal regulations must be satisfied.

Figure 1.4-6 shows an adjustable interference bond across an insulating (isolating) flange connecting a buried flowline to a cathodically protected well casing. Without the bond, interference current on the electrically isolated flowline would leave through the soil at point "A" causing corrosion on the flowline. To correct the problem, the current passing through the bond is adjusted to elevate the P/S potential of the flowline at point "A" to as close to the wellhead potential as possible while staying within the output (amperage) limit of the rectifier. The portion of the flowline closest to the wellhead is then under partial cathodic protection from the well casing c.p. system. If the extra current drain on the rectifier is greater than its allowable output (amperes) rating, the flange is left fully insulated (no bonding across it) and a magnesium anode is placed on the flowline at point "A" to drain off the interference current. Installation of the magnesium anode serves to increase the negative potential of flowline at point "A" to near that of the cathodically protected wellhead and thus eliminates interference corrosion on the flowline. The current output of the galvanic anode can be reduced with a variable resistor placed in the anode lead wire connecting to the flowline. By this means the P/S potential at point "A" can be kept no higher than that of the wellhead to prolong the anode life.

In the example of Figure 1.4-6, the insulating flange is being used to electrically isolate the casing from surface equipment and thus confine the c.p. current to the casing. Other devices used for electrical isolation of c.p. protected structures are plastic insulating joints, unions or couplings.

Special grounding devices are available to maintain the desired isolation while allowing voltage surges to bleed off to the earth.[7] Lightning strikes or induced AC current surges can ruin c.p. isolation devices by burning through the plastic gaskets and bolt sleeves. High temperature is another cause of failure. When isolation is lost, cathodic protection will also be lost if the c.p. current is not sufficient to maintain protective potentials on all the now electrically connected equipment. Isolating devices such as insulating flanges should be tested periodically to guard against shorting (electrical leakage).

Use of wellhead insulators to protect well casing from stray current corrosion is discussed in Section 1, Item 8.05.

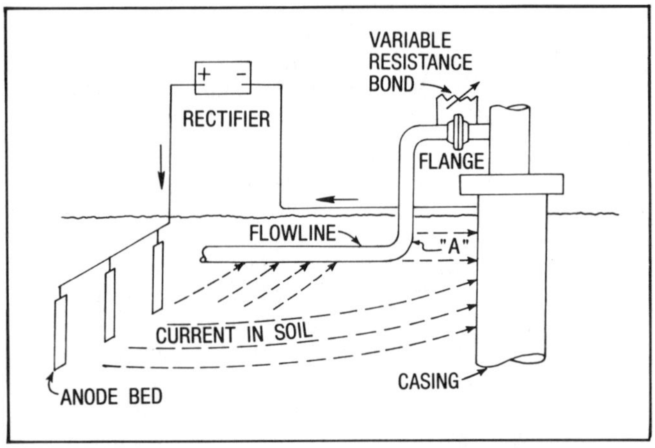

Figure 1.4-6. Interference bond on insulating flange at a cathodically protected well casing.

4.10 Offshore Structure Potential Surveys

Potentials along the leg jackets and braces of an offshore platform are measured in the same general way as that shown in Figure 1.4-4 except the reference half-cell is lowered with a long lead wire to measured depths. Care must be taken to position the reference half-cell the same distance away from the structure for each reading (usually 15-20 cm or 6-8 in.). In strong current areas, divers are used to position the half-cell. Silver/silver chloride reference half-cells are used and a −0.8 V (−800 mV) protection criteria is generally accepted. The 100 mV off potential criteria (see Item 4.07) is also gaining acceptance where impressed current c.p. is employed. Permanently mounted reference half-cells are sometimes employed at critical points on offshore structures to permit constant surveilance of structure to water potentials at those points.

4.11 Surface Potential Surveys

The pipe-to-soil (P/S) potential survey method shown in Figure 1.4-4 requires direct contact through a rod or test wire between the buried pipe and the negative (−) terminal of the high resistance voltmeter or potentiometer. The pipe may also be contacted at a riser location. To avoid having to contact the pipe for each reading, surface soil potential surveys are run as follows:

A first or base reading is made as indicated in Figure 1.4-4 using only one reference half-cell as shown. The initial base P/S reading is recorded, for example −0.65 volts vs. $Cu/CuSO_4$. Direct pipe contact is then eliminated by substituting a second half-cell for the contact bar. The two reference half-cells, connected one to each side of the voltmeter, are first tested by pressing their wet ends together and observing the meter needle deflection. If the needle deflection is less than 2 mv during this test, the two half-cells are matched and pipe testing can resume. A matched pair of half-cells can be found by testing several as required.

Half-cell 1 (used for the initial P/S reading during pipe contact) is left in position 1 and half-cell 2 is pressed to the soil directly over the pipeline at position 2.

If the planned spacing between P/S readings is 30 m (100

ft) apart, then position 2 is 60 m (200 ft) along the pipeline away from the point where the direct contact with the pipe was made. A second reading is made between the two half-cells now in the soil over the pipe. For example, the second reading shows a voltage difference between the half-cells of 0.10 volts (10 mv) with half-cell 2 (forward one) being positive (+). The voltage difference, ΔE, between the half-cells is numerically *added* to the initial P/S reading of −0.65 to obtain the second P/S reading of −0.75. In contrast, if the forward half-cell (No. 2) is negative (−) on the voltmeter, ΔE is *substracted* from the initial reading for a value of −0.55 for the second P/S S reading. Half-cell 2 is left in position 2 and half-cell 1 is moved forward 30 m (100 ft) to position 3 and ΔE is read. This reading is added or subtracted from the previous P/S value depending on the + or − polarity of the forward half-cell as described previously.

The series of values obtained are plotted to form a potential profile for the length of pipeline surveyed. Again, the points of highest negative P/S values on the plot indicate anodic hot spots if no stray or interference currents are present. It is good practice to rerun the P/S survey with closer spacings of 5 to 10 m (15 to 35 ft) between the half-cells over suspicious areas to pinpoint the hot spots.

The surface potential surveys using two half-cells as described previously are subject to cumulative error in the readings as variations in soils are traversed. Therefore, it is advisable to start over with a new base reading about every 300 to 400 m (1000 to 1300 ft) by recontacting the pipe. Starting over with a fresh base reading is especially important when resurveying suspected anodic hot spot areas.

4.12 IR Drop Method

These tests are made by using two contact bars spaced 30 to 150 m (100 to 500 ft) apart, as depicted in Figure 1.4-7. No reference cells are used. The voltage (IR) drop between the contact points is measured. Conventional

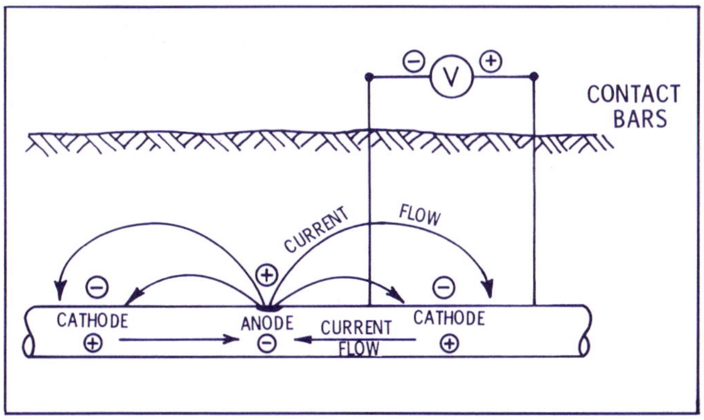

Figure 1.4-7. IR drop or current flow method for detecting anodic areas.

current flow (opposite of electron flow) in the pipe is assumed to be from positive to negative. When the anodic area is reached and passed, a reversal in current flow direction will be observed, indicating the location of the anode (on the pipe) where conventional current is discharging to the earth. Closer-spaced tests are repeated for better anode location definition.

This procedure is rarely used on deep buried pipe due to expense of multiple bell-holing for contact. However, basically the same procedure is used for detecting anodic areas on well casing by lowering a special casing potential profile tool down the hole on a wire line.[8] The tool has retractable contact knives set about 7.6 m (25 ft) apart. As the tool is moved up the hole from point to point, voltage differences between the contact knives are recorded from a potentiometer at the surface. The bottom contact is assigned a negative (−) polarity for reference. A plot of mV values vs. depth will show current flow reversals across the anodic zones. More discussion of this procedure is presented in Section 1, Item 8.07.

4.13 Effect of C.P. on Structure Potentials

After application of a properly designed c.p. system on a structure, a rerun of the potential survey will show that the potentials at various points have been elevated to a reasonably uniform high negative value. As mentioned earlier, all potentials should be −0.85 vs. $Cu/CuSO_4$ or higher (−0.8 or higher vs. Ag/AgCl).

All of the low values illustrated in Figure 1.4-5 will have been increased to or above the original potential of the original anodic areas. This is consistent with the theory discussed above under Item 4.07. The important point is that *potential differences* that originally existed on the structure have been essentially removed. Since the whole structure is now at a high uniform potential, there is little or no tendency for current flow (and corrosion) from one part of the structure to another. The major potential gradient is now between the whole structure and anode bed.

4.14 Soil Resistivity Measurements

This is an important technique for use with P/S surveys to detect anodic hot spots on a long line and for choosing locations of anode beds. The corrosivity of soils is nearly always inversely proportional to their resistivities, viz., low resistivity means a high probability of corrosion. Table 1.4-2 is a general guide to classification of soil corrosivities by measured resistivity values.

The unit of *resistivity* is the ohm-centimeter (ohm-cm), which is equal to the *resistance* in ohms of a cube of substance measuring 1 cm on each side. Resistivity is a discrete property of a substance and should not be confused with resistance in an electric circuit.

Table 1.4-2
Classification of Soil Corrosivity by Resistivity

Soil Resistivity in Ohm-Cm	*Corrosion Tendency*
<1,500	Very corrosive
1,500–5,000	Corrosive
5,000–10,000	Moderate
10,000–25,000	Mild
>25,000	Generally not corrosive if uniform

Their relationship is expressed in the following equation:

$$\rho = \frac{R}{\frac{L}{A}},$$

where ρ = resistivity in ohm-cm,
R = resistance in ohms,
L = length in cm,
A = area in cm^2

The most commonly used method of measuring soil resistivity utilizes an instrument having circuitry for impressing a measured current between two outer pins and measuring the change in voltage between two inner pins as shown in Figure 1.4-8. A d.c. current circuit is illustrated to demonstrate the principle, but modern instruments utilize a.c. current and the meter reads ohm-cm values directly.

Figure 1.4-9 illustrates a soil resistivity profile along a pipeline traversing both dry and marshy terrain. Points of low resistivity and thus probable anodic hot spots are indicated at A, B, and C. Closely spaced P/S potential surveys often indicate peaks of high potential coinciding with the low soil resistivity locations and confirm existence of the hot spots. For example, an anodic area would be likely where soil resistivity is less than 5000 ohm-cm and P/S potentials are higher than −0.6 volts vs. $Cu/CuSO_4$. Soil resistivity are often run initially, followed by pipe-to-soil surveys, at least through the zones of lowest soil resistivity.

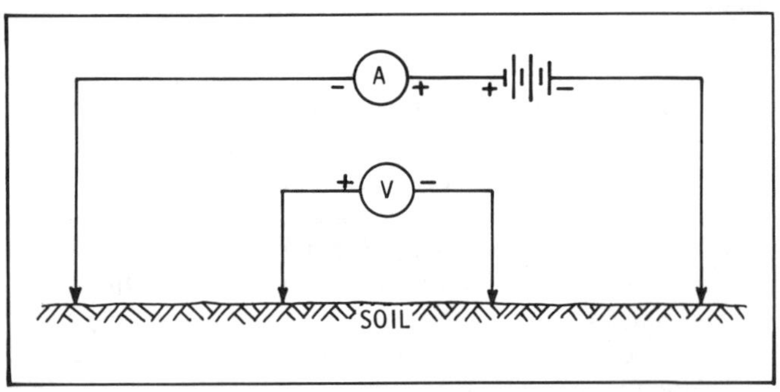

Figure 1.4-8. Four pin method for soil resistivity measurement.

4.15 Field Testing for Current Requirement

Hot spot protection of bare pipelines is done by simply attaching sacrificial anodes at the selected anodic locations and then resurveying potentials along the line. More anodes are added as needed to obtain a fairly uniform potential profile.

A more engineered approach, which is recommended where complete protection of a coated line is being considered, is to set up a temporary anode ground bed with a power supply. A good quality welding machine with variable amperage output is a convenient source of power. The temporary groundbed may consist of one or more pieces of clean pipe (five to ten cm or two to four inches in diameter) standing in augered holes that are kept filled with salt water.

The ground bed pipes are connected with a conductor wire to the positive (+) terminal of the power source and the negative terminal connects to the pipeline. Voltage required by the power source is not important. The ground bed is usually positioned 15 to 30 m (50 to 100 ft) to one side of the pipeline and the number of pipes in the groundbed can be varied to reduce or increase current flow. Total current flow (amperes) from the power source is adjusted until P/S potential on the pipeline reaches the protected level, viz., −0.85 vs. $Cu/CuSO_4$ reference.

A rule of thumb for estimating current requirement on a coated pipeline is:

$$2 \text{ mA} \times \text{total pipeline surface area in ft}^2 \times 0.02.$$

This calculation assumes that 2 percent of the pipeline area is bare or not adequately coated.

4.16 Sacrificial vs. Impressed Current Systems

On pipelines and well casings where total current requirement is more than two or three amperes and high driving voltages are desired to minimize the number of ground beds required, impressed current systems are generally preferred over sacrificial anodes.

Conversely, magnesium anodes may be best for protecting well coated pipelines in remote areas where electric power is not available and where thermoelectric or solar (photovoltaic) generators cannot be used. The anode beds are positioned along the pipeline at anodic areas located by the P/S potential surveys and in the lowest resistivity soils available.

Single magnesium anodes must be spaced at least 15 m (50 ft) apart to achieve full current output. Closely spaced multiple anodes are often grouped in beds (in low resistivity soil) but the output of each anode is reduced because of a mutual crowding effect. This is shown by

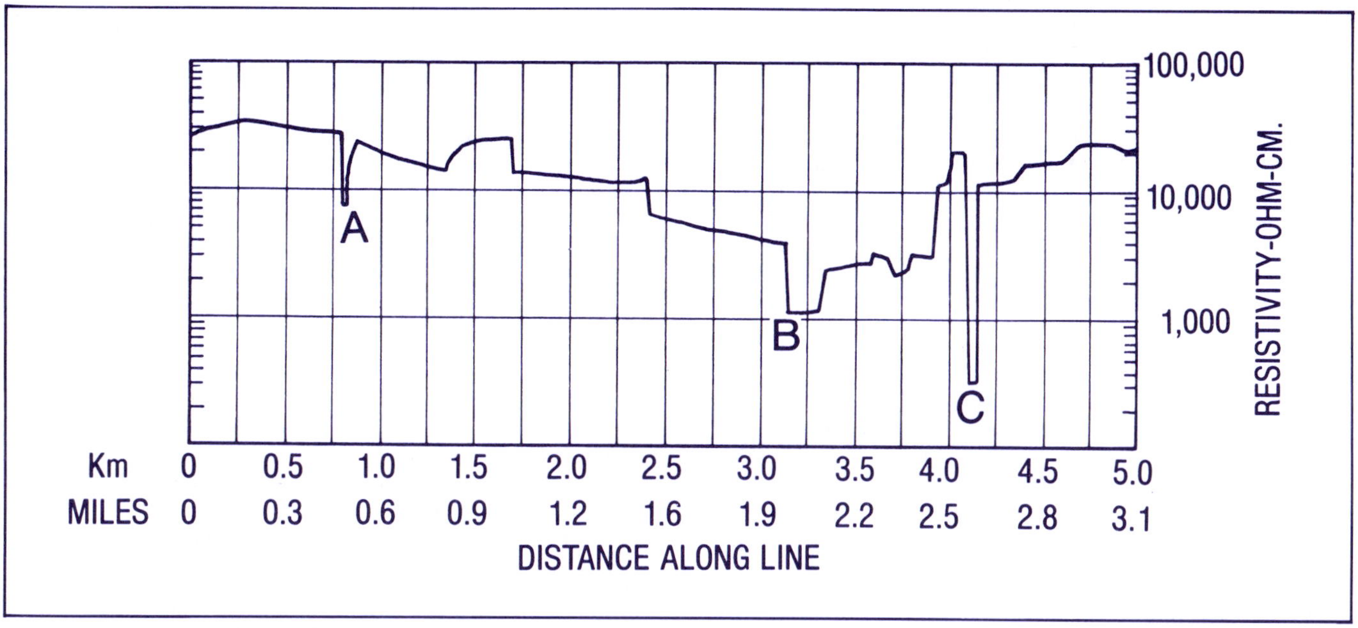

Figure 1.4-9. Soil resistivity profile of a pipeline.

the data of Table 1.4-3. It can be seen that the total current output of grouped anodes does not increase in proportion to the number of anodes in the bed. Cathodic protection service companies use more complete tables and charts of data for designing magnesium anode beds for a variety of soils and anode spacings. The size and weight of the anodes are primarily guided by the desired life of the bed before replacement is required. Accessability and cost of installation or replacement are also important factors in the design decisions.

As previously indicated, special aluminum alloys are generally used for sacrificial anode installations on offshore (marine) structures. This type installation has proven adequate and economical in relatively shallow waters such as the Gulf of Mexico. In deep water and where current density requirements exceed about 15 ma/ft^2 (160 ma/m^2) impressed current c.p. systems are generally preferred for longer life.

A computer aided cathodic protection design technique for complex offshore structures has been published.[9]

Table 1.4-3
Approximate Current Output of Magnesium Anodes

	Total Current Output[2]		mA
	Size: *17 lb (7.7 kg)*	*32 lb (14.5 kg)*	*50 lb (22.7 kg)*
No. of Anodes[1]			
1	100	116	130
5	320	373	416
10	480	557	624

[1]Commercial H-1 magnesium anodes spaced 3 m (10 ft) apart.
[2]Assumes soil resistivity in 1250 ohm-cm.
Output values are *reduced* by 50% if soil resistivity is 3000 ohm-cm.
Output values are *increased* by 125% if soil resistivity is 500 ohm-cm.

4.17 Protection of Tanks and Oil Treating Vessels

The bare underside of water tanks, oil storage tanks and similar vessels resting on the ground can normally be protected by sacrificial magnesium anodes. The total bottom area is calculated and a sufficient number of individual anodes (prepackaged or backfilled) are placed in the soil around the periphery of the vessel to supply 8 to 10 mA/ft^2 (86 to 107 mA/m^2). Connection to the vessel is made by a lead wire to each anode. Anode output is usually measured and adjusted by a variable resistor to prolong life. Copper coated rods for grounding lightning strikes should not be used because they create an unfavorable dissimilar metal couple (see Section 1, Item 1.13).

The interior of internally *coated* tanks can also be cathodically protected by bare magnesium anodes suspended in each compartment. The anode lead wires from individual magnesium anodes in large tanks are normally brought through but insulated from the upper tank wall or roof deck. The output of each is usually adjusted with an inline resistor to provide 2 to 3 mA/ft^2 (21 to 33 mA/m^2) of well coated wall surface.

Impressed current anodes fed by a rectifier are used for *uncoated* tanks and treating vessels. All metal surfaces contacted by a continuous body of salt water must be in direct line of sight of at least one anode. One or more anodes must be placed on each side of compartment walls or baffles to avoid shielding.

Current output from distributed impressed current anodes in uncoated compartmented tanks and treating vessels may be adjusted individually or the design may permit uniform current flow from each. The applied current density may be as much as 15 mA/ft^2 (160 mA/m^2) if the water in the vessel is acidic or contains H_2S. Details for c.p. systems in oil treating vessels such as heater treaters are given in NACE Standard RP-05-7.[10]

REFERENCES

(1) Gartland, P. O., Strommen, R., and Bardal, E.: "Current Density Requirements for Cathodic Protection of Steel Structures in the North Sea," *Materials Performance*, (NACE), June, 1983, p 40.

(2) "Design, Installation, Operation and Maintenance of Impressed Current Deep Groundbeds," NACE Standard RP-05-72.*

*NACE Standards (Recommended Practices) and Test Methods may be purchased from the National Association of Corrosion Engineers, NACE Publications Department, P. O. Box 218340, Houston, Texas (USA) 77218.

(3) Leavenworth, J.: "Solar Powered Cathodic Protection for Saudi Arabian Oilfields," *Materials Performance*, (NACE), December, 1984, p 21.

(4) "Control of External Corrosion on Underground or Submerged Metallic Piping Systems," NACE Recommended Practice RP-01-69 (1983 Revision).

(5) "Control of External Corrosion on Metallic Buried, Partially Buried or Submerged Liquid Storage Systems," NACE Recommended Practice RP-02-85.

(6) Barlo, T. J., and Berry, W. E.: "An Assessment of the Current Criteria for Cathodic Protection of Buried Steel Pipelines," *Materials Performance*, (NACE), vol 23, n 9, September, 1984.

(7) "Mitigation of Alternating Current and Lightning Effects on Metallic Structures and Corrosion Control Systems," NACE Standard RP-01-77 (1983 Revision).

(8) Allen, T. O., and Roberts, A. P.: *Production Operations, Vol 2*, OGCI, 1978, chap 10, p 187.

(9) DeCarlo, E. S.: "Computer Aided Cathodic Protection Design Technique for Complex Offshore Structures," *Materials Performance*, (NACE), vol 22, no 7, July, 1983.

(10) "Design, Installation, Operation and Maintenance of Internal Cathodic Protection Systems in Oil Treating Vessels," NACE Recommended Practice RP-05-75.

5 Protective Coatings and Plastics

Introduction

The technology of industrial plastics, protective coatings and coating application methods is enormous. Oil production personnel discover this fact rather quickly when they are faced with selecting a coating system for a particular job. Their ultimate decision is usually based on:

1. advice and recommendations of coatings suppliers,
2. past experience with specific coating systems, and
3. comparative results from testing laboratories.

With all these sources of information, some prior knowledge about coatings is helpful in the decision making process.

The purpose of this chapter is to acquaint the reader with:

1. types of plastics and coating materials used in oil field operations,
2. surface preparation methods,
3. application and inspection procedures,
4. types of coatings suitable for oil production operations, and
5. comparative testing and information sources.

Except for the intangible value of appearance, the only justification for coating equipment is to save money by preserving the equipment. It should be stressed that it is false economy to select a good coating system and then "save" money on inferior application.

The three essential steps in obtaining a good coating job are:

1. selection of a suitable coating system,
2. proper surface preparation, and
3. proper and careful application.

Each of these steps are equally important.

5.01 Common Generic Types of Coatings and Plastics

For this discussion, the types of coatings and plastics most widely used in the oil industry are grouped into six main classes. These are the materials that will be encounted most frequently for protection of tubular goods, tanks, and other vessels, and structures such as offshore platforms.

1. Hydrocarbon Base
   a. Natural
      Asphalt enamels and mastics
      Coal tar enamels and mastics
      Greases
   b. Synthetic
      Polyethylene
      Polypropylene
      Butadiene-styrene
      Acrylonitrile-butadiene-styrene (ABS)
2. Resins or Plastics
   a. Epoxies
   b. Acrylics
   c. Polyurethanes
   d. Vinyls
   e. Phenolics
   f. Polyesters
   g. Chlorinated rubber
   h. Drying Oils
3. Combinations
   a. Coal tar epoxy
4. Metallic
   a. Powdered zinc
      Inorganic binder
      Organic binder

   b. Fused, Plated or Chemically Deposited
      Aluminum
      Nickel
      Stainless steel
      Zinc (galvanized)
5. Cements and Concrete Sheathing
6. Composite Inorganics
   a. Ceramics—Glass
   b. Silicones

5.02 Characteristics of the Generic Types

The chemistry and compositional details of all the broad classes of plastics and coating materials are too broad to be covered here. However, one can develop a familiarity with their basic nature and properties by brief consideration of each of the various groups:

1. **Hydrocarbon Base—Natural**
As the name implies, these materials are primarily made up of repeating hydrocarbon chains:

$$-CH_2-CH_2-CH_2-CH_2- \text{ (alkyl)}$$

and

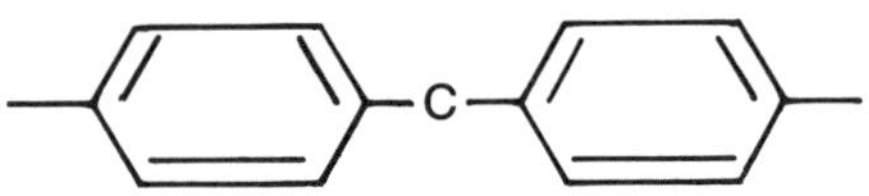

(Benzene rings $-C_6H_5$)

Asphaltic residues from petroleum distillation and coal tars from coal are familiar to most everyone. In addition to the high molecular weight hydrocarbons, these natural materials contain a variety of nitrogen, oxygen and sulfur groups that affect their solubility, melting point, resistance to oxidation and bacterial decomposition.

Asphalts and coal tars have a long history of use for protection of buried pipelines and tank bottoms. They are usually applied in a molten state and strengthened with wraps of fabrics, such as woven fiberglass. Some are deposited by solvent evaporation for thin film applications.

Coal tar base enamels are generally considered superior to ordinary asphalts for buried pipe coating due to greater resistance to moisture penetration and long term decomposition. However, asphalt has a proven record of long term satisfactory use in water immersion service including the interior of water storage tanks and on offshore pipelines. Submerged pipelines are typically coated with an asphalt primer and then alternate layers of hot applied asphalt and glass fiber tape wrap, building to a total thickness of 7-9 mm (275-350 mils). Concrete outer jacketing may be applied for weight. Cathodic protection on coated seawater lines is usually provided by zinc bracelet anodes.

A variety of heavy oil or grease base coatings are offered for temporary atmospheric protection.

2. **Hydrocarbon Base—Synthetic**
These are classed as synthetic plastics because they are manufactured and are not obtained directly from natural sources. Again, their major structure is that of repeating carbon-hydrogen groups that are caused to link together or polymerize by heat, pressure and catalysts. The starting monomers are generally either:

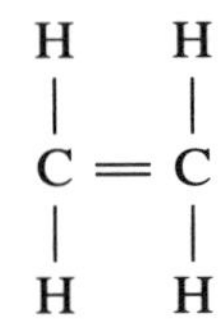

Ethylene

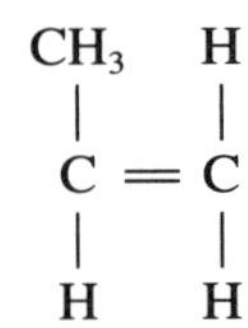

Propylene

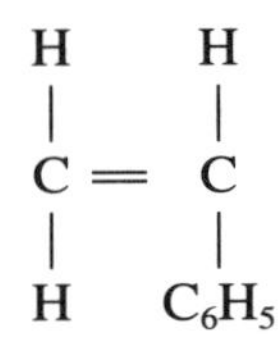

Styrene

The double bonds between the carbons split and connect to form the long hydrocarbon chains. The chains also cross link one to another to form three-dimensional networks and thus solidify.

Polyethylenes and *polypropylenes* (also called polyalkylenes or polyolefins) are often extruded into thin sheets or strips to make tapes. These are coated on one side with adhesive and wrapped mechanically onto pipe surfaces. The ease of application and convenience of wrap tapes have prompted their widespread use. However, tapes often disbond in time. Failure is usually due to poor quality adhesive or to improper pipe surface preparation prior to tape application.

An effective and long lasting pipe coating for temperatures up to 85°C (185°F) is made by extruding polyolefins into sleeves that are typically 0.6-0.8 mm (24-32 mils) thick.[1] The sleeves are bonded to the pipe exterior with butyl rubber mastic adhesives (8 mils or 0.2 mm minimum). An example is the Republic Steel X-Tru-Coat system. The pipe coating operation must be done under carefully controlled conditions in a coating plant from which the finished pipe is shipped.

3. **Resins or Complex Plastics**

A wide range of organic compositions are represented by the materials of this list. The vinyls and acrylics may be considered the most simple of this group with respect to chemical structure. Both polymerize by way of carbon to carbon double bonds, which are analogous to the polyalkylenes. Their properties and stability result from substituents on the carbons. For example, two common vinyl resins are polyvinyl chloride (PVC) and polyvinyl acetate (PVA), which are formed from their respective monomers:

$$\underset{\text{Vinyl Chloride}}{CH_2 = CHCl} \qquad \underset{\text{Vinyl Acetate}}{CH_2 = CHCO_2CH_3}$$

The *vinyls* and *acrylics* are usually pigmented and make good thin film coatings for atmospheric protection. Vinyls are especially popular as top coats over zinc-rich primers (see Metallic Coatings below). On marine structures and bridges, vinyl topcoats over zinc-rich primers (properly applied) commonly last five to eight years.

Vinyl and acrylic paints are usually dried by solvent evaporation. After drying they are odorless, nontoxic, nonflammable, and have good flexibility and abrasion resistance. Vinyls are also applied as plastisols to form heavy, tough coatings on instruments and specialty equipment.

PVC plastic, containing fillers such as clays or silica, is molded into pipe used for water service. The normal temperature limit for PVC plastic pipe is 60°C (140°F). However, chlorinated PVC (CPVC) can be used up to 80°C (180°F).

Phenolics are reaction products of phenols and aldehydes and are usually formulated with drying oils. Curing of non-solvent phenolics is done by polymerization of the oil. The films are tough and hard. They have poor resistance to alkalies and have poor wetting power. Phenolics are shop applied by baking and overcoated with epoxies for improved resistance to alkaline materials such as drilling muds.

Polyesters are made by reaction of polybasic acids and polyhydric alcohols. They are cured in place by catalysts to form hard, glossy coatings with good acid and solvent resistance. Like phenolics, their resistance to alkalies is poor. They can be built with glass fibers or chopped glass to form thick films. Thick, troweled polyester coatings are good for repair of tank bottoms and sides. Glass fiber reinforced polyester (FRP) resin is molded into pipe and sheets for tank walls. Use is primarily for water up to 60°C (140°F) but some types are suitable for hydrocarbons at up to 50°C (120°F).

Polyurethanes (or simply urethanes) are growing in use as coatings, especially for applications where wear resistance is important. They are often applied clear as topcoats. Chemical resistance is rated as good but they do show some susceptibility to attack by alkalies. They tend to chalk in the atmosphere but generally stay more flexible outside than epoxies.

Epoxies in various forms are perhaps the most widely used generic coatings and plastics in the oil industry. The basic resin structure is relatively complex, consisting of repeating hydrocarbon rings connected by oxygen links. The characterizing reactive group that splits to form the ether oxygen links is:

Epoxide group

Epoxy resins alone are themoplastic (melt when heated) and do not form good films. Therefore, they are always combined with cross-linking or curing agents such as amines, polyamides or anhydrides. Polyamide curing agents are most popular for epoxy paints since they increase pot life and are relatively nontoxic.

In general, epoxy coatings are so popular because they are tough and water resistant. They also resist alkalies, solvents and most chemicals.

Two-package coating materials are used, one containing epoxy resin and the other the curing agent. Pot life after mixing varies with curing agent, solvents and temperature. Low temperatures (0 to 5°C or 32 to 42°F) drastically slow curing and complete cure requires one to two weeks even at 25°C (76°F). Some shop applied epoxy coatings are force dried or baked at up to 85°C (185°F).

Special formulations for shop coating of pipe use combinations of epoxy and thermosetting (harden on heating) resins that are fused and cured directly onto the heated metal.

Epoxies react with constituents of coal tars to form a type of copolymer. These are the coal tar epoxies that have been widely used for pipe and tank coating. Their properties are variable due to variations in coal tars, and they are less popular now than epoxies and polyesters.

Epoxy plastic is also popular for making pipe and tanks, due to its chemical resistance. The pipe is usually reinforced with filament-wound glass fiber in the walls and a smooth layer of epoxy plastic on the interior. Good quality fiberglass reinforced epoxy pipe in nominal sizes of 50 to 100 mm (2 to 4 in.) can handle pressures up to 17.2 MPa (2500 psig) at temperatures up to 65°C (150°F). Some grades such as Ameron Bondstrand Blackgold are rated for temperatures up to 120°C (250°F) at pressures of 2 MPa (300 psig) for 50 mm (2 in.) pipe and 1.2 MPa (175 psig) for 100 mm (4 in.) pipe. Larger sizes are available.

Bisphenol-A resins are resistant to strong alkalis at temperatures up to 120°C (248°F). These are used for applications such as FRP (fiberglass reinforced plastic) vats, tanks and piping in refineries and chemical plants.

Isophthalic resins are widely used in filament wound FRP pipe, tanks and linings; especially in corrosive water service.

Furan resins have good resistance to oils and chemical solvents. They are used in FRP laminates for certain alkaline or acidic environments where isophthalic and other polyester plastics are not suitable.

4. **Metallic Coatings—Zinc Rich**

A unique class of coating meterials are those containing 60 to 75 weight percent zinc powder in the dried film. The zinc and binder act as a diffusion barrier to moisture vapor and salts in marine atmospheres. In addition, the zinc pigment corrodes sacrifically to protect underlying steel at pinholes (holidays) and scratches. For these reasons, zinc rich coatings are used worldwide for protection of marine structures such as offshore platforms.

The most popular mode of use is for prime coating over sand-blasted steel. Pigmented topcoats of vinyls, epoxies, polyurethanes or acrylics are then applied. Alkyd topcoats should not be used since they disbond from zinc primers.[2] Only limited use of zinc rich coatings is made for water-immersion service.

Two basic types of binders for the zinc powder are used:

a. inorganic or organic silicates, and
b. organic resins such as vinyls, chlorinated rubbers or epoxies.

Early inorganic silicate binders required post-curing of the coating, after application, with alcoholic phosphoric acid. Organic silicate binders are self curing by moisture adsorption. Organic binders such as chlorinated rubbers are self-curing by solvent evaporation. Table 1.5-1 lists selected characteristics for four types of zinc-rich primers.

Despite some claims to the contrary, zinc-rich coatings should always be applied over white or near-white sand or grit-blasted surfaces for best results. High pressure (20,000 psi or 138 MPa) water blasting also provides an acceptable surface.

A major cause of coating failure where zinc-rich primers are used on properly prepared surfaces is application of the topcoat too soon before the primer has had time to cure properly. One advantage of zinc-rich primers is that topcoating can be delayed several weeks or months.

Table 1.5-1
Types of Zinc-Rich Primers

SSPC* Type	Characteristics	
1-A	Inorganic silicate vehicles. Applied in one coat but requires chemical treatment after application for curing. Apply above 10°C (50°F).	
1-B	Inorganic self-curing silicate vehicles. Apply above 10°C (50°F). No undercutting or underfilm corrosion.	
1-C	Inorganic self-curing vehicles. Requires high humidity to cure. Can be applied at 0°C (32°F).	
2	Various organic vehicles	
	Epoxy Polyamide	—Generally best performing organic. Accepts epoxy or polyurethane topcoat in 16–24 hours.
	Vinyl	—Good flexibility and adhesion. Water immersion service with vinyl topcoat.
	Chlorinated Rubber	—Good flexibility. Easy brush application. Low film build. Poor abrasion resistance.

*Steel Structures Painting Council, Pittsburgh, PA

5. **Metallic Coatings—Fused, Plated or Chemically Deposited**

Hot dip galvanizing with molten zinc is used for atmospheric protection of certain prefabricated items. It is generally not suitable for immersion service in salt waters.

Some use is being made of chemically deposited metals such as *electroless* nickel. However, coating quality varies and severe galvanic corrosion of the steel substrate is always a possibility at breaks or scratches in the coating.

Molten or plasma spray-applied metals such as chromium-nickel (stainless) alloys are used for abrasive service such as sucker rod couplings. Again, local galvanic cell action is a problem. Stainless steel coated sucker rods are usually topcoated with epoxy for sealing.

Plated wear surfaces coated with stainless or high chromium alloys are used in pump parts where the cost is tolerable.

Aluminum metal coatings applied by proprietary methods (such as Alonizing) are used in heat exchangers and vessels exposed to hot gases containing sulfur compounds. Flame-sprayed zinc-aluminum mixtures can provide better protection than either metal alone. The most promising application is atmospheric exposure in marine environments but exposure experience is still quite limited.

6. **Cement Linings and Concrete Sheathing**

Properly applied cement linings are effective for internal protection of steel tubulars handling corrosive salt waters. The cement should be API Class B or C (high sulfate resistant) as specified in API Recommended Practice RP10E 1978. The lining is formed in individual pipe joints by centrifugal casting to uniform density and thickness. Recommended lining thickness varies from 3 to 30 mm (120 to 1200 mils) for pipe diameters of 60 mm (2-3/8 in.) to 508 mm (20 in.) respectively. Alkalinity from the cement helps protect exposed steel at cracks in the linings.

Concrete or cement sheathing is used for external protection of well casings, marine pilings and other buried structures. Water-saturated concrete or cement will conduct electricity and encased steel, including reinforcing bars, can suffer electrolytic corrosion. Cathodic protection of encased steel is recommended for critical structures. However, full length cementing is the best available process for external casing protection, with or without cathodic protection (see Section 8, Item 8.06).

7. **Composite Inorganics**

Included in this group are a variety of vitreous or ceramic coatings and refractory materials. Their most popular uses are for wear prevention and protection of steels at high temperature. Polymeric silicones are employed in high temperature paints for use on hot vent stacks.

8. **Conversion Coatings**

A final group of coating materials that should be mentioned are those designed for application on rusted steel surfaces. Both inorganic and organic compounds of various sorts have been offered for this service.

They are described as having the ability to combine with rust and incorporate it into the structure of the final dried coating film. Typical older compositions contain unsaturated oils, such as linseed or fish oils, which polymerize by air oxidation. Synthetic resins with rust bonding chemicals are now used. These materials are useful for temporary protection or for use where it is physically impossible to sandblast or otherwise properly prepare the surface. However, as a general rule, a conventional coating system properly applied over clean steel provides longer and more economical protection.

5.03 Coatings Systems and Nomenclature

Most suppliers of industrial protective coatings provide a variety of coating materials. Combinations in specified thicknesses and order of application are called coatings systems.

Two component systems consisting of a primer and top coat are common. Some systems employ an intermediate or tie coat to aid adhesion of the top coat. Two or three top coats may be used sequentially, thus building five or six component systems.

Following are examples of systems used effectively in marine atmospheres:

1. Zinc-rich primer—70 to 116 mm (3 to 5 mils),
   High-build epoxy—116 to 230 mm (5 to 10 mils),
2. Zinc-rich primer—70 to 116 mm (3 to 5 mils),
   High-build epoxy—116 to 230 mm (5 to 10 mils),
   Polyurethane overcoat—45 to 70 mm (2 to 3 mils).

In some systems, inhibitive primers (wash primers) are applied first as thin films to stabilize freshly cleaned metal surfaces against oxidation. There is considerable disagreement about the effectiveness of these for field applications. New types of wash primers that act as conversion coatings (see group 8 above) do extend coating life where vinyl topcoats must be applied over hand-brushed rusty surfaces. Vinyl wash primers or those containing phosphoric acid are used to treat galvanized steel prior to top coating. Weathering will accomplish the same thing.

The literature and container labels for coating systems prescribe safety precautions, surface preparation, application procedure and sequence, and wet and dry film thickness. Composition information is usually also given. The percent solids shown by the printed analysis includes weight of the resin or plastic plus pigment and fillers. Volatiles include the vehicle (solvent) and thinner. Lacquers are coatings formed by vehicle evaporation, e.g., vinyls and acrylics. Enamel is the traditional name for coatings formed by curing (polymerizing) in place.

5.04 Methods of Steel Surface Preparation Prior to Coating

Improper and incomplete surface cleaning is responsible for more failures of coating systems than poor quality coating materials. A good guideline is that anything worth painting is worthy of the cost of proper surface preparation.

Table 1.5-2 is a summary listing of the preparation methods most widely used. Both the National Association of Corrosion Engineers (NACE), Houston, Texas, and the Steel Structures Painting Council (SSPC), Pittsburgh, Pennsylvania, recognize the importance of proper pre-cleaning and have established code identifications for processes and results as indicated. NACE also publishes Test Method TM-01-75, "Visual Standard for Surfaces of New Steel Centrifugally Blast Cleaned with Steel Grit and Shot", and RP-01-72, "Surface Preparation of Steel and Other Hard Materials by Water Blasting Prior to Coating or Recoating."

5.05 Coating Application and Inspection

The person responsible for a coating job should approach the task with a professional engineering attitude. The exposure conditions should be studied and documented and the coating system should be selected for that service. The supervisor should also establish a step-

Table 1.5-2
Methods of Steel Surface Preparation

Code Number		Method and Results	
SSPC	*NACE*		
SP-1	—	Solvent Cleaning	Wiping or dipping with naptha, alcohol, mineral spirits, etc.
SP-2	—	Hand Cleaning	Use of hand tools, wire brushing and scraping. Tight mill scale is not removed by this method.
SP-3	—	Power Tool Cleaning	Power sanding, scraping or grinding.
SP-4	—	Flame Cleaning	Direct flame is continuously moved over surface to cause mill scale to pop off.
SP-5	No. 1	White Metal Blasting	Removes everything to bare metal to extent that further cleaning does not change appearance.
SP-6	No. 3	Commercial Blast Cleaning	Removes everything except the gray mill scale binder and surface rust stains leaving a streaky appearance.
SP-7	No. 4	Brush-Off Blast Cleaning	Removes only loose dirt, paint, rust and some corrosion products. Tight mill scale is not removed.
SP-8	—	Pickling	Immersion in dilute acids.
SP-9	—	Weathering and Cleaning	Surface allowed to rust heavily and then brushed or blasted.
SP-10	No. 2	Near-White Metal Blast Cleaning	All oil, grease, mill scale, paint, corrosion products, etc., are removed except for light shadows, streaks or slight oxide discolorations. At least 95% of surface is SP-5.

wise program of application requirements and inspection during and after the application. Following is a guideline list of things to observe:

1. The surface should be inspected to ensure it is properly prepared—no dirt, no grit, no rust, no mill scale are left on.
2. The surface must be free of dew or other moisture. Painting a wet surface must never be permitted; the threat of rain is the signal for stopping work. Another factor is the temperature at which application is made. The danger point is 4°C (40°F), especially if the temperature is dropping. Higher temperatures are desirable and necessary for some types of coatings, such as water base and epoxies.
3. The paint must be properly mixed and applied. If thinning is required, the thinner specified by the manufacturer must be used.
4. In the case of epoxies or other two-component materials, the contractor must watch the limited potlife and necessary sweating time once the two components have been mixed together. Applying materials that have been allowed to sit around for more than the alloted potlife will result in poor adhesion or complete failure.
5. The proper film thickness (wet and dry) must be achieved and the film must be free of holidays (pinholes) and other deficiencies. Sharp edges, bolts, rivets, etc., are difficult to protect and should therefore be given extra protection and close inspection.
6. The applicator must know what he is doing and, if not, must be instructed.

The supervisor should satisfy himself that the applicator is competent, especially if his charges are unusually low. For example, some inexperienced applicators leave thin spots by improper handling of the spray gun.

During application of coatings, the wet film thickness should be checked to insure the coverage is within specificiations. Dry film thickness must also be proper and uniform. Simple test gauges are available for these measurements.[3]

Finished coatings, either shop or field applied, should be tested for pinholes (holidays) and thin spots using a dielectric tester. Spark testers are available for flat surfaces and tubulars. Use of the latter are described in NACE RP-02-74, "High Voltage Electrical Inspection of Pipeline Coatings Prior to Installation."

A recommended procedure for detecting holidays in thin film internal tubular coatings, e.g., baked phenolics or fusion bonded epoxies, has also been published.[4] A schematic of the testing apparatus is shown in Figure 1.5-1. The power source imposes a voltage of up to 90 volts (usually 67.5V) across the coating by means of a water wet porous sponge held between metal plates on the end of a plastic tube (lance). As the wet sponge is pulled through a joint of internally coated pipe, an alarm in the detector instrument is activated at any spot (holiday) where the electrical resistance of the coating is less than 80,000 ohms. Holiday free coatings are usually specified for well tubing. Some companies allow one or two holidays per joint for surface flowlines in non-marine service.

A supervisor or inspector should also check shop applied coatings and the application techniques. Numerous failures of internal pipe coatings have resulted from poorly

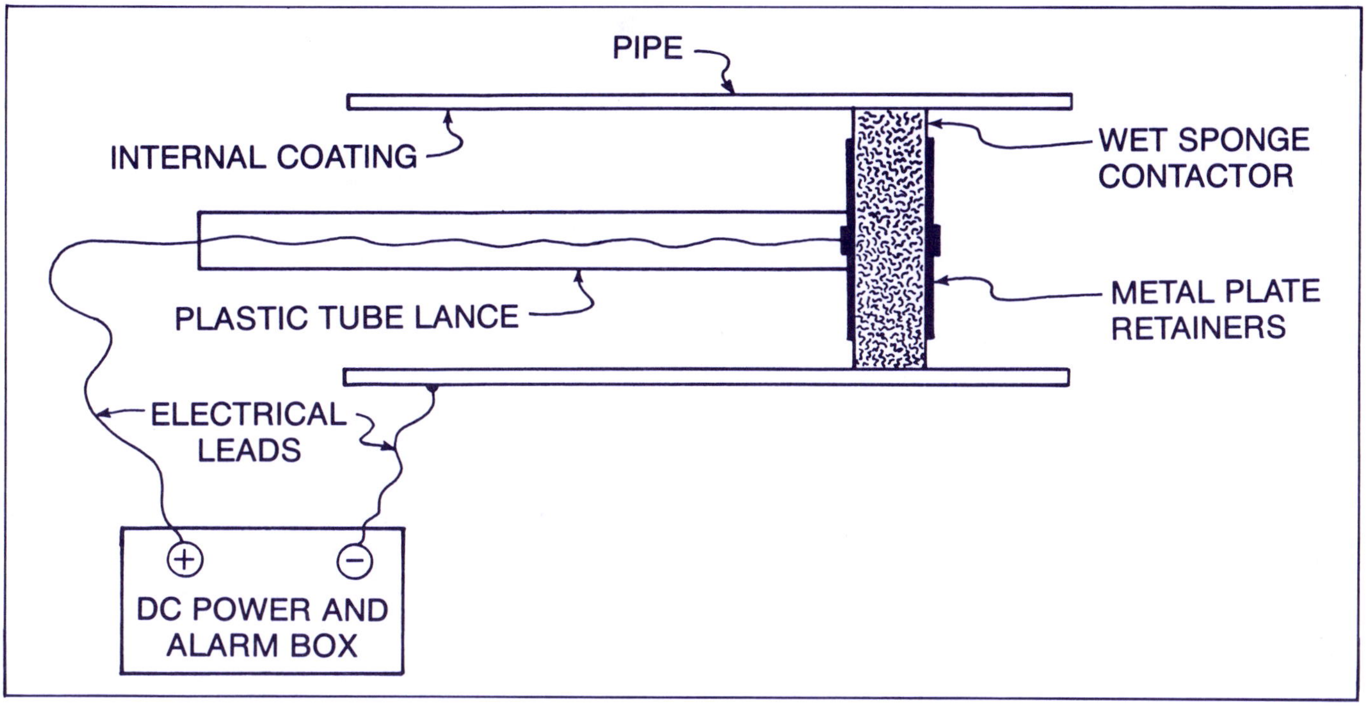

Figure 1.5-1. Holiday detector for thin film internal pipe coatings. (After NACE TM-03-84).

cleaned surfaces in the coatings application plants. On-site inspection of shop operations and holiday and film thickness checks of the finished product are worthwhile insurance, especially for coated well tubulars.

5.06 Performance Testing and Laboratory Inspection

Certain tests can be performed in the laboratory for (1) comparative evaluation of various coating systems under specific environments and (2) testing of coating properties. Such tests are beneficial for monitoring product quality, verifying supplier claims and for failure analysis.

Table 1.5-3 lists some important types of laboratory tests. Some of these are conducted routinely by several major oil producing companies and by testing laboratories. In addition, empirical tests are conducted to simulate field-exposure conditions. For example, short sections of internally coated pipe are connected in series in a closed loop with circulating salt water. Results indicate relative performance for use in flowline and injection well tubing. Short joints of internally coated tubing are also fabricated into test cells. These are filled with water and condensate saturated with produced gas and heated to well temperature as a test of downhole performance in gas wells. Several specimens can be tested simultaneously in a heated high-pressure autoclave.[5]

Close visual inspection for blistering and disbond-

Table 1.5-3
Laboratory Tests for Evaluation of Coatings Properties and Comparative Performance

Designation or Type	*Property Measured*
Cathodic Disbondment ASTM* G8 and G42	Loss of adhesion at holidays during exposure to salt water and electrical current flow.
Impact Resistance ASTM G14-72	Force required to rupture coating film.
Penetration Resistance ASTM G17-72	Relative resistance to indention.
Ultrasonic Film Thickness ASTM D1186	Film thickness.
Electric Spark Holiday Detector ASTM G9-77	Dielectric properties and pinholes.
Salt Spray	Resistance to salt-water fog (marine exposure).
Weatherometer	Resistance to sunlight (UV) and rain.
NACE** TM-03-75	Abrasion Resistance (by falling sand method).
ASTM G6-77	Abrasion Resistance (by rolling drum method).

*American Society for Testing Materials, 1916 Race St., Philadelphia, PA 19103

**National Association of Corrosion Engineers, P. O. Box 218340, Houston, TX 77218

ment, after exposure, is a good guide to relative performance.

5.07 Cathodic Disbondment and Soil Stress—External Pipe Coatings

ASTM test methods G8 and G42 (see Table 1.5-3) evaluate external pipe coatings for resistance to "voltage stress" when the coated pipe is under cathodic protection. The recommended practice for corrosion control on underground or underwater pipelines calls for coating supplemented by cathodic protection.[6] This practice recognizes that practically all coatings have or will develop pinholes or holidays despite careful application and inspection.

When coated pipe receives cathodic current, hydroxide (OH^-) alkalinity develops on bare areas of steel (see Section 1, Item 1.06). This happens even when the pipe potential is no greater than −0.85 volts vs. $Cu/CuSO_4$. High alkalinity at coating holidays has little if any effect on usual pipe coatings such as coal tar enamels, but can cause deterioration and disbondment of alkali susceptible materials such as phenolics or polyesters.

Another effect of cathodic protection that can cause localized loss of adhesion or disbondment of coatings at holidays is generation of gaseous molecular hydrogen (H_2). Recent studies[7] indicate H_2 evolution commences at *off* potentials (see Section 1, Item 4.08) of −1.12 volts vs. $Cu/CuSO_4$ and becomes vigorous at −1.17 to −1.22 volts. The on potential (potential measured with current being applied) is not considered a valid indicator of hydrogen evolution.[7] Based on available knowledge it appears advisable to avoid excessively high P/S potentials on coated pipe. The high potentials that can cause coating disbondment are most likely to occur near the rectifier or other source of impressed current.

Another source of coating stress that can cause disbondment is shrinking of high clay soils during dry periods. Most pipe coating systems utilize fabric wraps to reinforce the coating and adhesive additives are used to enhance bonding to the steel. Plastic tapes with no fabric reinforcement are susceptible to soil stress disbondment, especially if the adhesive is poor quality or too thin. Soil stress is, of course, not a problem in water-immersion service.

5.08 Suggested Types of Coatings for Oil Production Operations

Table 1.5-4 lists classes of coatings that have given good performance for various field applications. This list is intended solely for instruction purposes and it is not intended to be all inclusive nor exclusive of any company's products. The types of coating systems listed are the principal or identifying constituents and each are generally composed of several components such as primers, tie coats, top coats, etc., that vary with manufacturers. Due to variations in constituents and application procedures, any type listed may fail or give poor performance. However, the list is useful as a general guide to coatings for specific applications.

Table 1.5-4
Coatings for Specific Field Uses

Use Application	*Suggested Coating*
Internal Tubular Protection (Wells and Flowlines)	Baked Phenolics Fusion Bonded Epoxies
External Pipe Surfaces (Buried—Immersed)	Coal Tar Enamel Extruded Polyalkylenes Phenolic-Epoxy (Fusion Bonded) Concrete-Asphalt Mastic
Atmospheric Surface (Structures—Vessels)	Inorganic Zinc with Epoxy, Acrylic or Vinyl Topcoats Urethanes Polyesters
Internal Tank Surfaces	Polyester with Glass Mat over Epoxy Primer Hot-Applied Coal Tar Epoxies (100% solids)
Pipe Field Joints and Riser Pipes (Marine)	Preformed Heat-Shrinkable Polyalkylene Sleeves with Mastic Adhesive Hot-Applied Fabric Wrap Impregnated with Coal Tar and Resins Hot-Applied Asphalt

(Note: Additional information on coatings systems is given in Section 1, Chapters 8 and 9.)

REFERENCES

(1) "Extruded Polyolefin Resin Coating Systems for Underground or Submerged Pipe," NACE Recommended Practice, RP-01-85.

(2) Berger, D. M.: *Journal of Protective Coatings and Linings*, vol 1, no 2, July, 1984.

(3) KTA-Tator, Inc., 115 Technology Drive, Pittsburgh, PA 15275.
Elcometer, Inc., 118 E. Big Beaver Road, Troy, MI 48083 (Branch office in Manchester, England).

(4) "Holiday Detection of Internal Coatings of Less Than 10 Mil (0.25 mm) Dry Film Thickness," NACE Standard TM-03-84.

(5) "Evaluation of Internal Plastic Coatings for Corrosion Control of Tubular Goods by Autoclave Testing," NACE Standard TM-01-85.

(6) "Control of External Corrosion on Underground or Submerged Metallic Piping Systems," NACE Recommended Practice RP-01-69 (1983 Revision).

(7) Husock, Bernard, and Wilson, Richard: "Potentials and Hydrogen Evolution on Coated Pipe," *Materials Performance*, (NACE), vol 23, no 9, September, 1984.

6

Removal of Corrosive Gases—Mechanical and Chemical Methods

Introduction

Most of the corrosion of production equipment is caused by liquid water containing dissolved hydrogen sulfide (H_2S), carbon dioxide (CO_2), oxygen (O_2) or combinations of these corrosive gases.

In some situations, corrosion can be stopped or reduced to an acceptably low level by removing the corrosive gases that cause the trouble. This is a well-proven and cost-effective procedure for treatment of aerated surface waters prior to subsurface injection in waterfloods and for boiler feedwater. Deaeration or removal of dissolved oxygen is also a growing practice for drilling fluids.

Removal of the unwanted gases is accomplished by:

1. counter-current stripping with an inert gas (e.g., natural gas or flue gas),
2. vacuum deaeration, and
3. chemical treatment.

Theoretically any of the corrosive gases can be removed by some combination of the procedures just listed. Economics usually limits the removal method to dissolved oxygen since it is the most harmful, has low solubility in water and can be taken out relatively easy. However, there are exceptions and removal information for all three gases will be presented.

6.01 Hydrogen Sulfide

H_2S is roughly 200 times more soluble than oxygen and about three times more soluble than carbon dioxide in water at atmospheric temperature and pressure. When methane, nitrogen or some other inert gas is bubbled through sour produced water containing H_2S only part of the H_2S is stripped out. The reason is that only a part of the H_2S exists in solution as molecular (gaseous) H_2S and the remainder is non-volatile ^-SH ions. In neutral (non-acidic) field waters, typically about 50% of the H_2S is in the ionic form. Removal of dissolved CO_2 along with the H_2S increases water alkalinity and further reduces the amount of H_2S dissolved as a gas.

Some field installations have successfully removed H_2S from produced waters by stripping with CO_2 rich exhaust gas[1] or by adding a strong acid such as hydrochloric ahead of a natural gas stripper tower.[2] In both cases the acidity of the treated water is sufficiently high (pH less than 6.0) to convert most of the H_2S to the volatile gas form. Low concentrations of H_2S have been removed by sweeping hot refinery water streams with air.[3] All these processes leave the water very corrosive due to residual H_2S, CO_2 or O_2. H_2S stripping is not recommended for corrosion control in ordinary production operations.

Low concentrations of H_2S in domestic or boilerfeed water can be removed by treatment with chlorine (chlorination).

Chemical removal of H_2S in drilling fluids is a useful and practical corrosion control procedure (see Section 1, Item 8.02).

6.02 Carbon Dioxide

Dissolved CO_2 can be removed by gas stripping after the water is acid treated. The method is rarely used for that purpose alone since other methods of corrosion control are usually more economical.

In gas transmission lines where CO_2 is the primary corrosive agent, it is good practice to dehydrate the gas to the point that no water condenses at the lowest temperature and highest pressure encountered. Where complete dehydration is too expensive, excess water is trapped and drained downstream of the gas compressors and a CO_2 neutralizing chemical such as ammonia or a volatile amine is added to the gas stream. The amount of neutralizer added should be sufficient to maintain the pH of all condensed water in the system at 6.5 to 7.5. Higher pH values should be avoided if there is any entrained salt water that can deposit scale. Drip pots should be placed at low points and at the end of the line for water sampling and pH measurement.

6.03 Oxygen

Dissolved oxygen undoubtedly causes more corrosion in water injection systems than any other single factor. Contamination by even trace amounts of O_2 increases corrosiveness of sour produced water by as much as ten times. Oxygen is also the major corrosive agent in drilling fluids. Consequently, the highest priority for corrosion control in water systems is elimination of dissolved oxygen either by exclusion or removal.

Produced waters, as they come from the production wells, are usually free of dissolved oxygen. Properly designed and operated produced-water handling equipment (separators, free water knockouts, emulsion treaters and holding tanks) exclude air and thus avoid the problem of oxygen caused corrosion. However, the produced waters are often mixed with additional water from wells or a surface source such as rivers, lakes or oceans, to provide ample volume for reservoir waterflooding. In this situation, it is common practice to deaerate the source or make-up water prior to blending with the produced water stream.

Since both air exclusion and oxygen removal are equally important in production operations, both procedures will be discussed.

6.04 Oxygen Exclusion

Most water injection systems are of the closed type; that is, they are deliberately designed to avoid oxygen contamination by excluding air. Even when the flowlines and vessels are internally coated, oxygen exclusion is extremely important to avoid severe local damage at bare areas or at holidays in the coatings.

The two most likely points where air can enter closed injection systems are (1) through holding tank vent hatches and (2) at packing glands on the suction side of water pumps. The packing glands should be routinely checked for tightness. Use of water-sealed packing glands is also good practice, especially if the water being pumped contains H_2S.

Exclusion of air from holding tanks is best accomplished by gas blanketing the vapor spaces with oxygen-free gas under a small positive pressure, viz., 18-35 g/cm^2 (2-4 oz/in.^{2}). This is easily done, if oxygen free gas is available on location, by piping the gas to the top of the tank under sufficient constant pressure to prevent air from being drawn in as the water level in the tank rises and falls. The gas delivery pipe must be large enough to meet the volume demand, especially on small surge tanks where the water level may be drawn down rapidly.

Purchased nitrogen has been used for tank blanketing where oxygen-free natural gas is not available. An external expansion bladder is used to confine the nitrogen for recycling as the water level fluctuates. Due to high installation cost, nitrogen blanketing is not widely used although it is quite effective. The same can be said for floating roof tanks (no vapor space) using a pleated fabric seal around the periphery of the floating top.

Solubility of O_2 in ordinary crude oil is higher than the solubility in water. Therefore, floating oil blankets do not prevent oxygen contamination. A thick layer of weathered crude can slow the rate of oxygen entry. Floating layers (up to 1 m or 3 ft thick) of small plastic spheres (microballons) have been tried for blanketing oil-free water. However, the benefits normally do not justify the trouble and expense, especially in tanks that are pumped down frequently.

6.05 Oxygen Removal

Counter-current gas stripping and vacuum deaeration are the two physical methods most used for oxygen removal in production operations. Steam stripping is often used for pretreatment of boiler feedwater.

These removal processes operate by reducing the partial pressure of O_2 in the gas phase that is in equilibrium with the water. Dalton's law states that the total pressure of a mixture of gases is equal to the sum of the partial pressures of the individual gases in the mixture. Air contains approximately 20 mol percent O_2. Therefore, in air at atmospheric pressure (1 bar, 100 kPa, or 14.7 psi) the partial pressure of O_2 is about $0.2 \times 14.7 = 2.9$ psig or 0.2 bar. Henry's law states that the equilibrium concen-

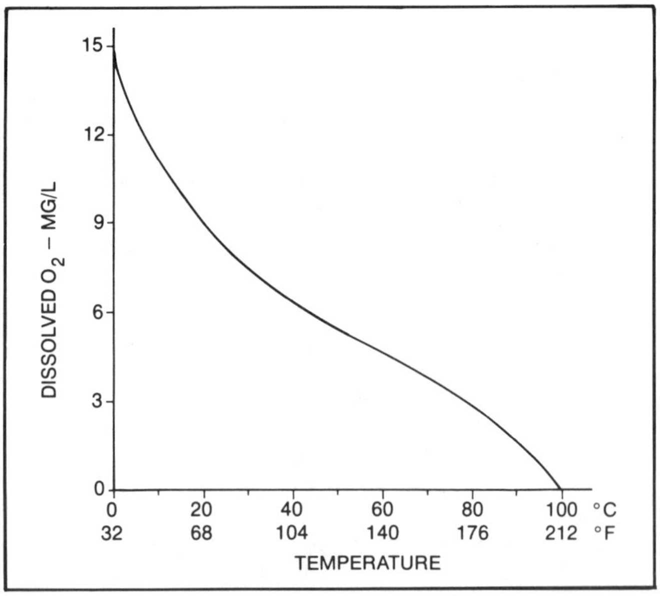

Figure 1.6-1. Concentration of dissolved oxygen in pure water in equilibrium with air at atmospheric pressure.

tration of O_2 dissolved in water is proportional to the partial pressure of O_2 in the gas in contact with the water at a given temperature. Consequently, the amount of dissolved oxygen in water can be reduced by replacing air with a gas that has a lower O_2 content (gas stripping) or by reducing total gas pressure (vacuum deaeration).

Figure 1.6-1 shows the amounts of oxygen present in water that is in contact with air at various temperatures. Note that the oxygen concentration (solubility) is zero at the boiling point of water under atmospheric pressure. Dissolved salts reduce O_2 solubility in water as was shown in Figure 1.1-9, Section 1.

6.06 Gas Stripping

Gas stripping is accomplished by passing water downward through a contact vessel while gas containing little or no oxygen is moving upward (counter-current) through the vessel or stripping tower as illustrated in Figure 1.6-2. Gas stripping vessels are available from oilfield equipment suppliers such as CE Natco. Standard design features and design criteria are published.[4] NACE Standard RP-02-78 states that gas stripping is most applicable (economically) to water systems handling over 1600 m^3/day (10,000 barrels per day) where the cost of natural gas is low or where the exhaust gas can be reused as a fuel. It should be noted that stripper exhaust gas contains oxygen removed from the water and is therefore not suitable for gas blanketing for air exclusion (see Item 6.04).

RP-02-78 further recommends that natural gas to be used for O_2 stripping must contain no more than 1 grain of $H_2S/100\ ft^3$ ($0.23\ mg/m^3$) and less O_2 than would be in equilibrium with the residual O_2 content allowed in the water.

The generally accepted upper limit for residual O_2 in water to be used in water injection operations is 0.05 mg/L (about 0.05 parts per million or 50 parts per billion). However, many operators prefer no more than 0.01 mg/L or essentially zero 0_2, especially in seawater used for injection offshore.

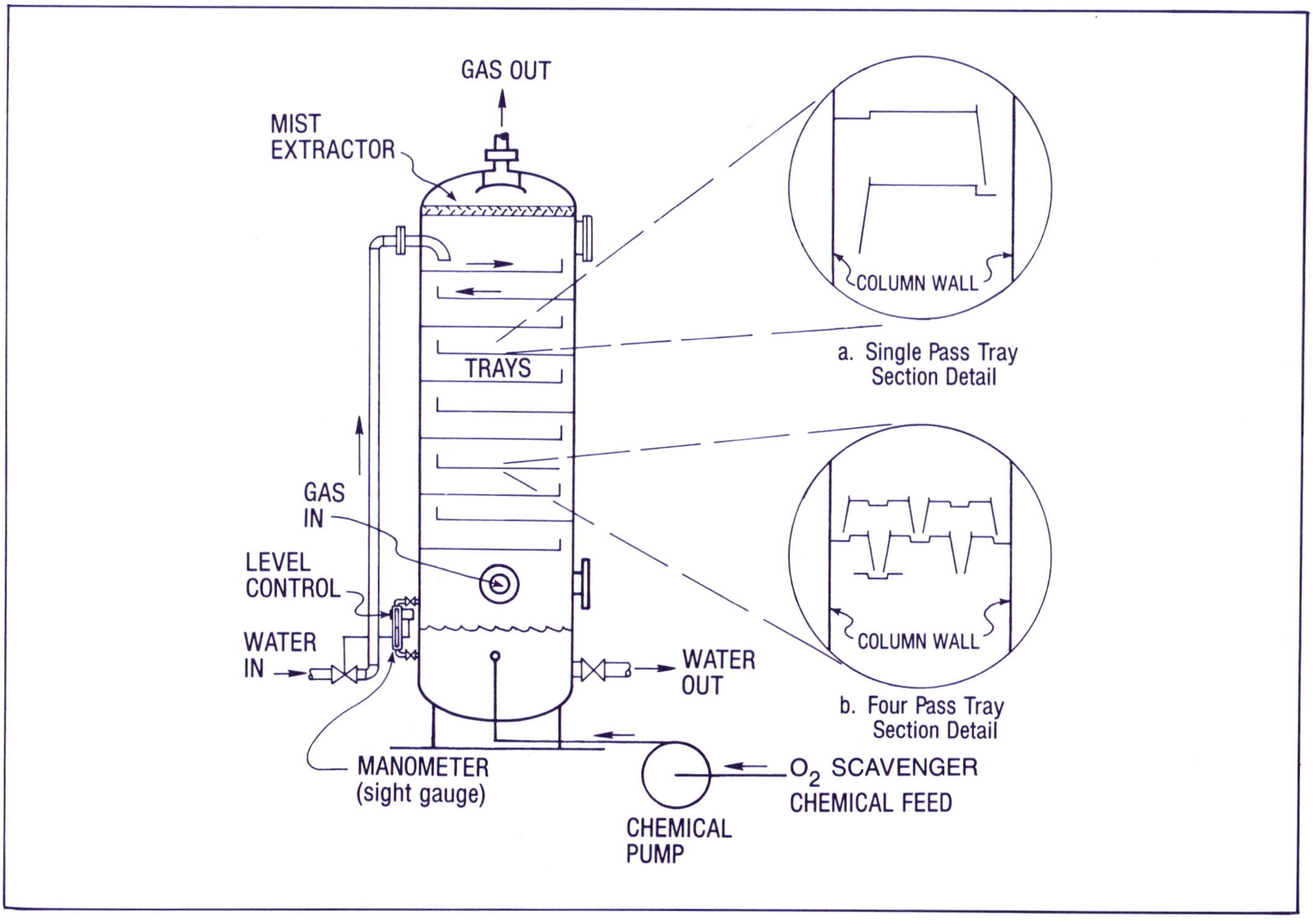

Figure 1.6-2. Counter-current gas stripping tower for removal of dissolved oxygen.

To achieve residual O_2 concentrations in these ranges by gas stripping with methane at 20°C (68°F), the maximum allowable O_2 content in the stripping gas is as follows:

Residual O_2 in Water mg/L	Maximum O_2 Content of Gas ppm or (Mol % × 20,000)*	
	Fresh Water	Seawater
0.01	227	297
0.05	1134	1491

*Assume gas is primarily methane

Stripper column diameters typically range from 1 to 4 m (3 to 12 ft) with 20 to 25 trays and gas to water ratios of 0.36 to 0.9 m^3/m^3 (2 to 5 scf/bbl) at pressures of 1.5 to 3 bar (20 to 45 psia).

Tray configurations vary. Both single and four pass trays are illustrated in Figure 1.6-2 (tray configuration after RP-02-78).

6.07 Vacuum Deaeration

Vacuum deaeration is carried out in downflow towers roughly similar in appearance to gas stripping columns. Tray columns are used but packed towers are generally more efficient. Typically, 2.5 cm (1 in.) Raschig rings are placed 1.5 to 2.5 m (5 to 8 ft) deep on support grids. Three layers of packing are often used with vacuum takeoffs above each layer (stage). Pressure at each stage is progressively reduced with lowest pressure (highest vacuum) at the bottom stage. Vacuum is drawn with one or more liquid seal vacuum pumps. Water flowing downward through the packing rings literally boils at ambient temperature, giving off the dissolved air and water vapor.

To produce boiling, the total pressure in the column must be reduced to a value equal to the sum of the partial pressures of dissolved gases (oxygen and nitrogen from air) plus the vapor pressure of water, all at the operating temperature. The data curve of Figure 1.6-3 shows the computed total (remaining) pressure as mm of Hg (mercury) at various temperatures when residual dissolved oxygen is 0.1 mg/L (or 0.5 mg/L as air). To compute the millimeters of Hg vacuum required, substract the total pressure at operating temperature (Figure 1.6-3) from prevailing atmospheric pressure (760 mm Hg at sea level).

6.08 Corrosion Protection in Deaerators

The materials of construction recommended by NACE RP-02-78 for gas stripping columns are summarized in Table 1.6-1. The same recommendations should generally be applicable to vacuum deaeration parts.

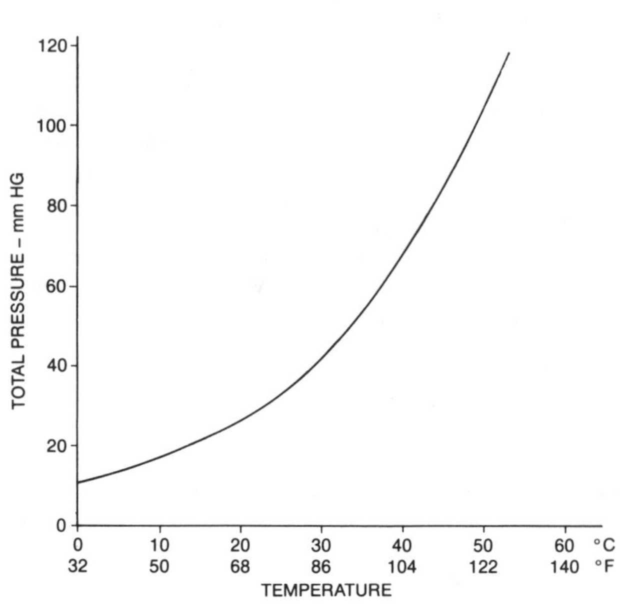

Figure 1.6-3. Vapor pressure of water with 0.5 mg/L of dissolved air (0.1 mg/L O_2).

6.09 Removal of Oxygen with Chemicals

Chemical removal of dissolved oxygen is accomplished by adding an oxygen scavenger (reducing agent) such as sulfite ion or hydrazine. These react with oxygen in water as follows:

(a) SO_3 (sulfite ion) + $\frac{1}{2}O_2$ (oxygen) → SO_4 (sulfate ion)

(b) N_2H_4 (hydrazine) + O_2 (oxygen) → N_2 (nitrogen) + $2H_2O$ (water)

The hydrazine reaction is slow except at temperatures near the boiling point of water or in very alkaline (caus-

Table 1.6-1
Materials of Construction Recommended for Gas Stripping Columns

Item	*Fresh Water*	*Saltwater (Brine)*
Column shells and support rings	Mild Steel, uncoated with 0.16 cm (1/16 in.) corrosion allowance	Same, but coated with holiday-free epoxy, fiberglass-reinforced epoxy or polyester
Trays[1] and bubble caps	AISI 304 stainless steel	AISI 316 stainless steel trays with Monel 400 bubble caps

[1]For all columns, stainless steel trays must be electrically insulated from mild steel support rings to prevent galvanic corrosion of the latter.

tic) solutions at pH 10 or higher. Consequently, use of hydrazine is primarily in boilers but it has been added to alkaline drilling muds left behind well casings.

Sulfite ion is applied by adding sodium sulfite (Na_2SO_3) in water solution (about 10 to 12 wt %). However, a more popular form is ammonium bisulfite (NH_4HSO_3) which is supplied by service companies in a premixed water solution containing about 65 wt % of the ammonium bisulfite.

Sodium sulfite is usually supplied as a dry powder or crystals that often have a cobalt salt added to increase reaction rate. Concentrated ammonium bisulfite solution without an added catalyst is more stable than catalyzed sodium sulfite solutions and does not lose strength as rapidly when exposed to air in chemical feed tanks. Traces of iron present in most waters are apparently adequate to catalyze (speed-up) the reaction rate of the ammonium bisulfite.

Both sodium sulfite (dry basis) and ammonium bisulfite (65% solution) are fed at a rate of 10 ppm (mg/L) for each mg/L of dissolved oxygen. The sulfite scavenger solution is commonly injected directly into the water sump at the bottom of a gas stripping tower, as indicated in Figure 1.6-2, to remove residual oxygen. For proper treatment control, it is essential that the water be tested frequently for oxygen content. Consequently, sampling taps should be placed in the incoming water line and at an appropriate location in the deaerated water line downstream of the tower. Further discussion on oxygen control tests is provided in Section 1, Item 8.15.

Sulfite chemical can be used economically as the sole oxygen removal treatment (no gas stripping or vacuum deaeration) in small water systems, especially where oxygen content is low. Again, good control requires periodic testing for O_2 content and close surveillance of the chemical injection equipment. If the water is to be chlorinated for bacterial control, the chlorine must be added at a separate point downstream of the sulfite injection where the oxygen has been reacted out. Chlorine and other oxidizing agents react with and consume the sulfite. This should also be noted in the treatment of seawater where chlorine (hypochlorite) is generated electrically from the seawater upstream of deaerators. Any hypochlorite residual will react with sulfite downstream of the deaerator towers, thus increasing total sulfite requirement.

Attempts have been made to remove dissolved oxygen from injection or flood waters by treatment with sulfur dioxide (SO_2) generated on site by controlled burning of elemental sulfur. Although theoretically feasible and economical, the process has not gained general acceptance because of operational problems in burning the sulfur and intense corrosion of equipment.

REFERENCES

(1) Weeter, R. F.: "Exhaust Gases Strip H2S from Sour Flood Water," *Petroleum Engineering*, vol 35, no 5, 1963.

(2) Hart, W. J., and Wingate, R. G.: "Removal of Sulfides From Brine by Aeration with Exhaust Gas from Submerged Combustion," AIME Paper No 515G, 1955.

(3) Martin, J. D., and Levanas, L. D.: "Air Oxidation of Sulfide in Process Water," *Oil and Gas Journal*, vol 60, no 24, 1962.

(4) "Design and Operation of Stripping Columns for Removal of Oxygen from Water," NACE Standard RP-02-78, National Association of Corrosion Engineers, Houston, TX, U.S.A.

7 Metals Selection, Design and Handling Considerations

Introduction

The major properties of a metal that make it suitable for a specific application in production operations are: (1) adequate strength, (2) resistance to general and localized corrosion and (3) resistance to failure by cracking. High cost and lack of availability exclude many high alloy steels and non-ferrous metals that might otherwise meet these three basic requirements. Fortunately, the degree of corrosion resistance needed in most applications can be minimized by use of chemical inhibitors, coatings, cathodic protection, or removal of corrosive agents from the metal environment. Strength requirements can be met by adjusting design thickness or by controlled changes in the metal's chemical and physical properties.

The approach of adding strength by increasing thickness is limited by the maximum weight that can be sustained especially in drill pipe, casing and tubing strings, and marine structures. Extra thickness also increases susceptability to defects and hinders uniform heat treatment.

Increasing strength by alloying and heat treatment is the common practice for ferrous (iron base) metals. A major constraint for these techniques is increased susceptability to sudden cracking above certain strength levels.

The importance of cracking as a failure mechanism has increased with deeper drilling and deserves special attention by production personnel. Corrosion related cracking processes are most common and are consequently within the scope of this book. However, certain mechanical properties of metal must be understood before cracking and metals selection can be adequately discussed.

METAL PROPERTIES AND CRACKING FAILURES

7.01 Elastic Limit—Ductility—Yield and Tensile Strength

When a piece of metal with definite dimensions is pulled at each end, it is under tensile stress. Stress is the load or force per unit of cross-sectional area. Total elongation or the extent of stretching compared to original length of a portion of the specimen is called the strain. Within the elastic limit, strain is not permanent and dimensions return to normal when stress is removed. At stresses beyond the elastic limit, permanent plastic deformation occurs and continues with increasing stress until the metal fractures and breaks. Ductility or toughness is a measure of plastic deformation prior to fracturing.

The stress being applied at the elastic limit or onset of plastic deformation is called the yield point or yield strength. The exact point is hard to define and the 0.2% offset yield point is commonly used. This is the stress load when a marked portion of the test specimen has been permanently stretched 0.2% of original length. Tensile strength (ultimate) is the maximum load just prior to fracture divided by cross-sectional area.

7.02 Hardness

Hardness is a measure of a metals resistance to indention or scratching. It is an indicator of ductility.

Rockwell hardness testing is applicable for oil field steels and is done by pressing a cone-shaped diamond point into the steel with a 150 kg load. Depth of penetration of a 0.2 mm point at this loading is translated to a reference scale called Rockwell C (HRC). Low strength API grade steels such as H-40 or J-55 usually have an HRC of less than 20. Maximum hardness attainable in carbon steels is HRC 65.

Other hardness scales are used, a common one being the Brinell. Rockwell C 22 is approximately equal to Brinell 237 hardness.

7.03 Fatigue Strength and Corrosion Fatigue

Fatigue strength and endurance limit are synonymous terms used to designate the unit stress that a metal can withstand for an indefinite but large number of repetitions or cycles of stress in an inert environment. In practical tests, stress levels are often expressed as a percent of yield strength for the metal being tested.

The endurance or fatigue limit for steels is roughly 50 to 75% of their tensile strength. However, hard, high-strength steels have relatively lower fatigue strength (as a percentage of yield strength) than softer, more ductile steels.

When a corrosive agent such as oxygen in water is present, the fatigue limit is progressively lowered in proportion to corrosion intensity. The effect is illustrated for a low-alloy steel by the data curves of Figure 1.7-1. The combined effect of stress and corrosion is termed *corrosion fatigue* and is more damaging than either contributor alone.

Corrosion fatigue is the major mode of failure of drill pipe and sucker rods.

7.04 Heat Treatment of Steels

Physical properties of steel and some other metals are strongly affected by their heat treatment, which largely determines their grain structure.

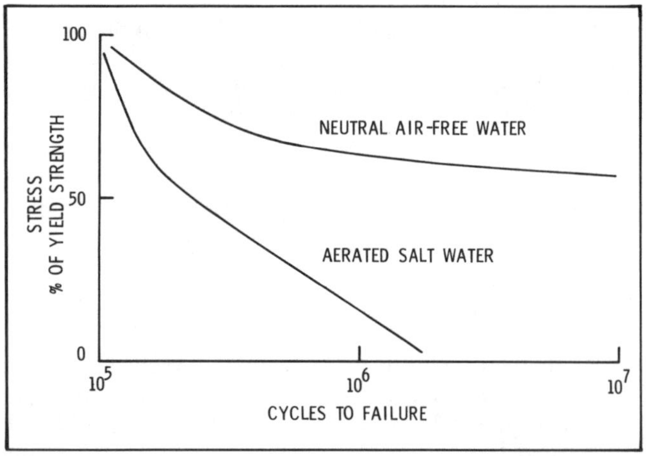

Figure 1.7-1. Environment vs. endurance limit of a low alloy steel.

Hardening is achieved by heating to a critical temperature and then cooling. The grain structure after heating is called *austenite*. With rapid cooling (quenching), austenite is converted to martensite, which is a brittle, hard needle-like grain structure.

To reduce the amount of martensite and achieve a more uniform hardness without internal stresses, the quenched steel is reheated to a temperature below 732°C (1360°F) for a time and then cooled at a controlled rate. Very slow cooling in the furnace produces a soft annealed steel (partial annealing is used for stress relief in fabricated or welded parts). Slow cooling in still air is called normalizing. Tempering is a process of intermediate heating and controlled, uniform cooling to maintain desired martensitic hardness properties.

Other processes for controlling hardness and strength levels alter the grain structure by a combination of alloy additives and heat treatment. Some alloys produce a relatively soft metal on rapid quenching and are called age or precipitation-hardened (PH) steel.

Stainless steels (300 series) retain austenitic microstructure and are non-hardenable by heat treatment.

7.05 Sulfide Stress Cracking (SSC)

The metallurgical properties that control strength, hardness, ductility, etc., are responsible for susceptability to various forms of cracking. The most important cracking mechanism in oilfield equipment is sulfide stress cracking (SSC).

SSC occurs when a susceptable metal is under tensile stress and exposed to water containing hydrogen sulfide or certain other sulfur compounds; generally under air-free conditions. The amount of water present may be only a thin film such as moisture condensed from sour gas or the metal may be totally immersed. Corrosion cells on the metal generate iron sulfide, FeS, and atomic hydrogen, H^o (see Chapter 1, Item 1.17 (1) and (2). The amount of metal loss is usually minimal and the surface layer of FeS very thin when SSC occurs. The surface layer of FeS promotes movement of H^o into the metal lattice. Penetration of H^o into the body of the metal reduces ductility. The H^o may also accumulate at imperfections and generate tremendous pressure by converting to molecular or gaseous hydrogen, H_2, in pockets or voids. Both loss of ductility and local internal stresses created by H_2 are believed to be involved during SSC.

When the metal is a hard, high-strength steel for example, the combination of ductility loss and local internal pressure superimposed on the tensile stress loading causes the metal to break and crack. Cracking failure is often sudden and, if tensile stress is sufficiently high, the crack propagates completely through the metal.

The time to failure by SSC is shortest at low temperatures (20 to 40°C or 75 to 105°F). For tubular-grade, low-alloy steels with hardnesses up to HRC 26, SSC generally does not occur as long as the temperature exceeds about 80°C (175°F). However, cracking of steels in the hardness range of HRC 22 and above can occur rapidly if the steel is cooled in the presence of H_2S even for brief periods.

A generally accepted rule for resistance of carbon and low alloy steels to SSC is that uniform hardness be maintained no higher than HRC 22.[1] Certain non-ferrous metals such as Monel K500 are acceptable up to HRC 35.

The AISI 4100 series steels (e.g., 4140 such as used in drill pipe) are acceptable up to HRC 26.[2] The H grade steels (such as 4140 H) are manufactured with controlled hardness that is uniform through the wall thickness to improve SSC resistance.

The preferred heat treatment for SSC resistant steels is to quench and temper. This is preferable to normalizing and tempering. Grades C-75 (Type 2), L-80 and C-90 tubing and casing steels are quenched and tempered with controlled hardnesses. API specification 5AC[3] sets maximum hardness for L-80 at HRC 23. However, a recent report by Ciaraldi[4] of an L-80 casing failure under severely corrosive conditions at high stress levels, has suggested that a hardness limitation of HRC 22 as recommended by NACE MR-01-75 should be considered.

The MR-01-75 document recommends using SSC resistant metals when the hydrogen sulfide partial pressure is equal to or greater than 0.005 psi (0.345 kPa). At a total gas pressure of 10,000 psi (68.95 MPa) this partial pressure of H_2S represents only 5 ppm in the gas phase. Tests by Watkins and Vaughn[5] indicate N-80 and C-95 grade steels may be suitable for sour service at H_2S partial pressures of 0.75 psi (5 kPa) and 1.5 psi (10 kPa) respectively. This would still put a maximum limit for H_2S in produced gas at only 150 ppm at 10,000 psi (68.95 MPa) or roughly 0.01 mol %.

7.06 Hydrogen Blistering (Internal Fracturing)

Penetration of atomic (nascent) hydrogen, H^o, into relatively soft low-strength steels can produce internal fractures or blisters. The mechanism is similar to that described previously for SSC where H^o inside the steel changes to the molecular H_2 form and generates high pressure. Conversion to H_2 occurs below the surface at discontinuities or defects such as laminations in rolled sheet steels. In soft steels such as that used for storage tank walls, the internal fracture commonly expands and follows discontinuities running parallel to the surface. The metal is literally pried apart by the pocket of H_2 gas and the surrounding walls are bowed outward in the form of a blister (Figure 1.7-2). The blisters are most often

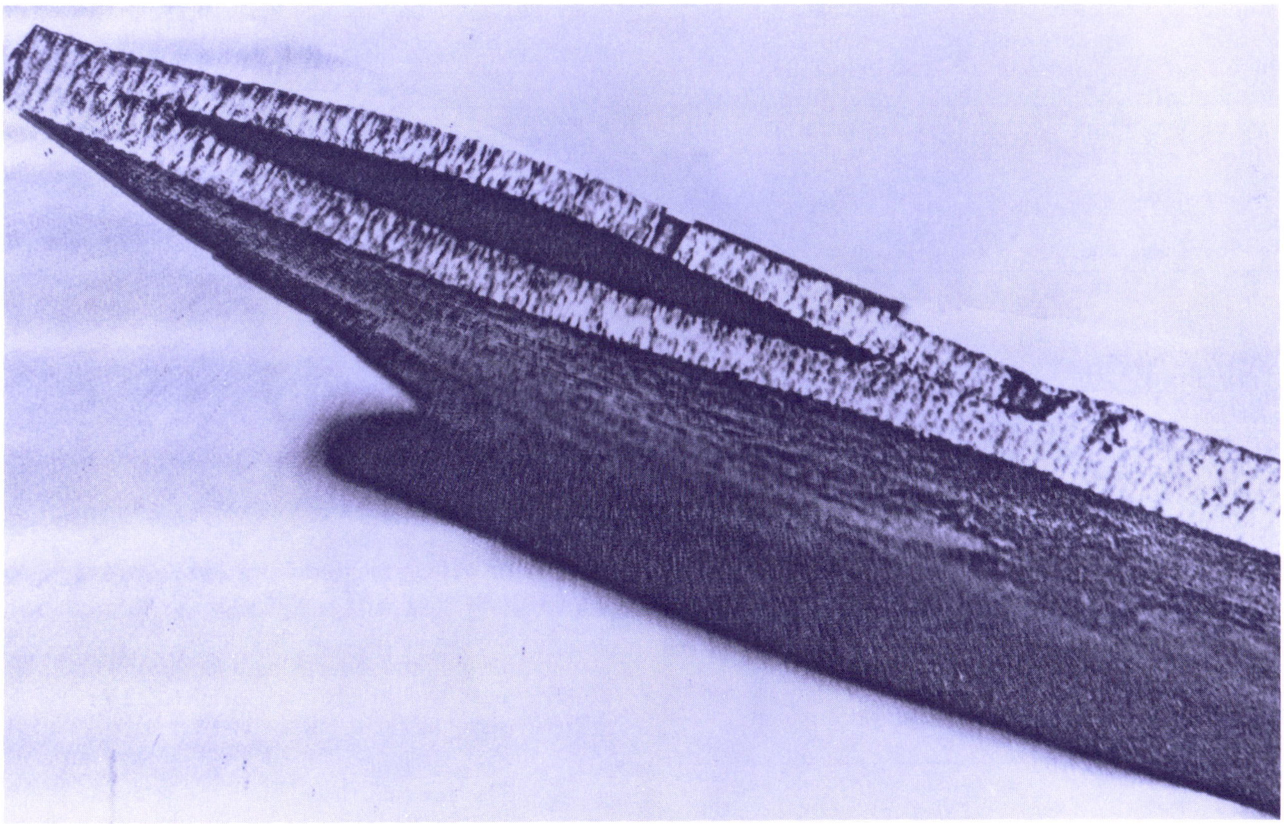

Figure 1.7-2. Cross section of a hydrogen blister in the wall (0.25 in. or 0.64 cm thickness) of a sour oil storage tank.

found on the walls of tanks or vessels holding sour water or hydrocarbons and can be as large as one m (three ft) in diameter. Chemical inhibition and use of internal coatings helps alleviate the problem. In severe cases such as sweetening vessels in gas plants, use of higher quality (cleaner) steel grades has been required.

7.07 Chloride Stress Cracking (CSC)

Austenitic (300 series) stainless steels can fail by cracking when exposed to water containing chlorides at temperatures above about 95°C (200°F) while under tensile or hoop stress. The latter is actually tensile stress on the walls of pipes or vessels containing fluids under pressure. The ferritic 13% Cr stainless grades (no Ni) such as 410 are not affected by CSC but are susceptable to SSC. Nickel-base alloys (very high nickel content) such as Incoloy 825 and Hastelloy G are generally immune to both SSC and CSC. (See Table 1.7-1.)

7.08 Stress Corrosion Cracking (SCC)

The chloride stress cracking of austenitic stainless steels just described is a specific example of a broad range of cracking failures described under the general name of stress corrosion cracking. SCC is broadly defined as corrosion which is accelerated by tensile stress.[6] Typically, the failure occurs suddenly (after an incubation period) with little apparent corrosion having occurred.[7] SCC is distinguished from corrosion fatigue (Item 7.03) in that the latter occurs under cyclic stress, which involves repetitions of both tensile and compressive forces. It is similar to corrosion fatigue in that it (SCC) is always initiated at a surface notch, which is most commonly a pit.[2] The pit often starts by dissolution of some component of the metal by a specific type of corroding agent (e.g., chloride attack of austenitic stainless steels). An example of intergranular SCC is shown in Figure 1.7-3. The metal in this case is high manganese stainless steel from a non-magnetic drill collar that failed in service during drilling of a hot (> 100°C or 212°F) well with high chloride content drilling mud. Pitting occurred at the surface (top of photo) and cracks progressed into the metal, primarily along the grain boundaries. Cracking resistant, non-magnetic alloys are available (Section 1, Item 8.02).

Low alloy (mild) steel suffers SCC when exposed to hot, concentrated caustic or nitrate solutions while being stressed near or beyond the elastic limit.[8] Under similar conditions, brass can be cracked by ammonia and certain aluminum and titanium alloys are cracked by chlorides, bromides and iodides. Many other specific combinations of metal, environment and tensile stress have been identified.

Some investigators feel that sulfide stress cracking (Item 7.05) should be considered a type of SCC. Unlike most other forms of SCC, sulfide stress cracking involves hydrogen embrittlement that occurs within the body of the metal and does not necessarily initiate at a surface pit or notch. As a practical matter, however, surface damage such as pits, scratches and tool marks act as stress concentrators and can promote cracking failure by any of the three basic modes, viz., corrosion fatigue, SSC or SCC.

Table 1.7-1
Some Prospective Alloys for Hot Well Tubulars

Alloy	Major Alloying Constituents wt% Cr	Ni	Mo	General Corrosion Resistance[1]	Cracking Resistance SSC	CSC
304SS	19	9	—	F	S	S
316SS	17	12	3	F	S	S
410SS	13	—	—	S	S	R
430SS	18	—	—	F	S	R
Duplex SS	24	6	3	G	S	G
Carpenter 20 Cb	20	29	3	G	R	R
Incoloy 825[2]	22	42	3	G	R	R
Inconel 625[2]	19	65	9	R	R	R
Haynes 20 Mod	22	26	5	G	R	R
Hastelloy G[3]	22	43	7	G	R	R
Hastelloy C-276[3]	16	59	16	R	R	R
MP 35 N[4]	20	35	10 (+35 CO)	R	R	R

[1]To CO_2—Organic acid attack in saltwater at 130°C (266°F)
[2]International Nickel Company
[3]Cabot Corporation
[4]Latrobe Steel Company
G = Good F = Fair S = Susceptible R = Resistant

DESIGN AND HANDLING CONSIDERATIONS

7.09 Metals Selection

Equipment design features are generally governed by physical parameters such as pressure containment and flow capacity and by standard mechanical specifications, i.e., API Specifications. However, attention should always be given to design factors and metal selection that can minimize corrosion or make treatment for corrosion control easier.

1. **Well Installations**
   a. *Tubing and Casing Selection*

      Following is a guideline list for highly corrosive sour wells (after Bradburn and Kalra[9]) that is consistent with NACE MR-01-75:[1]

      —Select tubulars with a low sulfur content that have homogeneous hardness throughout. Hardness should not exceed HRC 22 or use 4100 series steels, quenched and tempered, up to HRC 26.

      —Use internal flush joints (no recess).

      —Select tubulars that have been properly stress relieved (partially annealed) after upsetting and straightening operations.

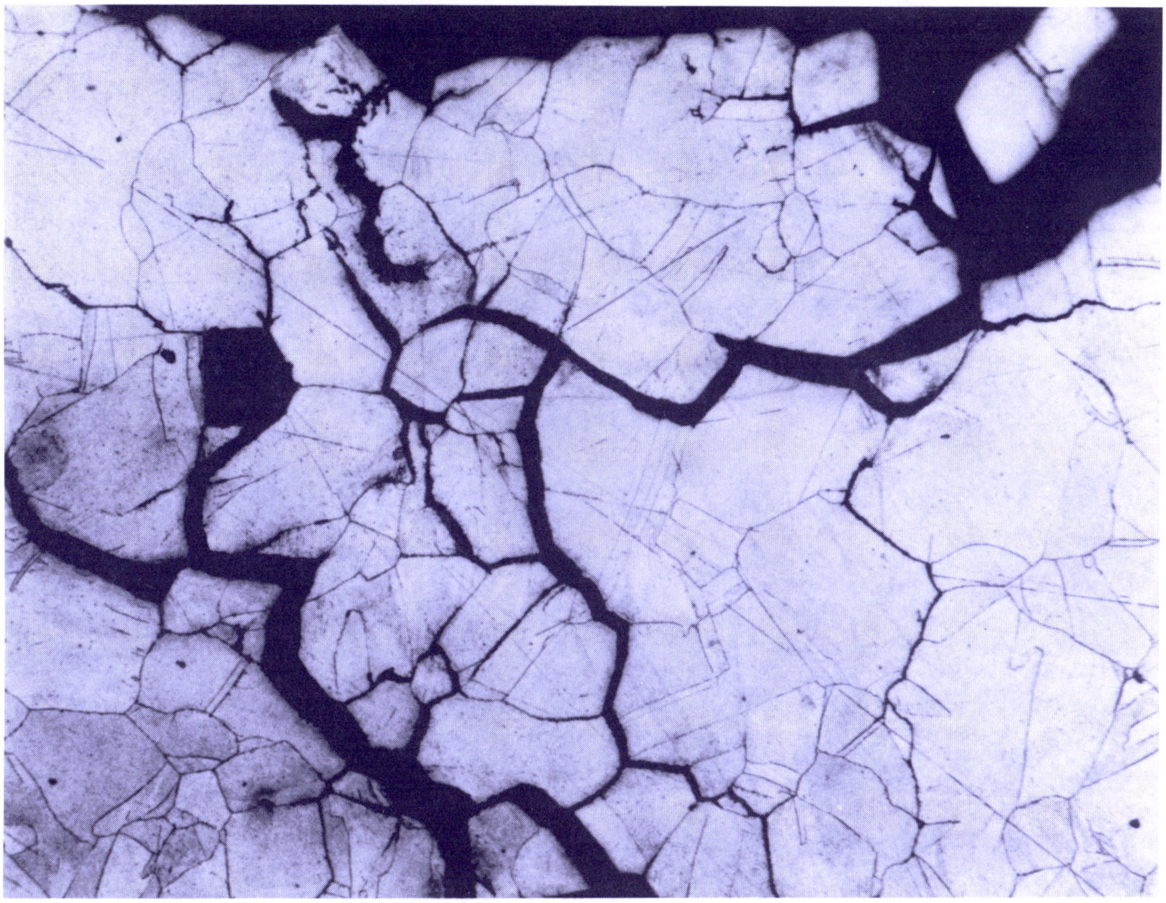

Figure 1.7-3. Stress corrosion cracking (SCC) in a stainless steel drill collar. Approx. 300x.

—Avoid austenitic stainless steels (300 series).

—Although the austenitic stainless steels, e.g., AISI 304 and 316, are generally not suitable for tubulars in wells producing chloride brines containing H_2S, these alloys are highly useful for pump and valve trim (see Item 7.10). They are particularly useful in applications where oxygen is present and in low temperature CO_2 service (see Section 1, Chapter 9).

Considerable research and study is being devoted to corrosion resistant metals, especially for tubing in ultra-hot wells. Table 1.7-1 lists some prospective alloys and their comparative corrosion resistances. The list is intended solely for study purposes and should not be considered as a recommendation for or against use of any product. Many other alloys are available. The 13% chromium alloy represented by 410 SS is of special interest since it is being used in many offshore wells. This alloy resists acidic corrosion by CO_2 and organic acids up to about 120°C (250°F) but rapidly becomes susceptible to general attack and pitting at higher temperatures. Note that it resists CSC better than the austenitic nickel-containing stainless grades (304 and 316) but is susceptible to SSC. There is evidence that 13% Cr alloy may be susceptible to pitting corrosion by salt water containing combinations of H_2S, CO_2 and O_2 at temperatures as low as 43°C (110°F).[12]

The duplex stainless steels are more resistant to CSC than austenitic (300 series) types. These typically contain 22 to 25% Cr, 3 to 5% Mo, and 5 to 15% Ni. They are resistant to general pitting corrosion at high concentrations of CO_2 at temperatures up to 200%C (392%F) but are susceptible to cracking by H_2S at high temperatures.[13]

Ferritic stainless steels containing 18 to 20% Cr and 1-3% Mo with little or no Ni are suitable for many severely corrosive situations such as CO_2 EOR projects (see Section 1, Item 9.01). Increased resistance to both general pitting attack and cracking with increasing major alloy constituents is reflected by the data of Table 1.7-1.

b. *Completion Design*

Design the wellhead and tubing arrangement as required to permit inhibitor injection (see Section 1, Item 8.08, Inhibitor Treatment Methods, including deep hot wells).

c. *Sucker Rods*

For rod-pumped wells, the rod metallurgy and properties are prescribed by API Specification 11B. In sour production, hardness should be controlled to HRC 22 (23 maximum). Alloy rods are normally not required when the well is properly inhibited. Rod loading should not exceed specifications. Joints should be tightened properly but not over torqued.

Load limits are listed in API RP 11BR.

2. **Wellheads and Flowlines**

Steels should be selected for SSC resistance (HRC 22 maximum, quenched and tempered) unless it is certain that H_2S will not be present. Type 410 stainless is suitable up to 120°C (250°F) if H_2S is absent. The same or Monel K500 is used for gates and seats and valve stems.[1]

Christmas trees internally clad with a thin layer of Inconel 625 have been installed for trial use in very corrosive production (H_2S, CO_2 at 300°F).[10] Cameron Iron Works was the manufacturer using hot isostatic pressing (HIP) to bond the high alloy layer over the base steel. Weld overlaying by metal inert gas (MIG) welding of high nickel-chrome alloys is performed but the HIP process appears preferable.[2]
Flowline metallurgy should be SSC resistant[1] for sour production. The design should provide inhibitor injection points if needed. Sharp bends should be eliminated especially in gas production with entrained sand or other solids. A general guideline is that the bend radius should be no less than five times the pipe diameter.

ERW (electic resistance welded) pipe is often used for flowlines due to lower cost compared to seamless. ERW pipe should be thoroughly inspected at the mill to avoid incomplete welds. The weld seam must be properly heat treated to eliminate hard spots and internal stresses. Figure 1.7-4 shows internal corrosion of an ERW flowline. The hole occurred where the weld seam was not completely fused.

3. **Tanks and Vessels**

Crevice corrosion is initiated where two surfaces are in contact.[2] These may be gaskets, metal touching metal (overlaps or incomplete welds) or deposits that settle in tanks or other vessels. When oxygen is present, differential concentration cells are formed in crevices or under deposits and cause deep pitting (see Section 1, Item 1.16).

Drain lines should extend to the tank bottom for sludge removal. Vessels should be designed to eliminate low flow dead spaces where solids can accumulate. No partially welded overlaps should be permitted. Tank and vessel interiors should be as free as possible of braces, baffles and irregular support members.

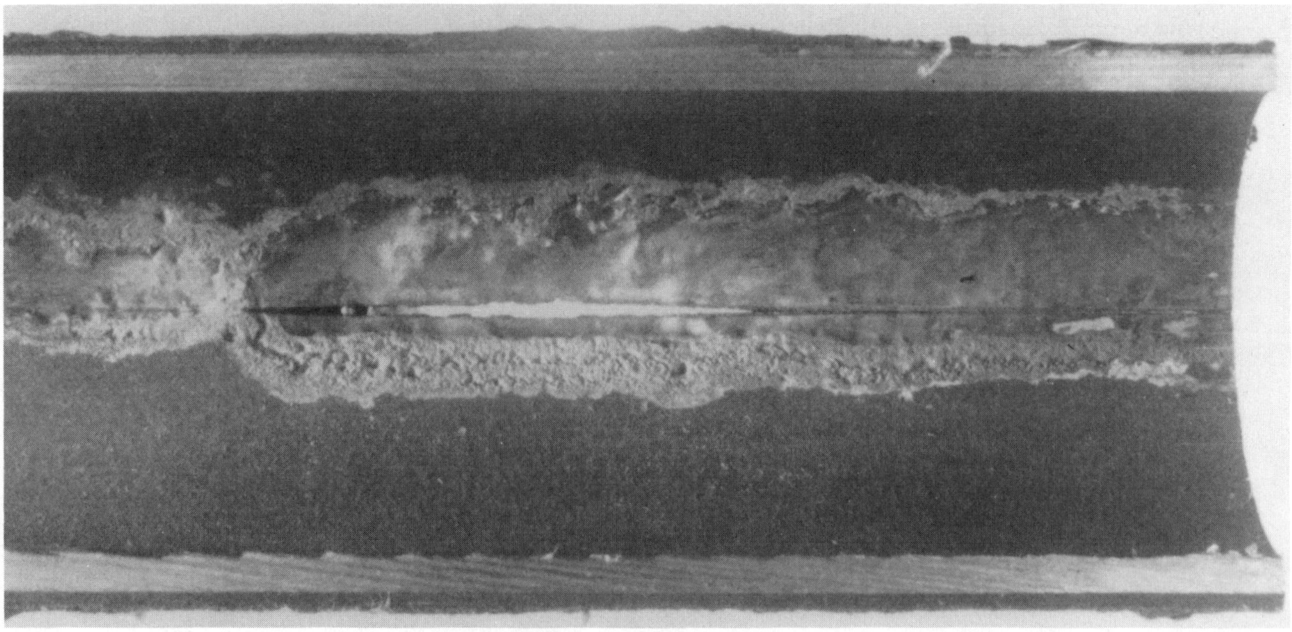

Figure 1.7-4. Internal corrosion along the weld seam in ERW pipe.

7.10 Designing with Dissimilar Metals

In the preceding discussion, the use of certain stainless steels and non-ferrous alloys such as Monel K500 was mentioned, especially for wellhead valve gates and stems. It is also common practice to chrome plate subsurface pump barrels and use alloys such as Colmonoy No. 6 (chromium boride and chromium carbide in a nickel base) for wear-resisting pump plungers. In waterflood surface pumps, chromium-plated impeller cages and aluminum bronze pump heads are used. These are a few examples where corrosion and wear-resistant alloys are used in specialty applications. Pump and valve manufacturers generally have adequate experience to guide proper use of these materials.

Most of the corrosion-resistant alloys contain high concentrations of chromium (Cr), nickel (Ni), and copper (Cu) and some contain molybdenum (Mo), cobalt (Co) and aluminum (Al) plus various amounts of iron, manganese, silicon, and carbon.[11] Chromium and Cr-Ni (stainless) alloys ($>$ 22% Cr) are generally resistant to corrosion by waters containing oxygen but are corroded by non-aerated sour waters. They are also susceptable to SSC and CSC. Nickel and alloys containing more than about 40% Ni are resistant to slightly acidic, non-oxidizing solutions such as sour brines. The Inconels and Incoloys (International Nickel Co.), Hastelloys (Cabot Corporation) Chlorimet (Duriron Co.) and the Monels are very corrosion resistant and contain 44% or more Ni. Copper and alloys with 70% or more Cu are used in valves, heat exchanger tubes and piping-handling aerated salt waters. Examples are aluminum bronze (7% Al) and cupronickel (30% Ni).

As a general rule the corrosion resistant alloys, such as 304 and 316 stainless, are cathodic to steel and cause galvanic (dissimilar metal) corrosion of steel when the two are coupled in salt water (see Section 1, Item 1.13). Their satisfactory use in valves and pumps is usually the result of the relatively small size of the alloys parts. In this case, the surrounding steel corrodes sacrificially and provides extra protection to the alloy part. With a small cathode, corrosion rate on the larger (and usually thicker) steel anode is tolerable.

On chrome or nickel-plated parts where the plating is cracked or broken, intense local attack occurs on the underlying steel. Here the cathode area is large and the exposed steel anode is small. Therefore, corrosion is localized and intense. For successful use, the plating must be uniform, non-porous and undamaged.

7.11 Equipment Handling Precautions

The most important precaution is to avoid surface damage during installation. Gouges, nicks, scratches and dents on tubing, drill pipe and sucker rods initiate localized corrosion cells (see Item 1.14). Surface damage (gouges, etc.) can also reduce the threshold stress for SSC by as much as 50%. This is the tolerable tensile stress above which the frequency of sulfide stress cracking increases drastically.

As previously mentioned, all joints and couplings should be properly tightened but not over torqued.

A more obvious precaution is to avoid permanent bending of pipe, rods and tubing joints and kinking of highly stressed items such as wirelines. Many companies issue printed handling instructions to employees and contract crews on proper handling procedures.

7.12 Velocity Effects

Low (near stagnant) liquid flow velocity should generally be avoided since this promotes build-up of slimes and sludges and thus promotes isolated pitting. As velocity increases through tubulars the overall *untreated* corrosion rate tends to increase proportionally even when suspended solids are not present. This effect is illustrated by the data of Figure 1.3-2 (Section 1, Item 3.04). In these tests involving impingement of liquid water droplets on a steel surface, the overall corrosion rate increased nearly linearly up to velocities of 30-40 ft/sec (9-12 m/sec). The corrosive gases present in these tests were H_2S and CO_2 at 107°C (225°F). Other tests have shown corrosion rates of steel exposed to salt water containing dissolved oxygen at 25°C, increased 2.5 times when flow velocity was increased from 3.3 to 13 ft/sec (1 to 4 m/sec).

Because untreated corrosion rates do generally increase with increasing velocity, attempts are being made to establish design criteria for an "erosional" or "limiting" velocity that should not be exceeded for flow of corrosive water in tubulars such as well tubing. The calculation usually involves the following equation:

$$V = \frac{C}{\sqrt{\rho}}$$

where V is the erosional or limiting velocity as ft/sec, ρ is computed density of total fluids (water, oil and gas) as lb/ft^3 under system conditions, and C is an empirical constant. The value of C is still being debated and may fall between 100 and 500.

At present it appears that any value of V calculated for untreated systems with no entrained solids would not be applicable where solids are present or where filming amine inhibitors are being used (see Figure 1.3-2, Section 1).

The most damaging condition that should be avoided is high velocity impingement of entrained solids, such as at sharp bends or elbows mentioned under Item 7.08.

7.13 Cavitation

A form of erosion-corrosion that involves both liquid velocity and pressure is cavitation caused by repeated formation and collapse of gas or vapor bubbles. Metal loss due to cavitation can be localized and intense. In production equipment, cavitation often occurs in water injection pumps where there is inadequate pressure on the suction feed line and is occasionally seen in flowlines and tubing handling water with dissolved gas at turbulent velocities. The best solution to the problem of cavitation in water pumps is to increase pressure on the suction side to the point that bubbles of gas or water vapor are not created at the lowest point on the pressure cycle.

REFERENCES

(1) "Sulfide Stress Cracking Resistant Metallic Material for Oil Field Equipment," NACE Standard MR-01-75 (latest revision), National Association of Corrosion Engineers (NACE), 1440 South Creek, Houston, TX 77084 USA.

(2) Craig, B. D.: *Practical Oil Field Metallurgy*, PennWell Publishing Co., Tulsa, OK, 1984.

(3) "Specification for Restricted Yield Strength Casing and Tubing," API Specification 5AC, March, 1982.

(4) Ciaraldi, S. W.: "Materials Failures in Sour Gas Service," *Materials Performance* (NACE), vol 25, no 1, January, 1986, p 9.

(5) Watkins, M., and Vaughn, G. A., "Effects of H_2S Partial Pressure on the Sulfide Stress Cracking Resistance of Steel," *Materials Performance* (NACE), vol 25, no 1, January, 1986, p 44.

(6) *Corrosion Basics—An Introduction*, (NACE), 1984.

(7) Atkinson, J. T. N., and VanDroffelaar, H.: *Corrosion and Its Control*, (NACE), 1982.

(8) Uhlig, H. H.: *Corrosion and Corrosion Control*, John Wiley and Sons, New York, 1971.

(9) Bradburn, J. B., and Kalra, S. K.: "Corrosion Mitigation—A Critical Facet of Well Completion Design," SPE Paper 11185, September, 1982.

(10) *Ocean Industry*, February, 1984, p 92.

(11) "Selection of Metallic Materials to be Used in All Phases of Water Handling for Injection into Oil Bearing Formations," NACE Recommended Practice RP-04-75, (NACE).

(12) Craig, B. D., "Field Experience With Alloy Clad Grade L-80 Tubing," *Materials Performance* (NACE), vol 25, no 6, June, 1986, p 48.

(13) Miyuki, H., et al., "Localized Corrosion of Duplex Stainless Steels in $CO_2-H_2S-Cl^-$ Environments at Elevated Temperatures," *Advances in CO_2 Corrosion* (NACE), vol 2, 1985, p 96.

8 Specific Corrosion Control Procedures in Oil Production Operations

Introduction

In this chapter, practical aspects of corrosion prevention in the industry will be considered. Most of the techniques have come into widespread use since about 1950 and are undergoing constant change along with other areas of production technology.

The amount of private and published information is obviously enormous and no set of cookbook instructions can possibly cover all situations. The author's approach will be to examine the causes of corrosion on various types of equipment and the control methods that have proven most suitable. To provide some continuity to the overview, the discussion starts with drilling operations and progresses from subsurface equipment to water injection wells. Additional items covered are gas processing equipment and offshore platforms. Corrosion in enhanced recovery (EOR) processes is covered in Section 1, Chapter 9.

As appropriate, both the causes and remedies are related back to the basics already covered and additional details are provided on the control procedures introduced in Section 1, Chapter 2.

8.01 Drilling Equipment Corrosion

The major *cause* of most drill stem corrosion is dissolved oxygen. Pitting by dissolved oxygen leads to corrosion fatigue (Section 1, Item 7.03). Traces of H_2S from sour formation zones increase the corrosivity of aerated drilling fluids by as much as ten times. H_2S can cause both general corrosion and sulfide stress cracking (Section 1, Item 7.05).

The *mode*, or pattern, of attack and ultimate failure of drill pipe is strongly influenced by a multitude of factors such as:

1. Tensile and bending stresses,
2. Metal properties (imperfections, scratches, hardness, etc.),
3. Erosion by solids, and
4. Improperly made-up joints.

Racked pipe left with wet solids on the inside surfaces suffers oxidation corrosion that initiates pitting. In the hole, the pits continue to deepen and effective wall thickness at these spots is proportionately reduced. The metal at the bottom of the pits is actually subjected to higher mechanical stress than that of the total pipe. Local stress magnitude is influenced by pit geometry and is more concentrated at sharp bottom or V-shaped pits. These sharp angle pits tend to form more often on inside surfaces. External pitting is minimized by rubbing of the pipe against the hole walls.

Metallurgical properties also have a strong influence. The hard, high strength steels often required for drilling generally have a lower relative fatigue strength, or endurance limit, than softer, more ductile steels (see Section 1, Item 7.03). Continued application of cyclic stressing in the corrosive aerated drilling mud causes the metal to "fatigue" and crack. Cracking usually starts internally at the sharp bottom pits discussed above and then propagates through the pipe wall. Once initiated, cracking through the wall can occur rapidly in hard ($>$ HRC 25-30) steels. In hard steels the crack also tends to spread and enlarge circumferentially. When remaining uncracked metal is unable to withstand the stress load, a "twist-off" occurs and the pipe separates.

In more ductile steels, a through-the-wall crack may simply enlarge due to fluid jetting and a "washout" hole is formed. A washout causes a rapid drop in pump pressure and the pipe must be pulled to replace the perforated joint. A clever method for pinpointing the location of a washout in a drill string was published in the *Oil and Gas Journal*:[1]

After a drillpipe washout, common practice is to pump a piece of rope (softline) down the drill pipe to partially plug the perforation and increase pressure. Using the following Bering-Sharbutt formula, the pipe joints (stand) with the washout can be located:

$$S = \frac{TV}{L}$$

where S = stand number containing washout
T = pumping time in minutes (min) until pressure increase
V = feet/minute velocity = $\frac{\text{bbl/min pump output}}{\text{bbl/ft pipe capacity}}$
L = length of average stand in feet

Example: A piece of softline was pumped in at 7 bbl/min down 4.5 in. drill pipe (capacity of 0.01422 bbl/ft) and pressure increased after 6 minutes pumping time. Average L = 93 ft.

$$S = \frac{6 \times \frac{7}{0.01422}}{93} = 31.76$$ (washout should be found near the 32nd stand)

8.02 Drillpipe Corrosion Control Methods

Assuming the pipe design and steel properties conform to recommended standards, the joints are made up and lubricated properly, and reasonable care is taken to avoid mechanical damage; the principle means for minimizing corrosion and cracking of drill pipe is control of the exposure environment.

Four proven procedures for reducing drillpipe corrosion are:

1. Removal of dissolved oxygen from the drilling fluid,
2. Use of chemicals to react with and inactivate hydrogen sulfide,
3. Flushing out mud deposits prior to racking, and
4. Application of filming inhibitors.

In addition, considerable use is made of internal pipe coatings. Special drill pipe coatings using baked phenolics top coated with epoxy are available from AMF Tuboscope or Baker Tubular Services (both in Houston, Texas).

Continuous deaeration of drilling fluids is a growing practice (see Section 1, Item 6.09). Sodium sulfite and ammonium bisulfite are two oxygen removal chemicals that have been widely used. Treatment rates are 8-10 milligrams/liter (or ppm) of the sulfite for each ppm of O_2 present, plus 10-30% excess. A reaction catalyst such as a trace amount of a cobalt salt may be added to speed up the reaction:

$$\underset{\text{(sulfite)}}{SO_3^{=}} + {}^1\!/_2 O_2 \xrightarrow{Co++} \underset{\text{(sulfate)}}{SO_4^{=}}$$

Concentrated liquid solutions of a sulfite salt such as ammonium bisulfite are convenient and can be used without added catalyst.

Mechanical deaeration by gas stripping or by vacuum (Section 1, Items 6.06 and 6.07) in packed towers is employed in some cases to remove oxygen down to about 0.1 milligram/liter residual. For best results, remaining O_2 is taken out by subsequent chemical (sulfite) treatment.

As previously mentioned, hydrogen sulfide (H_2S) greatly intensifies corrosion when O_2 is present. Even when all oxygen has been removed, contamination of the drilling fluid by H_2S from a sour zone can cause serious damage. High strength steels that are harder than RC 22 tend to crack rapidly under tensile stress when exposed to H_2S (see Section 1, Item 7.08). NACE Standard MR-01-75 (see Chapter 7, Reference 1) recommends that when drilling operations require use of high hardness materials and equipment components, the drilling environment should be controlled by use of one or more of the following:

1. Maintain drilling fluid density and hydrostatic head sufficient to minimize in-flow of formation fluid,
2. Maintain drilling fluid at pH 10 or higher,
3. Use chemical sulfide scavengers, or
4. Use drilling fluid in which oil is the continuous phase.

A widely used hydrogen sulfide scavenger is basic zinc carbonate ($ZnCO_3 \cdot Zn(OH)_2$), which reacts with H_2S as follows:

$$H_2S \rightarrow HS^- \rightarrow S^= + Zn^{++} \rightarrow \underset{\text{(Insoluble)}}{ZnS \downarrow}.$$

Other sulfide scavengers are: chelated zinc that is soluble in drilling fluids, porous iron oxide, and copper salts. The copper salts are not now widely used since they can deposit metallic copper onto steel and cause localized pitting by the dissimilar metal couple effect (see Section 1, Item 1.13).

Water-soluble organic corrosion inhibitors are, in general, not effective for control of oxygen corrosion. In the absence of oxygen, they can be used effectively for reducing corrosion pitting caused by H_2S. Intermittent high concentration slugs of oil-soluble organic inhibitors that film the pipe every 8-12 hours have been successfully used. However, these are inconvenient and are strongly adsorbed to drill cuttings and clays. Sulfide stress cracking (SSC) of hard steels can occur even when overall metal loss is minimized by application of inhibitors.

API drillpipe specifications require full length, heat-treated seamless pipe with limited phosphorus (0.040%) and sulfur (0.060% maximum). Quenched and tempered uniform hardness steel grades such as AISI 4145H are preferred for cracking resistance. These grades are also used for tool joints and drill collars. Nonmagnetic drill collars made of Monel K500 or 316 L (low carbon) stainless steel are used for drilling directional holes. The austenitic (typically 18% Cr, 8% Ni) stainless collars are subject to stress corrosion cracking (see Section 7, Item 7.08). High nickel stainless such as Hughes Nipponel 280 has recently been introduced to minimize SCC cracking of the nonmagnetic collars.

8.03 Drilling Corrosion Test Methods

When oxygen removal is being practiced to reduce drill pipe attack, it is important to monitor dissolved O_2 residuals to insure control. Some operators simply monitor the concentration of the sulfite oxygen scavenger and assume that O_2 is absent if excess $SO_3^=$ is present. It is better practice to check both $SO_3^=$ and O_2 levels every few hours. The latter tests are made with commercial oxygen meters or with an oxygen test kit (see Item 8.15). Chemical test kits for H_2S are also readily available (see Item 8.15).

A simple and recommended technique for monitoring corrosiveness of drilling fluids downhole is the placement of metal test rings in the tool joint box recess.[2] Test rings are commercially available that are mounted in plastic as illustrated in Figure 1.8-1.

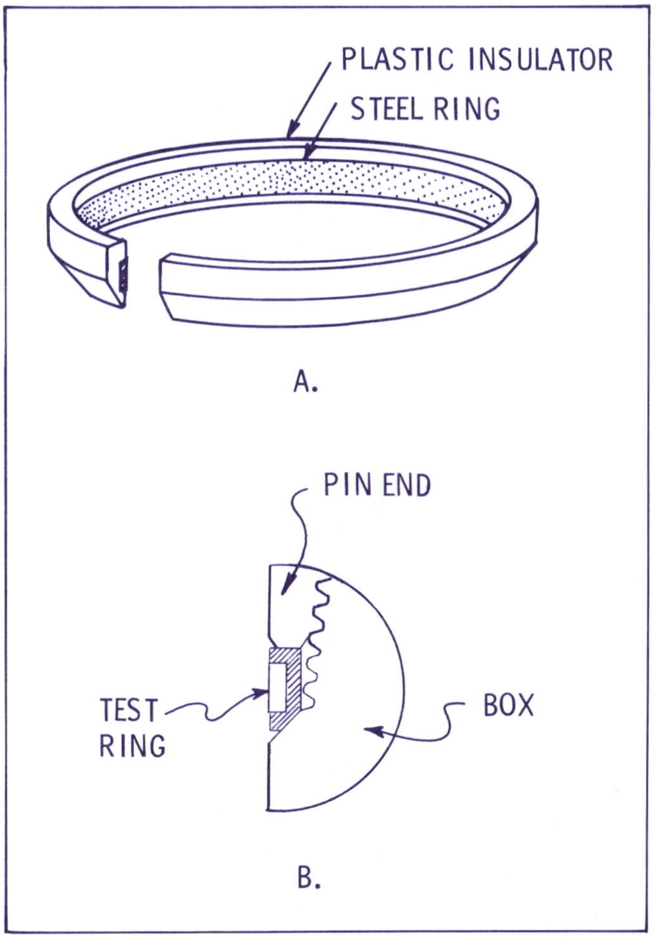

Figure 1.8-1. Drill string corrosion test ring. A. *Plastic-mounted ring.* B. *Ring positioned in tool joint recess.*

Two rings (or coupons) are generally used; one near the top of the string and one near the bottom. The test rings are normally installed at the beginning of operations and are changed on a weekly basis until the hole is completed. Spot analysis of the corrosion film and visual examination of the coupon's surface will evaluate the type of corrosion and severity of the attack. Weight loss can be determined and calculated to corrosion rate (see API RP 13B). Service companies that sell oxygen scavengers and corrosion inhibitors normally monitor the corrosion test rings. Preweighed, drill pipe corrosion rings are also available from Metal Samples Co., Route 1, Box 152, Munford, Alabama 36268, U.S.A.

8.04 External Casing Corrosion

Post-mortem examinations of failed casing reveal four primary causes of external casing corrosion:

1. electrical currents,
2. attack by formation (connate) water,
3. bacterially generated H_2S, and
4. corrosion by completion fluids.

Excessive amounts of thread lubricant, containing metallic copper can cause severe localized attack around the outside edges of casing collars.

Given enough exposure time, any of the corrosive conditions cited can cause serious casing damage but electrolytic corrosion is usually the major contributor. Historically, the most rapid and severe attack in individual fields has often been found to occur at from one to three formation zones or depths. These "hot spots" are commonly at formations that have high salt content water.

Electrolytic corrosion, is the common name applied to metal loss resulting from discharge of conventional d.c. current to the earth (see Section 1, Item 1.21). Current flow may originate from either:

1. potential gradients between formations traversed by the casing, or
2. current entering the well from connecting flowlines or electrical grounding systems.

Potential gradients of greater than one volt can be generated between surface equipment, such as long flowlines and well casings, by differential oxygen concentration cell action (see Section 1, Item 1.16). Oxygen available to the flowline in surface soil favors the cathodic reduction reaction.

$$\text{(a) } H_2O + {}^1/_2O_2 + 2e \rightarrow 2OH^-,$$

while iron goes into solution from the anodic areas on the casing where no oxygen is present:

$$\text{(b) } Fe^o \rightarrow Fe^{++} + 2e.$$

Electrons flow up the casing while conventional electric current is considered to flow in the opposite direction down the casing (see Section 1, Item 1.09).

Stray electrical currents can also be induced onto flowlines from high-voltage power transmission lines, cathodic protection groundbeds and electrical grounding systems. In many cases the exact source or path of the stray currents cannot be readily determined but tests with a voltmeter across open flowline-to-wellhead flanges will indicate that electrical current is entering the well, viz., positive (+) pole of voltmeter on flowline and negative (−) pole of voltmeter on casing side of the open flange deflects the meter needle to the right, indicating current flow from + to −.

As mentioned previously, potential gradients between earth formations can produce subsurface currents between different portions of a single casing. This current cannot be detected by voltage measurement tests at the wellhead.

In all cases of electrolytic corrosion on casings, conventional electrical current discharges from the metal to the earth at the anodic corroding areas (see Section 1, Chapter 1, Item 1.21). Presence of salt water increases metal dissolution and other anodic reactions. The high electrical conductivity (low electrical resistivity) promotes current transfer between casing and earth at salt zones. Up to 9.1 kg (20.1 lb) of casing steel is lost for each ampere of current flow during one year. The rate of casing wall penetration by electrolytic corrosion is, therefore, directly proportional to the amount (amperes) of current flowing and inversely proportional to the area of current discharge.

In the oxygen concentration cell situation discussed previously, where no strong extraneous electrical currents are involved and metal dissolution is the primary anodic oxidation reaction, the corroding away of 9.1 kilograms of casing steel over a period of one year will produce a continuous current flow of one ampere.

8.05 Wellhead Insulators

Reduction in the frequency of casing leak occurrence has been achieved in several fields by electrical insulation of well casings from the flowlines. This stops current flow down the casing from the surface (and electron flow up the casing).

Dielectric insulating materials for both screw and flange joints are commercially available. It is good practice to insulate *all* wells connecting to a *single* treating battery. Plans for all new wells should include insulating joints. Cathodically protected casings should also be insulated (see Section 1, Item 4.09).

Three problems that can be encountered with casing/flowline insulators are:

1. Difficulty in testing for shorting due to current leakage around the joint through low resistivity earth. However, made up joints can be readily tested for shorts with commercial insulation testers utilizing radio frequency energy. An excellent insulation tester of this type is made by Gas Electronics Company and sold by cathodic protection equipment suppliers.
2. Interference corrosion on the surface casing when the flowline is at high potential due to cathodic protection. In these cases the insulating joints may be partially shunted, or the wellhead potential is elevated by attaching a sacrificial anode. Interference on insulated flowlines due to casing c.p. is remedied by the same procedure (Section 1, Item 4.09).
3. Failure of insulation materials at high (>120°C or 250°F) temperature. Heat resistant materials should be selected for hot, high pressure gas wells.

8.06 Casing Protection by Cement and Completion Fluids

In addition to wellhead insulation, the best available procedure for reducing casing failure due to external corrosion is the placement of a uniform cement sheath opposite all corrosive zones. Full-length cementing of surface pipe and the production casing (at least from the intermediate string down) is recommended for deep wells, especially offshore. Only sulfate resistant (API Class B or C) cement should be used.

Casing that is not cemented should be protected with oxygen-free (deaerated) high pH completion fluid (preferably thixotropic or gelable). Completion fluids left behind casings should be treated with an oxygen scavenger such as hydrazine or sodium sulfite and brought to a pH of 9.5 - 11.0 before placement. Residual dissolved oxygen initiates corrosion pitting and promotes subsequent bacterial growth. Oil-base completion fluids can prevent these problems as long as they remain stable, but they are expensive.

8.07 Cathodic Protection of Casing

In addition to cement and non-corrosive completion fluids, impressed current cathodic protection is widely used to mitigate external casing corrosion. The major technical difficulty with this procedure is that truly uniform current distribution is nearly impossible to achieve. Rectifier driven anode beds at the ground surface generally cannot effectively project current below about 2500 m (8000 ft). Deep anode beds (see Section 1, Chapter 4, Reference 4) are recommended to improve current distribution to deep zones when casing leaks occur at depths greater than about 300 m (1000 ft). As a general guide, the depth of deep anode bed boreholes should be about 5% of the deepest zone depth where corrosion failures are encountered.

Retrievable prepackaged deep anode systems are available from Matcor, Inc., Doylestown, Pennsylvania; Cathodic Engineering Equipment Co., Hattiesburg, Mississippi, and major cathodic protection installers. Graphite, high silicon cast iron (such as Durichlor 51), and platinum clad anodes are most often used in deep bed installations. Theoretically, very deep, remote ground beds can protect to nearly any depth but cost is usually prohibitive.

Despite its limitations, cathodic protection of casings is a viable and cost effective precedure in many locations. It should be considered especially for long-life wells and for protection of the upper portions of deep high pressure wells.

The amount of cathodic current supplied to *each* well in a group from the rectifier/ground bed system usually ranges from 3 to 30 amps depending on depth and corrosion severity. The current input to each well is adjusted to eliminate all major anodic areas on the casing (see Item 8.09). Final current input is further adjusted (balanced) so that the wellhead to earth potential at each well in a field is close to that of all surrounding wells. Wellhead potential balancing within groups of wells is necessary to avoid large potential differences that can lead to accelerated corrosion by the interference effect discussed in Section 1, Item 4.09. Computer controlled rectifiers are available which automatically balance wellhead potentials (against a permanent reference half cell) for each well receiving its current from that rectifier.

Cathodic protection engineering and installation services for well casing is available worldwide from numerous companies such as Cathodic Protection Services (CPS) Co., Houston, Texas; Harco Corp., Medina, Ohio; Cathodic Protection Co. Ltd., Grantham Lincolnshire, U.K.; Caproco Corrosion Prevention Ltd., Edmonton, Alberta, Canada, and others.

8.08 Measuring External Casing Corrosion

Several types of instruments are available for detecting and measuring casing corrosion. Some utilize a combination of inside mechanical calipers and electromagnetic casing wall thickness measurement. The combination logs are interpreted by difference to indicate the locations where external metal loss is occurring. Advanced inspection tools and services are provided by several companies such as NL-McCullough, Dresser-Atlas, Welex and Schlumberger. Properly employed, these devices are excellent for diagnostic purposes, but they only detect corrosion *after it has occurred.*

Probably the most commonly used means for determining that a casing cathodic protection program is economically justified is a graphical plot of casing leak history in each field. A graph is prepared by plotting the total (cumulative) number of leaks vs. time (years) on semi-log paper. The vertical scale (ordinate) is logarithmic and the bottom scale for years is linear. Each leak repair for each well is added to the previous total and plotted as one point until a substantially straight line is clearly established. The line is then projected (extrapolated) to provide a number for probable future total leaks for the group of wells. By assigning a total expense for c.p. installation on all wells, the economics can be established. From overall prior experience, a reduction in leak frequency of at least about 75% can be expected from properly installed c.p. systems.

8.09 Casing Potential Profiles

External anodic areas where serious corrosion damage may result can be detected early with a downhole casing potential profile tool. As discussed in Section 1, Item 4.12, the procedure measures the IR drop across a length (25 ft or 7.6 m) of casing between two contact knives. Figure 1.8-2 depicts the features of the tool and a data plot.

Leads from each contactor are brought up through the supporting cable on which the tool is run and are connected to a micro-voltmeter on the surface. A voltage (IR drop) reading is obtained between the two contactors. Logging is done from the bottom to the top, at intervals equal to the spacing of the knife contactors. A plot of the voltage (IR) drops versus depth yields the "casing potential profile."

Current is assumed to flow from + to − (conventional current) and the bottom contact is used as the reference. Data curves or plots are drawn (as indicated in Figure 1.8-2) with negative (−) values on the left of zero and positive (+) values on the right. Readings on the left (−) side of zero then indicate current flow down the pipe while positive (+) values indicate flow is upward. Thus, a curve sloping toward the left from the bottom indicates an anodic or corroding zone where current is leaving the casing.

By application of an adequate amount of c.p. current, all downward current flow (negative polarity on bottom contact) is stopped, as indicated by the dashed curve in Figure 1.8-2. This occurs since the impressed c.p. current flows to the casing from the earth and uniformly upward.

Since $E = IR$ and assuming resistance is uniform for the specific size casing, then the current flow, I, can be computed because both E and R are known. Approximate values of R for various weight casing can be computed with the equations of Table 1.8-1.

Cathodic protection of well casings has been achieved (all anodic areas removed) with application of as little as 6 mA/m^2 (0.5-0.6 mA/ft^2). Current requirement can be determined with a temporary ground bed setup (Section 1, Chapter 4, Item 4.15). A native state potential profile is run, and additional profiles are then run with various levels of applied current (usually 3 to 30 amperes total). Potential survey service is provided by Cathodic Protection Services Company, Houston, Texas (Corrintec in Great Britain).

8.10 Surface E Log I Tests for Estimating Casing Current Requirement

As an alternate to the downhole casing potential profile procedure, attempts have been made to estimate current requirement by measuring the pipe to soil potential

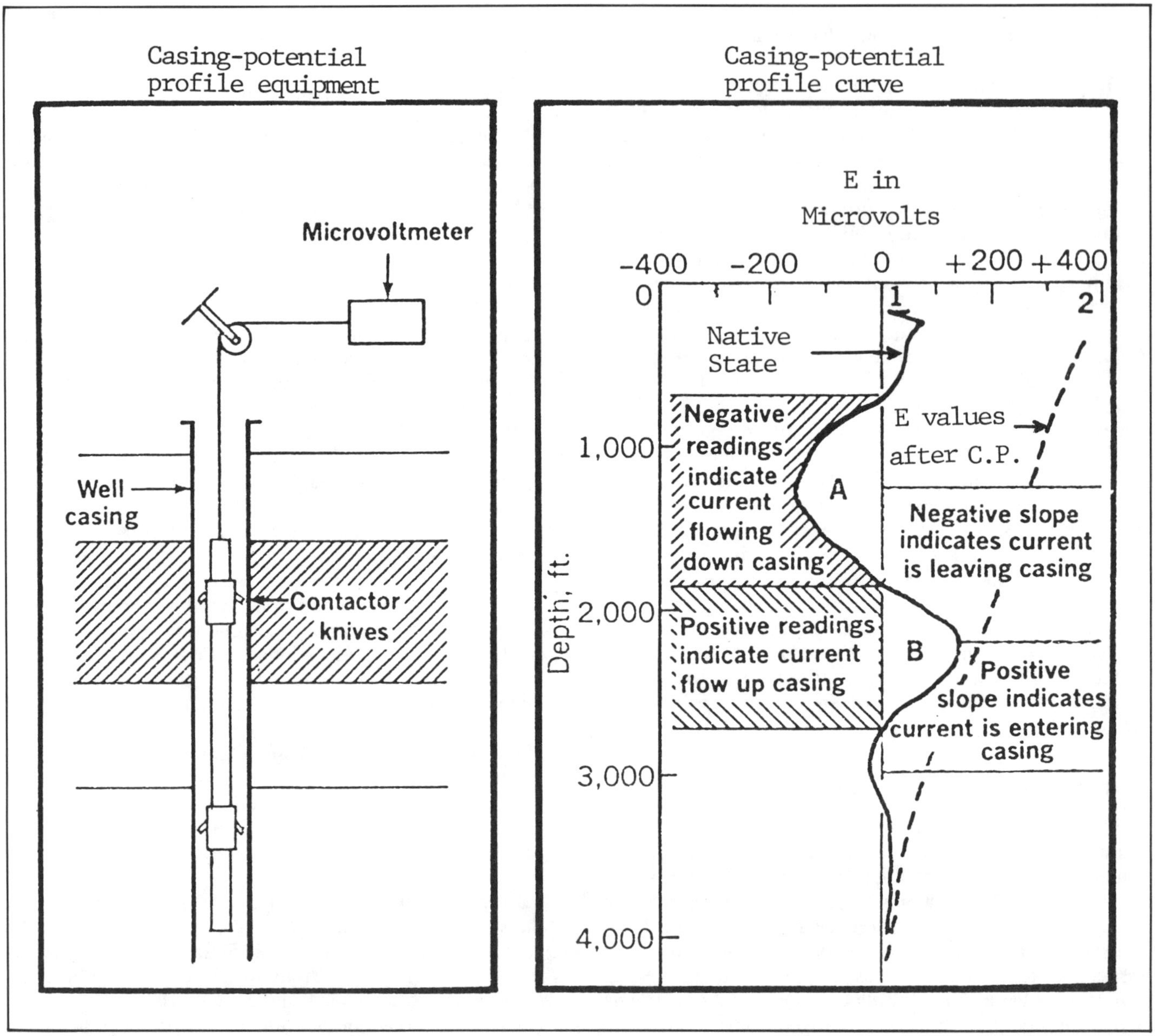

Figure 1.8-2. Casing potential profile test equipment and sample data plot.

of the wellhead using a $Cu/CuSO_4$ reference. The reference cell is positioned about 30 m (100 ft) away from the well and potential readings are taken by the same method described in Section 1, Item 4.08. Various levels of current are supplied to the casing with a generator or batteries and a temporary ground bed. At each level of current, wellhead potential (E) is plotted against the log of total current (I) being applied (on a semi-log plot) to produce an E log I curve (see Appendix, Item 1A.09). Improved results have been reported for tests made using a current interrupter and the instant off procedure for the wellhead potential readings (see Section 1, Item 4.07).

When using the E log I method, some operators conduct a series of downhole casing potential profiles in at least one of a group of wells to establish a correlation with the wellhead E log I data. Subsequent E log I tests on surrounding wells are then presumed to be more reliable for estimating c.p. current requirements.

The impetus for use of surface test methods such as the E log I procedure is to eliminate the expense of downhole logging and the costs associated with removing the well from service. Since E log I surface tests are generally not accepted as reliable, the Pipeline Research Committee of the American Gas Association is spon-

Table 1.8-1
Calculation of Electrical Resistance of Casing

Metric (S.I.) Units:

$$R = \frac{1411.7}{w}$$

where: R = resistance as micro-ohms (10^{-6} ohms)/m of length
w = weight as kg/m

U.S. Units:

$$R = \frac{289.08}{w}$$

where: R = resistance as micro-ohms (10^{-6} ohms)/ft of length
w = weight as lb/ft

soring a research program to develop a predictive computer model for cathodic protection of field well casings. The program is being designed to predict downhole casing-to-soil potentials, c.p. current distribution, interference between wells and to establish the type of surface measurements required to assess casing protection levels. A summary of progress to date has been published by Dabkowski.[3]

8.11 Internal Subsurface Corrosion

Corrosion of well internals is caused for the most part by (1) stimulation, completion, workover or lift fluids entering from the top or (2) entry of formation fluids, generally from near the bottom. While this statement may seem trivial, its intent is to encourage consideration of the producing well as a dynamic system amenable to engineered corrosion control.

An unaccountable amount of subsurface corrosion damage is done during initial well operations, which does not cause immediate equipment failure and thus goes undetected. Corroded tubing that is pulled after a well has produced one to three years may still show evidence of attack by (1) stimulation acid that was left in the hole too long or (2) by improperly prepared saltwater completion fluid. When an unexpected failure does occur, a common practice is to initiate a hastily devised control program that will, hopefully, prevent a recurrence.

Some comparatively simple precautionary steps can be taken during planning, completion, and early production of any well that will help eliminate the guesswork and lead to a more rational control program. These steps are listed in Table 1.8-2.

1. **Will Corrosion Be a Problem?**

   Steps 1, 2, and 3 of Table 1.8-2 pertain to this question. Unfortunately, many wells are drilled and completed before any real effort is made to assess the probability of serious corrosion or to incorporate efficient means for prevention and treatment. In contrast, an exemplary case history of pre-planning for corrosion control and monitoring in producing wells and surface facilities in the remote Prudhoe Bay Field, Alaska, was published by Byars and Galbraith.[4]

   Referring back to Section 1, Item 1.17, it is established that the dissolved gases (O_2, CO_2 and H_2S) are the major promoters of corrosion in oil production operations. Subsurface equipment in producing oil and gas wells is affected most by the CO_2 and H_2S. In some reservoirs water soluble organic acids (acetic and propionic) also contribute significantly. However, the produced gas analysis (presence of CO_2 and H_2S) is the best first indicator of potential corrosion trouble.

   Any measurable trace of H_2S in the gas produced from a well is considered justification for a corrosion monitoring program when water is also present. In laboratory tests, it can be demonstrated that water containing as little as 3 mg/L (ppm) of H_2S can cause blistering and ultimate pitting of carbon steel. The iron sulfide corrosion product is highly insoluble and is also cathodic to the steel.

   According to the guidelines set out in NACE Standard MR-01-75 (Section 1, Chapter 7, Reference 1) use of control measures to prevent sulfide stress cracking (SSC) should be considered if the partial pressure of H_2S in the gas is greater than 0.05 psi

Table 1.8-2
Elements of a Planned Corrosion Prevention Program for New Wells

1. Determine most probable composition of the gas, especially CO_2 and H_2S content.
2. Unless prior experience proves otherwise (rare), expect sufficient water to support corrosion.
3. If the gas phase contains H_2S or if CO_2 partial pressure exceeds 0.05 MPa (7 psi), include means for corrosion control and monitoring in the well equipment design.
4. Steels and other materials that may be exposed to H_2S should meet requirements set out in the NACE Standard MR-01-75 (Section 1, Chapter 7, Reference 1).
5. Determine expected temperature, pressure and mode of production (flowing, gas or pump lift).
6. Make certain that stimulation acid is adequately inhibited and do not exceed maximum allowable equipment exposure time for the expected temperature. Remove spent acid completely and as rapidly as possible.
7. Add oxygen scavenger chemical to saltwater or other aqueous fluid used for pressure control during completion immediately prior to use. If fluid is to be left in the hole, replace saltwater with inhibited oil if possible.
8. If H_2S or CO_2 are present in the produced gas, institute a corrosion rate monitoring program with one or more types of wellhead (flowline) monitors (see Section 1, Chapter 10, Item 10.02 and 10.03).

(345 Pa) when total pressure is 65 psi (0.45 MPa) or higher.

Partial pressure of any gas constituent can be computed from the gas analysis as follows:

partial pressure = fractional mol % × total pressure

Example: Gas analysis shows CO_2 content of produced gas is 2.0 mol percent
Reservoir pressure is 42 MPa (6000 psi or 6 ksi)
CO_2 partial pressure (P_{CO_2}) = 0.02 × 42 = 0.84 MPa (122 psi)

Early studies[5] indicated that when P_{CO_2} is less than 7 psi (0.05 MPa or 0.5 bar) corrosion by CO_2 is generally not a problem in gas condensate wells. This guideline is still useful and wells producing gas with higher CO_2 content should be considered potentially corrosive.

Corrosion severity tends to increase with increasing CO_2 partial pressure and temperature.[6] However, above about 30 psi (0.2 MPa) P_{CO_2} and 60°C, there are no simple rules for predicting localized corrosion rate due to formation of more or less protective layers of iron carbonate ($FeCO_3$) corrosion product.[7] In general, the corrosion rate is dependent on the permeability of the $FeCO_3$ scale and a combination of scale dissolution and fluid flow rate.[8] Increasing bicarbonate (HCO_3^-) content in the produced water tends to decrease the protective scale permeability (increases protectiveness) whereas increasing calcium content tends to increase permeability. Experience in gas condensate wells has shown that corrosion rates of mild steels may be as much as 7 mm/yr (280mpy) in high turbulence areas such as choke bodies, ells, and trees.[7,9] Attempts are being made to classify the corrosiveness of fields producing CO_2 rich gas (prior to well completion) according to water chemistry[10] but at present, producing experience coupled with a vigorous program of corrosion rate monitoring and iron content analysis (see Chapter 10, Item 10.04) is the most widely used method.

2. **Factors Affecting Corrosion Rate and Chances of Failure**

Many wells produce enough H_2S and CO_2 during a few weeks to completely dissolve the tubing string if all the H_2S and CO_2 could react.

Generalized or overall solution corrosion of metal is limited by surface films including millscale, iron sulfides and carbonates, mineral scales and by polarization effects and hydrogen overvoltage (Section 1, Items 1.10 and 1.16).

Localization, in the form of pitting and ultimate perforation, is usually caused by one or more of the following:

a. Breaks in surface films (millscale, iron carbonate, etc.),
b. Metal inclusions,
c. Metal grain structure differences,
d. Local stress, and/or
e. Abrasion by solids.

The appearance of pitting caused by CO_2 attack is illustrated in Figure 1.8-3.

After an area starts to corrode and local attack is initiated, it tends to continue at that spot. This is one reason why even moderate corrosion started by stimulation acid or completion fluids can be so detrimental during the early life of a well.

The unpredictable local factors listed make it practically impossible to prejudge with precision the time a well will produce before a failure occurs. The major task of the corrosion engineer is to assess the probability of corrosion occurring at a serious rate and to plan a control program such as that outlined in Table 1.8-2. Special emphasis should be given to equipment materials selection and to designing the well completion for inhibitor injection if required. Metals selection and design considerations are discussed in Section 1, Chapter 7.

8.12 Inhibitor Treatment Methods for Wells

Use of chemical inhibitors for control of subsurface well corrosion is sufficiently important to justify special attention (see also Section 1, Chapter 3). Inhibitor suppliers are generally knowledgeable about the application methods and their recommendations should always be obtained. The consumer is well advised to compare products, services and specifics of treating procedures from several suppliers.

Table 1.8-3 is an abbreviated list of the many procedures that are used to treat different type wells. These

Table 1.8-3
Summary of Proven Inhibitor Application Methods for Producing Wells

Type of Well	*Treating Method*
(1) Open annulus —pumped —high fluid level	Intermittent batch with circulation
(2) Open annulus —low fluid level	Intermittent batch with flush
(3) Closed annulus —pumped or natural flow	Periodic batch down tubing Tubing displacement Formation squeeze Downhole injector valve (annular feed)
(4) Closed annulus —gas lift	Formation squeeze supplemented with continuous gas treatment
(5) Flowing gas or gas-condensate	Same as (3)
(6) High temperature sour gas (>150°C)	Continuous via dual string or special inhibitor injection line

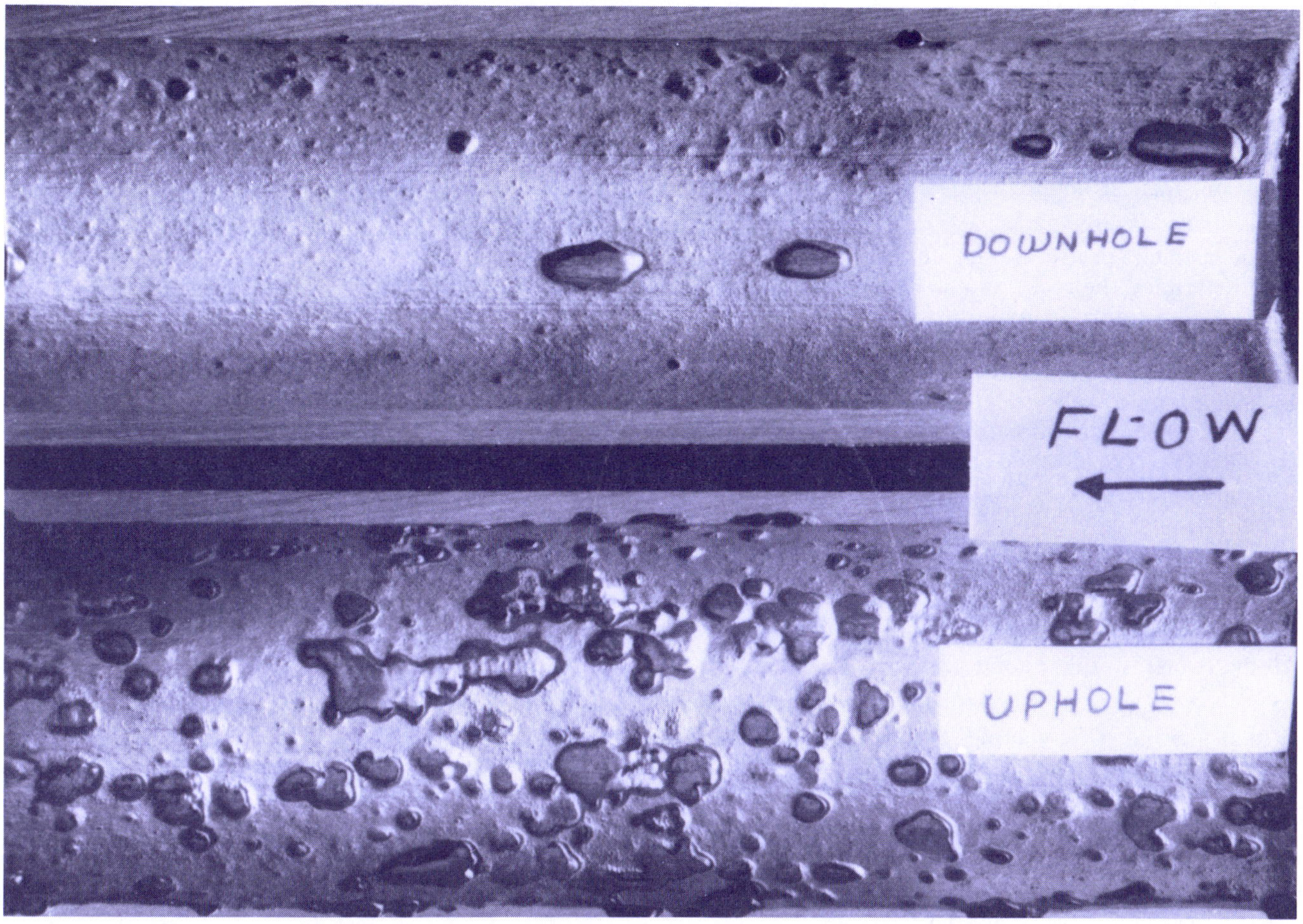

Figure 1.8-3. Gas well tubing corroded by CO_2. Intense corrosion uphole was promoted by water condensation.

were selected for Table 1.8-3 on the basis of proven effectiveness and are not intended to be totally comprehensive. More detail on each of the listed procedures is given as follows:

1. **Intermittent Batch Treatments Down the Casing-Tubing Annulus**

   This procedure is employed in wells without packers and relies largely on film persistency of oily amine-based inhibitors.

   Oil soluble inhibitors that are also water dispersible, appear to work best. Batch volume and injection frequencies vary but typically are 4 to 8 L (1 to 2 gal) injected every two to ten days, depending on corrosion severity. Treatment is always initiated with at least one high concentration (0.3 to 0.5 vol %) inhibitor slug (see Section 1, Item 3.07).

   In wells that pump off or have low fluid levels in the annulus, protection is largely dependent on tenacity of the periodically applied film. Each slug, or batch, is flushed with enough production or water from a truck to carry the inhibitor down the casing. When there is a high standing fluid level through which the inhibitor must pass, it is advisable to circulate each slug down with a volume of produced oil and water (from the tubing) equal to a least one-half the annular volume. In this case, the annular fluid may continue to feed inhibitor down between treatments if the fluid level oscillates.

2. **Treatment of Wells with Casing-Tubing Packers**

   This group includes single tubing completions, multiple completions (tubing and casing production) and tubingless completions.

   With the latter two types, the technique of *formation squeezing* is generally favored. Squeezing or forcing a batch of inhibitor out into the formation, is also used (as indicated in Table 1.8-3) in other wells where frequent batch treatments (see 1 above) cannot be employed.

   In the squeeze technique, a concentrated (5 to 10 vol %) solution of inhibitor in oil is injected into the producing interval. The well is then left shut-in for a period of time (12 to 24 hours) to permit adsorption of inhibitor onto the formation rock. Exposed well equipment is filmed as the high concentration solution goes down. Upon returning the well to production, the inhibitor desorbs from the formation into the produced fluid and maintains the inhibitor film on tubing and other subsurface equipment. The inhibitor and solution concentration are selected on the basis of effectiveness and by preliminary fluid mixing tests to insure that stable emulsions are not formed when the diluted inhibitor mixes with produced water.

Some operators design squeeze treatments by core tests that determine the rock volume required to adsorb a calculated volume of inhibitor. Inhibitor volume is taken to be that amount that will provide 15-25 mg/L of produced fluids over a three to nine month period, plus an allowance for irreversible adsorption. The most common procedure, however, is to simply inject 0.2m^3 (one drum) of as-purchased inhibitor for formations up to 30 m (100 ft) thick and proportionately larger amounts for thicker zones. The as-purchased inhibitor is prediluted to 5 to 10 vol % strength with lease crude or diesel oil prior to injection and this slug is displaced (overflushed) with oil containing 0.5% inhibitor. Addition of a selected demulsifier to the overflush oil is good practice.

Recommended overflush volume is one-half the injected inhibitor slug volume. Excessive overflushing increases irreversible adsorption losses in the formation. Subsequent squeezes with similar volumes usually give longer life than the first one, due to satisfaction of the rock adsorption capacity.

Tubing displacement treatments are performed similarly to squeeze jobs except the 5 to 10 vol % inhibitor solution is simply displaced to the bottom and is *not* squeezed out into the formation. This method has given good results in relatively dry gas production, where inhibitor film life can be six months or longer.

Single tubing gas wells producing at less than 34.5 MPa (5000 psig) pressure and 120°C (250°F) can be treated by batching a 5 to 10 % slug of inhibitor in oil into the tubing and letting it fall. Slug volume is 10 to 20% of total tubing volume. Fall rate varies with inhibitor viscosity and is slowed by chokes or other constrictions. About 250 m (820 ft) per hour is a typical fall rate for deep wells. In high pressure/temperature wells, the slug will sometimes not fall since the oil carrier is miscible with gas in the tubing and is held at the top like a piston. In this case, the regular tubing displacement method, described previously is used.

Wells with packers can also be treated by the *casing displacement* or annular feed method. A pressure-actuated injector valve (side pocket mandrel valve) is placed in the tubing string just above the packer to permit flow of inhibitor solution into the tubing from the casing-tubing annulus. The casing is kept full of inhibitor solution at all times; usually 5 to 10 vol % in oil. Periodic batch treatment is made by means of a surface pump that forces a volume of inhibitor into the casing and through the injector valve into the tubing. Continuous treatment with a water-soluble inhibitor in water solution can also be used. Dispersions of water dispersible inhibitors in water are not used because they are not sufficiently stable. The major problems with this method are (1) only the tubing above the injector valve is treated and (2) the valve often plugs with millscale from the casing and from sediment carried in with the inhibitor solution. Thorough cleaning of the casing prior to start-up and filtration of the inhibitor solution helps avoid valve plugging.

An alternate casing displacement method has recently been described which eliminates the side pocket mandrel valve.[11] The tubing is simply perforated with several holes just above the packer to allow entry of inhibitor solution from the casing annulus. The annulus is filled with inhibitor solution up to a level sufficient to balance bottom hole pressure in the tubing. Treatment is carried out continuously (after initial heavy slug filming) by pumping additional inhibitor solution into the casing at the top of the well.

3. **Gas-Lifted Wells**

These are considered separately due to their special problems. Gas lifting is one common production operation where oxygen can be a cause of subsurface corrosion in addition to H_2S and CO_2. Contamination of lift gas with O_2 has caused rapid gas-line failures and plugged lift valves. The best remedy is to totally exclude air from the lift gas.

Even in the absence of oxygen, entry of gas into the production stream encourages corrosion and tends to shorten inhibitor film life by creating turbulence. Attempts to treat by injecting concentrated as-purchased inhibitor into the gas lines at the surface have failed due to solvent evaporation and sludging. This can be minimized by diluting the inhibitor with oil to a concentration of no more than 10 vol %. Some operators use atomizing nozzles for inhibitor injection into the gas lines to achieve better distribution. Inhibitor treating volume is adjusted to provide 15-25 mg/L of as-purchased inhibitor in the total produced liquid.

Since treatment of the lift gas protects only from the top operating valve upward, supplementary protection for lower equipment is provided by periodic formation squeezes.

4. **Inhibitor Protection of Deep Hot Wells**

These are characteristically gas producers from formations deeper than about 4000 m (13,000 ft). Bottom hole temperatures range from about 150°C (300°F) upward and the gas often contains 5 to 20 mol % CO_2 and from traces to 35 mol % H_2S. Pressure may range from 35 MPa (5000 psia) to 172 MPa (25,000 psia).

Initial water production may be essentially zero but increases in most cases to significant amounts.

When the flowing well temperatures are above 150°C (300°F), use of internal tubing coatings is risky at best and most operators rely on chemical inhibition for protection. Well tubing made of special alloys is also being used to some extent (see Section 1, Item 7.07).

a. *Inhibition Problems*

Attempts have been made to treat deep, hot sour gas producers by continuous circulation of organic inhibitors (in a heavy oil carrier) down the casing when the annulus does not have a packer.

Unfortunately, the produced gas is often undersaturated at well temperature and pressure with hydrocarbons heavier than propane (C_3). As a result, the inhibitor oil carrier is miscible (becomes single phase) with produced gas and literally evaporates leaving a heavy sludge deposit.

The inhibitor evaporation problem can be overcome by diluting with oil and continuously injecting a sufficiently large volume of the solution to saturate the produced gas stream with heavy hydrocarbons.

The degree of undersaturation and consequently, the amount of heavy oil required to maintain the inhibitor in liquid form depends on:

Gas composition and volume
Temperature and pressure
Carrier oil composition

A detailed gas analysis and PVT calculations for the conditions of each well must be made to determine if inhibitor treatment using an oil carrier is feasible. Some wells are being treated by *continuous* application of up to 48 m^3 (300 bbl) of a 1 to 2 % solution of inhibitor each day. The carrier oil properties and volume are selected by PVT testing so that enough solvent oil is left to reach the top of the well in liquid form. Oils with an ASTM boiling range of up to 210°C (410°F) initial with 70% recovery point at 480°C (896°F) have been used. The oil was recovered and recirculated. Most often, however, a diesel oil cut is used and is not recovered for recirculation.

Water solutions of inhibitors have also been applied continuously. At temperatures higher than about 160°C (320°F), many of the water soluble inhibitors have proven ineffective. However, new and more effective water soluble (or permanently water dispersible) products are being developed by chemical manufacturers. Their attractiveness is largely due to the fact that water remains in the liquid state at the high well temperatures and pressures and does not evaporate as oil carriers often do.

b. *Other Inhibitor Application Techniques for Deep Wells*

Formation squeeze treatment of deep, hot zones is not generally practiced due to:

Expense of frequent retreatment,
Fear of formation damage, and
Lack of control over return inhibitor concentration.

The most advanced technology of application is use of special injection tubing for carrying the inhibitor solution directly to the well bottom. Figure 1.8-4 depicts two tubular configurations that have been used for this purpose. These are simply dual or concentric tubing strings, one for inhibitor solution and one for production. The dual string configuration is favored.

An alternate scheme that is growing in popularity uses small diameter inhibitor injection lines. These are similar to subsurface hydraulic control lines. Figure 1.8-5 illustrates one type supplied by Hydril. The nitrile rubber-base jacketing material can be used at up to 163°C (325°F) depending on the nature of the completion fluid left in the casing.

Another type, supplied by NL Sperry-Sun, utilizes a 316 L stainless steel capillary line that is not jacketed. An installed NL Sperry-Sun Chemical Transmission System (CTS) is illustrated in Figure 1.8-6. The capillary line has a wall thickness of 0.5 mm (0.02 in.) and an internal diameter of 1.4 mm (0.054 in.). Flow capacity is 19 L/day (20 qt/day) at 83 MPa (12 ksi) for a 6 cp viscosity fluid. Excellent results with the CTS have been reported.

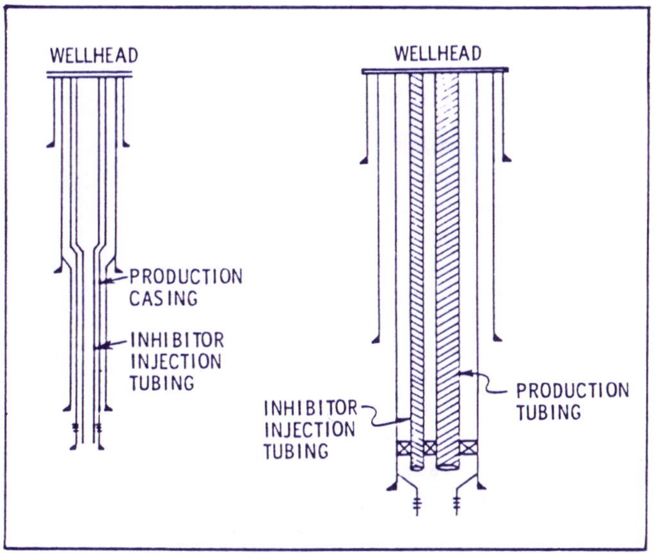

Figure 1.8-4. Two types of dual string completions for inhibitor treatment of deep hot wells.

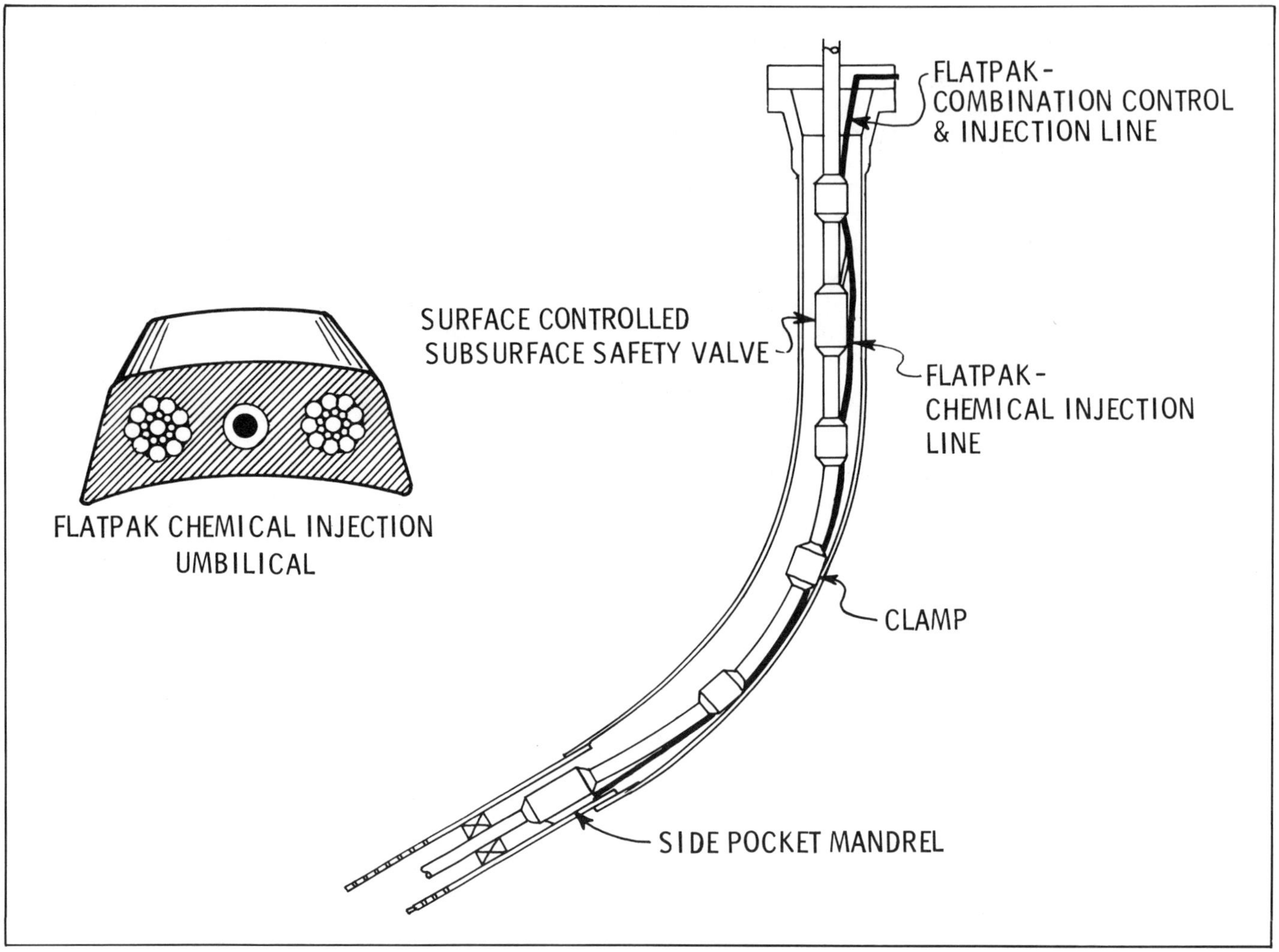

Figure 1.8-5. Cross section and installation of Hydril Flatpak injection line. (Courtesy Hydril Advanced Ocean Systems Division).

8.13 Protection of Surface Production Equipment

Carryover inhibitor from treatments of producing wells is commonly relied on for protection of flowlines and phase separation equipment (gas separators, freewater knockouts, emulsion treaters, etc.)

Additional inhibitor treatment is applied at the surface to high-pressure sour gas lines. Need for additional treatment to supplement carryover inhibitor from treated wells often results from gas cooling and subsequent water condensation. Corrosion can be particularly severe where gas flowlines slope upward and low flow velocity allows the bottom water to move sluggishly up the incline and repeatedly flow back down. Corrosion rates at these points should be monitored regularly with an ultrasonic thickness guage (see Section 1, Chapter 10, Item 10.05).

Holding tanks and similar vessels are coated on the bottom and a distance up the sides where water collects (see Section 1, Chapter 5). Cathodic protection, generally by impressed current anodes, is used to protect heater treaters and similar equipment. NACE Standard RP-05-75 provides guidelines for c.p. installations in these vessels (Section 1, Chapter 4, Reference 9).

Internal flowline coatings such as fusion bonded epoxies are usually cost effective, especially for thin walled ERW pipe handling corrosive production.

A frequent cause of failure in flowlines is concentrated corrosion at welded joints. All welds should conform to published welding standards and should be *post-weld heat treated* for stress relief and uniform metal grain structure.

8.14 Protection of Water Injection Systems

Severe corrosion in water injection (waterflood) systems can most often be traced to contamination by dissolved oxygen. Traces of oxygen in systems handling sour produced water greatly intensify the attack as previously indicated. Carbon dioxide alone is usually not a serious problem because of low temperatures and low gas pressure in most water systems. However, supply water from deep wells is sometimes accompanied by CO_2 rich gas that breaks out of solution and causes cavitation corrosion along the top of flowlines (see Section 1, Item 7.13).

Use of all plastic or either internally plastic coated or cement lined pipe is standard practice for handling waters containing H_2S, O_2 or both. Baked phenolic and fusion-bonded epoxy linings have a good success record for protection of sour water injection flowlines and well tubing.

Steel holding tanks for sour produced water are usually coated internally with polyester or epoxy, or they are cathodically protected. In *all* cases where H_2S is present, the tanks should be fitted with means to exclude air (Section 1, Item 6.04). The most effective procedure for this is to blanket the tank vapor space with air-free gas under positive pressure at all times. Gas supply should be sufficient to hold pressure as the liquid level drops. Ordinary oil is not adequate as an air-excluding barrier.

Waterflood supply waters from surface sources or shallow wells generally contain dissolved oxygen, which must be removed. In neutral low salinity waters, overall corrosion rates are roughly proportional to dissolved O_2 content. However, any concentration above about 0.05 mg/L (50 ppb) can cause pitting and also promotes bacterial growth. Consequently, the best corrosion control procedure is total deaeration. Either vacuum or gas stripping can be employed and properly designed stripping towers can reduce O_2 levels to about 0.1 mg/L (Section 1, Items 6.06 and 6.07).

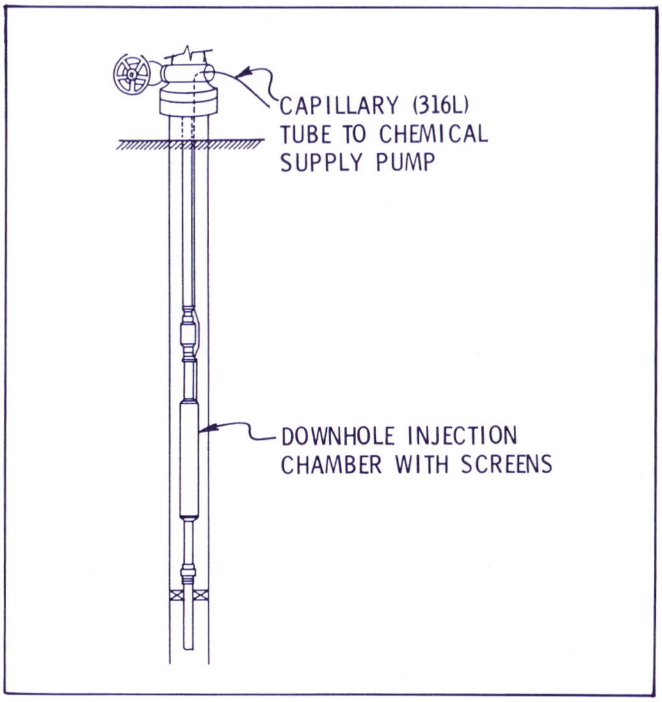

Figure 1.8-6. Installed downhole capillary (0.094" O.D. × 0.054 I.D.) Chemical Transmission System. (Courtesy NL Sperry-Sun).

Residual dissolved oxygen, not removed by stripping, should be reduced to essentially zero by treatment with an oxygen scavenger (see Section 1, Item 6.09). This is especially recommended in offshore operations using seawater for flooding and where deaerated supply water is to be subsequently mixed with sour produced water prior to injection.

Systems handling *air-free* sour water can be adequately inhibited with either water-soluble or water-dispersible organic inhibitors.

Aerated waters, especially where H_2S is also present, are difficult to inhibit. Periodic, high concentration slugs of organic inhibitors can control the attack but such treatments are cumbersome and not generally used.

Deaerated injection waters, such as seawater, which contain soluble sulfate ($SO_4^{=}$) support growth of sulfate reducing bacteria. The organisms metabolize the $SO_4^{=}$ and generate H_2S. Although the sulfate reducers are classed as anaerobic microbes (grow only in the absence of dissolved oxygen), they are most troublesome when traces of O_2 are present, even intermittently. The reason is that the sulfate reducing organisms find a favorable environment on the steel surfaces under aerobic slimes. The H_2S they generate causes both pitting and general corrosion. Treatment with biocides is standard practice when the organisms are determined to be active (see Section 2, Chapter 3).

Periodic cleaning of water lines with scraper pigs is highly beneficial for eliminating slime and solid deposits that promote localized corrosion.

Auxiliary equipment such as pumps and valves in corrosive water systems are constructed of resistant materials including 316 stainless, Monel K500 and Inconels. Grade D aluminum bronze and nickel aluminum bronze are widely used in water pumps. Ceramic-trimmed aluminum bronze gate valves are satisfactory. Additional materials selection information for water systems is given in Section 1, Chapter 7 and NACE Standards RP-04-75 and MR-01-75 (see Section 1, Chapter 7, References 1 and 11).

8.15 Corrosion Monitoring in Water Injection Systems

1. **Chemical Tests**

   Complete mineral analyses, including determinations of iron, H_2S and O_2 content should be run initially on all individual water streams and mixtures. Periodic repeat analyses, especially for H_2S and O_2, should be run during the life of the operation.

Convenient test kits for field measurement of dissolved iron, H_2S and O_2 are supplied by the Hach Company, P. O. Box 389, Loveland, Colorado 80539, U.S.A. Another supplier is Hellige, Inc., 877 Stewart Ave., Garden City, New York 11530, U.S.A.

The need for careful and repeated testing for dissolved O_2 (D.O.) cannot be overemphasized. These tests can be run quite easily in the field with commercial oxygen meters or with the prepackaged test kits. A popular type of oxygen test kit is Chemet™, (supplied by Chemetrics, Inc., Mill Run Drive, Warrenton, Virginia 22186 U.S.A.). Figure 1.8-7 illustrates measurement of dissolved oxygen (D.O.) in a flowing water stream with a Chemet™ ampoule. Color developed in the ampoule (intensity proportional to O_2 concentration) is measured by a comparator that is supplied. Kits covering four concentration ranges, from zero to five ppm O_2, are available. Water salinity and presence of H_2S do *not* interfere with the test.

Several good quality oxygen meters are available for field use. For example, Orbisphere Laboratories, (20902 South Brookhurst, Huntington Beach, California 92646,U.S.A.) supplies a range of oxygen measurement systems. The Orbisphere Microprocessor Oxygen Indicator features digital readout, autocalibration and automatic compensation for temperature and salinity effects. A special sensor is available for use with the Orbisphere D.O. meters when H_2S is present. Most polarographic (membrane) oxygen sensors lose accuracy in sour waters.

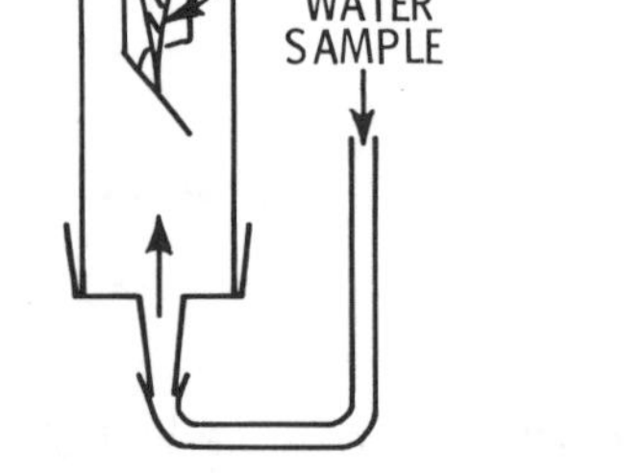

Figure 1.8-7. Measuring D.O. in a flowing stream with a Chemet® Ampoule. (Chemet is a registered trademark of Chemetrics, Inc.)

2. **Corrosion Rate Tests**

A program of corrosion rate monitoring should be initiated *at the start* of injection operations (especially in long life waterfloods) and periodically thereafter. It is especially important to measure corrosion rates at selected test points to detect changes throughout the system over time. Recommended test points in an exemplary system are shown in Figure 1.8-8. Corrosion rate plus oxygen or H_2S concentration tests run periodically at these locations provide data for a continuing corrosion profile of the system. Samples for periodic bacteria counts should also be collected at the same points (see Section 2, Chapter 3 for information on bacteria testing and control).

If desired, commercial corrosion rate meters such as the Magna Corrosometer™ (supplied by Rohrback Instruments, Santa Fe Springs, California 90670, U.S.A.) may be used for corrosion rate monitoring in water injection systems. However, simple weight-loss coupons are generally adequate and have the advantage of showing deposits, bacterial slimes, and pitting. It is definitely advisable to periodically run coupon tests as a double check when the electronic test meters are being used for routine corrosion rate testing. Test coupons are available commercially from chemical suppliers or can be user fabricated. Details on corrosion rate testing with meters, coupons and pipe nipples (spools) are given in Section 1, Chapter 10.

8.16 Gas Processing Equipment

In normal field gas handling operations, corrosion severity is influenced by the same factors previously discussed, viz., partial pressure of CO_2, presence of H_2S, temperature, metal properties and erosion by solids. In compression systems, liquid water forms and corrosion rates are increased at points of water impingement such as in scrubbers and discharge piping. Inhibitors have proven effective for this service, especially compounds which are oil soluble/water dispersible and have a volatile component such as diethylamine. Neutralization by aqueous ammonia injection to pH 7-8 in the condensed water phase is also an effective control procedure.

Non-volatile oily inhibitors can cause fouling of gas dehydrators, especially the molecular sieve solid desiccant type. Ammonia (used as an acid gas neutralizer) tends to accelerate degradation of the molecular sieve desiccants. However, satisfactory corrosion control has been obtained in inlet gas separators, heat exchangers and desiccant bed regeneration equipment by injecting commercial diethylamine* into the gas stream. The inhibitor injection rate is controlled by corrosion rate mon-

*Do not confuse diethylamine, $(C_2H_5)_2NH$, with diethanolamine, $(C_2H_4OH)_2NH$, used for sour gas sweeting (DEA).

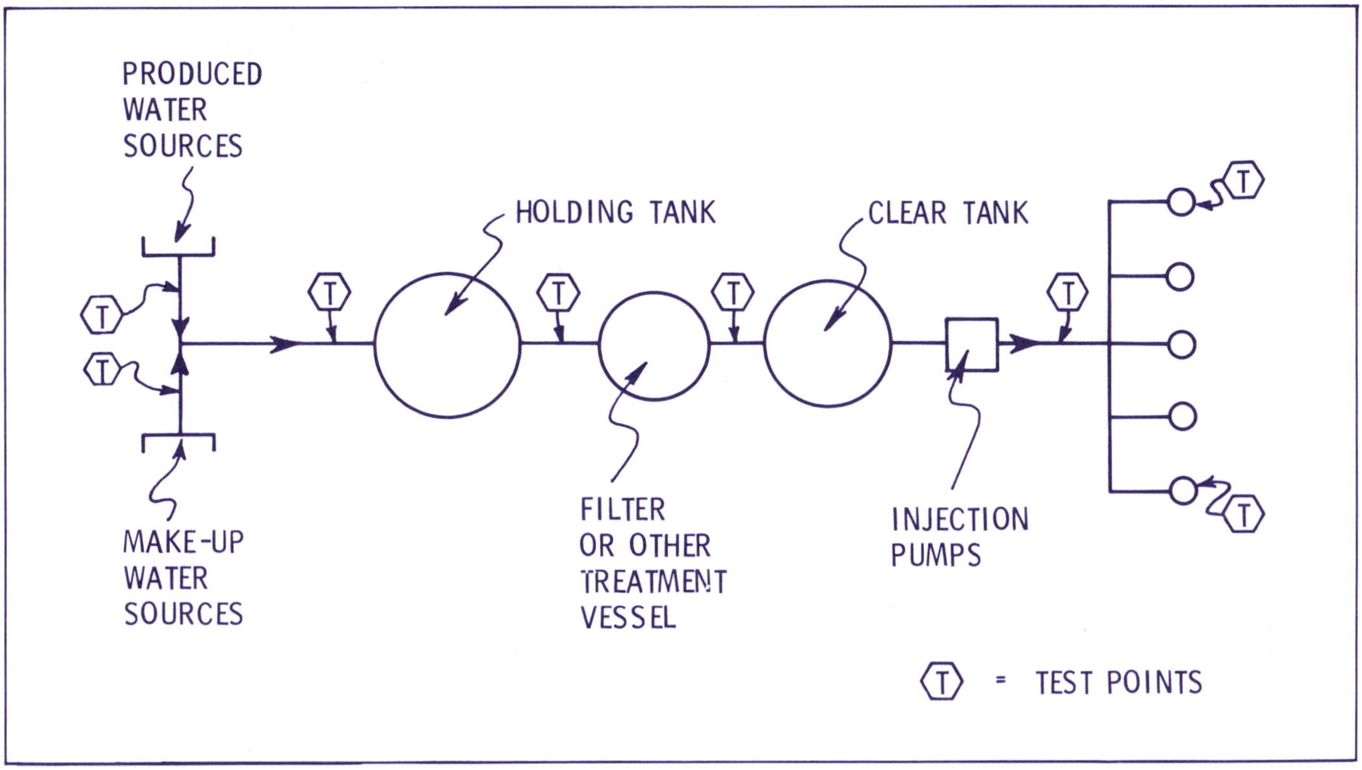

Figure 1.8-8. Recommended test points in an injection system.

itoring. The inhibitor addition rate is usually enough to increase the water phase pH value to 6.5-7.0.

Glycol dehydration equipment can be severely corroded if the glycol solution is allowed to become acidic (low pH) and contaminated with oxygen. The solution pH should be kept in the range of 7.5 to 8.5. Control of pH with a volatile amine neutralizer such as monoethanolamine (MEA) or morpholine helps to retard corrosion in the regenerator overhead. Oxygen corrosion can be reduced by adding 0.5 wt % of sodium MBT inhibitor to the glycol (see Section 1, Chapter 3).

A variety of *acid gas* (H_2S and CO_2) removal (sweetening) processes are used, each with their own characteristics. The process supplier/designer should be requested to provide corrosion control recommendations. The following general guidelines should also be observed:

1. Avoid contamination by oxygen.
2. Keep fluid velocities and temperatures as low as possible.
3. All steels and materials should conform to NACE Standard MR-01-75 to minimize sulfide stress cracking (see Section 1, Item 7.05).
4. Use filtration as appropriate to minimize abrasive particle concentration.

In sweetening systems using MEA or DEA, solution loading should not exceed 0.3-0.4 mols of acid gas/mol of amine. DEA is generally more stable and less likely to cause serious corrosion than MEA.

8.17 Offshore Drilling—Production Operations

Offshore well and production equipment are subjected to the same corrosion problems discussed in previous chapters and essentially similar control procedures are applied. An important difference, however, is in the economics. Repair of an offshore failure generally costs considerably more. In addition, offshore wells are typically high volume producers and erosion-corrosion (caused by a combination of corrosive fluids and entrained solids) is often severe. Special completion techniques involving formation sand consolidation and use of screens and gravel packs are employed to reduce solids erosion in the wells and surface lines. Flowlines are preferably designed to avoid small radius bends (see Section 1, Item 7.09), and a regular program of monitoring metal loss at selected points should be followed (see Section 1, Item 10.05). Steel grade and special alloys for well tubulars should be selected by the guidelines set out in Chapter 7.

Protection of the platform structures from corrosion by the marine environment is accomplished primarily by use of protective coatings (Chapter 5) and cathodic protection (Chapter 4). As an aid in planning a modular protection program, exposure conditions on offshore structures are grouped into five major categories: (1) Atmosphere, (2) Splash zone, (3) Tidal zone, (4) Submerged, and (5) Mud line.

The atmospheric and splash zone surfaces are generally coated, and cathodic protection is employed for the tidal zone and below. Combination coating/cathodic protection systems on submerged areas are being tested with favorable results.[12] Future use is anticipated, especially in deep water. NACE Standard RP-01-76, "Control of Corrosion on Steel, Fixed Offshore Platforms Associated with Petroleum Production," gives information on both coatings and cathodic protection practices. Another good source of pertinent information is API RP-14E, "Design and Installation of Offshore Production Platform Piping Systems."

1. **Coatings**

The most widely used coatings systems for atmospheric exposure consist of zinc-rich primers applied over a near white metal blast-cleaned surface (SP-10 or NACE No. 2) and one or more topcoats of polyamide epoxy, vinyl or acrylic. A polyurethane overcoat over a high-build epoxy topcoat is used for extra protection, especially on surfaces subject to some abrasion.

Detailed information on coatings systems and their application for marine use is available from major suppliers such as Ameron, Carboline, Cook, Ceilcote, Devoe, Glidden, Hempel, and International Paint, to name a few examples.

Some proven splash zone coatings systems are:

a. *Glass flake reinforced epoxies* such as International Paint Intergard No. 4490/4489 Series and Ceilcote Flakeline 600 Series.

The Intergard Series is preferably applied at the shipyard; in two coats over near white metal, to a dry film thickness (DFT) of 0.6 to 0.8 mm (25 to 30 mils). Application temperatures should be +5°C (40°F) or higher for rapid cure. Touchup of damaged areas is feasible with additional material or application of epoxy mastic following solvent cleaning or spot-blasting.

The Flakeline 600 Series is especially useful for low temperature (−12°C or +10°F) application offshore at a DFT of O.2 mm (8 mils).

b. *Glass flake reinforced polyesters* such as Carboline Carboglas 1687, Ceilcote Flakeline 200 and 300 and Napko Polyester Splash Zone Coating No. 9-4110.

The polyesters have excellent bond strength applied over near white metal blasted surfaces but are more brittle than the epoxies, viz., tend to crack if flexed significantly. A disadvantage in cold climates is their requirement of +10°C (+50°F) or higher application temperature. Field repair is feasible and resistance to impact and abrasion is very good. Resistance of polyesters to cathodic disbondment (see Section 1, Item 5.07) is also rated as good.

Recommended DFT (2 coats) ranges from 0.5 mm (20 mils) for Ceilcote Flakeline 300 to as much as 6.4 mm (0.25 in.) for Napko Splash Zone Coating No. 9-4110 applied in one coat.

c. *Silica filled epoxy* such as Ameron Tideguard 171. Applied over near white metal blasted surfaces to a DFT up to 6.4 mm (0.25 in.), the Tideguard 171 has excellent abrasion and impact resistance. However, adhesion strength is less than that of polyesters.

Many other good splash-zone systems are available and the previous proprietary types are listed solely as examples.

Experience with coal tar epoxy coatings in the splash zone has shown mixed results. They have proven generally acceptable in warm climates such as in the Middle East but numerous failures have occurred in the North Sea areas, possibly due to low temperature application. Coal tar epoxy has reportedly given good service on totally submerged portions of riser pipe after warm temperature application to near white metal blasted surfaces.

Other riser pipe external coatings include neoprene rubber (11-13 mm or 0.5 in. thick) in the splash zone, extruded polyalkylene sleeves (polyolefin or polyethylene-propylene) over butyl rubber mastic, fusion bonded epoxy, and hot applied asphalt. For the field joints, heat shrinkable polyalkylene sleeves with butyl rubber adhesive have been used extensively. Heavy duty cold applied polyalkylene tapes and hot applied asphalt tapes have also been used for field joints with varying success.

Fusion bonded epoxies and hot applied asphalts are probably the two most widely used external coatings for corrosion protection of submerged pipe, in conjunction with cathodic protection by zinc anodes.

2. **Cathodic Protection**

Most of the existing offshore fixed platforms are protected below the water line with aluminum alloy sacrificial (galvanic) anodes (see Section 1, Item 4.04). Special aluminum alloys containing zinc and indium are available for high current output and long service life.

Impressed current c.p. systems are used on some structures where current requirements are high and in deep water where the total weight of galvanic anodes is detrimental. A popular type of impressed current anode material is titanium or proprietary inert alloy that is externally clad with platinum.

The most generally accepted criteria of adequate protection is that all submerged surfaces exhibit structure-to-seawater potentials of at least −0.8 volts (−800 mV) vs. Ag/AgCl (silver/silver chloride) reference electrodes (see Section 1, Item 4.10). Potentials higher than about −1.1 volt should generally be avoided to reduce the risk of cathodic disbondment of coatings and possible hydrogen embrittlement of high strength steels.

Historically, offshore c.p. systems have been conservatively designed to provide a uniform average current density (mA per unit area) to all parts of the structure. The applied current density is designed to be sufficiently high to maintain the protected potential of −0.8 volts vs. Ag/AgCl. A majority of the installations are apparently successful. However, it has been recognized for at least 25 years that optimum current densities vary greatly from location to location and also from top to bottom on individual structures because of differences in the exposure environment.

The general influence of environment on c.p. current density requirements are discussed in Section 1, Item 4.02. For offshore structures, the major factors that influence current density requirements are seawater velocity, oxygen content, temperature and metal surface deposits such as rust and calcareous mineral scales. The latter form as a result of cathodic protection. Seawater salinity is fairly uniform around the world but this can be another variable for close-to-shore installations.

It has been found that best protection and c.p. efficiency are achieved when adequate c.p. current reaches all bare metal to build and maintain a thin but dense and uniform layer of the cathodic calcareous scale. For example, Gartland, et al., (Chapter 4, Reference 1) ran long-term cathodic polarization tests on steel specimens mounted at different depths on a platform in the North Sea Frigg Field. Only minor sensitivity to depth and seasonal variations were found as long as the calcareous deposits on the specimens remain intact. However, loss of scale, particularly at shallow depths during winter storms, strongly affected average current densities required for maintaining protection. The observed current densities required ranged from a summer low of 62 mA/m^2 (6 mA/ft^2) to a winter high of 340 mA/m^2 (32 mA/ft^2). Based on the test data, an initial current density of 160 mA/m^2 (15 mA/ft^2) appeared to be adequate at medium depths under average conditions. Det norske Veritas (Norway) recommends use of 130 and 160 mA/m^2, as initial current densities for the southern and northern parts of the North Sea respectively.

An example of a location where current density requirements are very high is the Cook Inlet in Alaska. Tidal velocities there can exceed 1.5 m/sec (5 ft/sec) and the water contains glacial silt which scours the platform legs and cross braces, exposing the surfaces to oxygen saturated salt water. Laboratory model tests and field experience determined that current densities of at least about 860 mA/m^2 (80 mA/ft^2) are required for protection. These high current levels were attained with impressed current systems using d.c. generators (on the decks) and platinum clad anodes mounted on concrete sleds resting on the sea floor around and under each platform. Protected structure-to-water potentials were not attained until current densities were elevated sufficiently to produce the white calcareous scale on the upper leg jackets and braces.

Long term experience indicates current densities of 86 to 107 mA/m^2 (8 to 10 mA/ft^2) are generally sufficient in relatively warm and quiet seawater such as in the Gulf of Mexico. High efficiency aluminum alloy anodes perform well in both the Gulf of Mexico and the North Sea but the installations must be designed carefully to achieve uniform protection.

Current density requirement estimates are used to compute total current in amperes needed for incremental sections of the platform structures. High efficiency aluminum alloy anodes produce about 2756 to 2844 ampere hours per kilogram (1250 to 1290 ampere hours per pound). Using the rated anode output, the total current requirement and desired life (typically 10-20 years), the total number and size of anodes required are calculated. The anodes are mounted with stand-off brackets at strategic locations on the platform braces and leg jackets. Positioning of individual anodes is guided by a master plan drawn up by the cathodic protection designer and installation is usually made during the latter phases of platform construction.

Ideally the anodes are positioned so that all structure surfaces are in direct line of sight of at least one anode. This is particularly important for protection of congested brace nodes where mechanical stress is highest and corrosion cracking or pitting is least tolerable.

The problem of anode placement to achieve optimum distribution is still being studied and will become even more important in the future. The impetus is the escalating cost of cathode protection as the complexity and size of offshore structures increase with deep water drilling. A computer current density (CCD) technique for predicting the distribution of c.p. current was recently published by DeCarlo (Section 1, Chapter 4, Reference 8).

Portions of the platform leg jackets and support piling at and below the mud line (sea floor) are usually protected by the c.p. system used for the submerged parts. These areas should be taken into account in computing the total submerged area to be protected, especially in locations where water motion can cause the sea floor to shift and move during storms.

REFERENCES

(1) *Oil and Gas Journal*, November 4, 1985.

(2) API RP 13B, "Standard Procedure for Testing Drilling Fluids," February, 1971.

(3) Dabkowski, John, "Cathodic Protection of Storage Field Well Casings," *Materials Performance* (NACE), vol 25, January, 1986, p 17.

(4) Byars, H. G., and Galbraith, J. M., *Advances in CO_2 Corrosion* (NACE), vol 1, p 103, 1984.

(5) "Corrosion of Oil and Gas Well Equipment," NACE-API (published by API, 1958).

(6) DeWaard, C, and Milliams, D. E., *Corrosion*, vol 32, no 5, p 177, 1975.

(7) Schmitt, G., "Fundamental Aspects of CO_2 Corrosion," *Advances in in CO_2 Corrosion* (NACE), vol 1, p 10, 1984.

(8) Burke, P. A., and Hausler, R. H., *Advances in CO_2 Corrosion* (NACE), vol 2, p 133, 1985.

(9) Bacon, T. X., *Oil and Gas Journal*, vol 44, no 52, p 119, 1946.

(10) Crolet, J. L., and Bonis, M. R., *Advances in CO_2 Corrosion* (NACE), vol 2, p 23, 1985.

(11) Hausler, R. H., and Weeks, S. G., "Low Cost, Low Volume, Continuous Inhibition of Gas Production Tubulars," *Materials Performance* (NACE), vol 25, no 6, June, 1986.

(12) Thomason, W. H., Pape, S. E., and Evans, S., "The Use of Coatings to Supplement Cathodic Protection of Offshore Structures," *Materials Performance* (NACE), vol 26, no 11, November, 1987, p 22.

9 Corrosion Control in Enhanced Recovery (EOR) Projects

9.01 Carbon Dioxide (CO_2) EOR Systems

Operations utilizing CO_2 injection to increase oil production are growing in number around the world. The SACROC CO_2 project in southwest Texas is one of the oldest, having been initiated in 1972. Corrosion continues to be a serious problem even in new systems that have operated less than two years. Although considerable research and field testing of corrosion resistant materials are still going on, enough information has accumulated to establish general guidelines for corrosion control.

Several schemes for utilizing CO_2 for increasing ultimate oil recovery have been proposed and tried. In general, the recovery procedures depend on reduction of reservoir oil viscosity by dissolved CO_2 gas and miscible displacement of the oil by a CO_2 rich liquid. In one popular process, CO_2 gas injection is initially alternated with water injection to improve sweep efficiency (the water-alternating-gas or WAG process) followed by ordinary water or carbonated water as a drive fluid.

CO_2 gas is obtained from gas processing plants and from wells that produce high purity carbon dioxide. The CO_2 gas is compressed to the supercritical state (about twice the critical pressure, e.g., about 15 MPa or 2177 psi) prior to delivery to a distribution pipeline and ultimately to the individual injection wells.

1. **Corrosion Severity and Mechanisms**

   Item 1.18 in Section 1, discusses the relatively mild corrosivity of CO_2 (in H_2O) to carbon steel at low CO_2 concentrations and low temperatures. However, as discussed in Item 8.11a, Section 1, corrosion by CO_2 can be intense at high partial pressures and high temperatures. The attack tends to be localized in the form of deep pits (see Figure 1.8-3, Chapter 8) but the pits sometimes grow together to form channels or large "washed out" areas. At temperatures higher than 60°C (140°F) patches of iron carbonate corrosion product often shield and partially protect areas of metal while adjacent areas are corroding vigorously, leaving islands of metal. Corrosion rates in the range of 3 to 6 mm/yr (120 to 240 mpy) have been observed at actively corroding sites, both in the laboratory and under field conditions.

   A poorly understood fact is that carbonic acid (H_2CO_3 which is CO_2 in H_2O) can be more corrosive than a stronger acid (such as HCl) at the same level of acidity (pH). Corrosion resulting in hydrogen evolution may occur at pH 6 with carbon dioxide, whereas pH 4 is required for hydrogen evolution with highly ionized hydrochloric acid.[1] One study[2] has suggested that the carbonic acid molecule (H_2CO_3) is electrochemically reduced (see Items 1.06 and 1.07, Section 1) directly at cathode sites without being ionized first, e.g.:

   $$\underset{\text{carbonic acid}}{2(H_2CO_3)} + 2e^- \rightarrow \underset{\text{hydrogen gas}}{H_2\uparrow} + \underset{\text{bicarbonate ion}}{2HCO_3^-}$$

   CO_2 dissolved in water which is in contact with essentially pure CO_2 gas, exists in solution primarily in the form of molecular CO_2 or hydrated carbon dioxide ($CO_2{\cdot}H_2O_x$). Rapid cathodic depolarization by direct reduction of molecular carbonic acid (instead of reduction of hydrogen ions, H^+) may be a major reason why high concentrations of CO_2 are extremely corrosive.

   Numerous publications have described corrosion by CO_2 in gas producing wells[3] and in the SACROC CO_2 Project.[4,5,6]

2. **Corrosion Control Procedures in CO_2 Environments**

   The following discussion is drawn largely from the excellent papers by Newton, Gill, et al.,[4,5,6] this author's experience, contributions by various individuals and equipment and chemical suppliers.

a. *CO_2 Gas Supply Wells*

These are generally single tubing completions with tubing-casing packers. Tubing is typically J-55, internally plastic coated with 6 to 8 mils (0.18 mm) of epoxy modified phenolic over a phenolic primer (examples: Baker Tubular Service Plasticap 400 and AMF Tuboscope TK 69).

Packer fluids consist of salt water treated with liquid sulfite oxygen scavenger (200 ppm) and a water soluble inhibitor (500 ppm).

b. *CO_2 Gas Supply Wellheads*

Wellheads for CO_2 gas wells producing very little water are generally constructed of internally plastic coated carbon steel, preferably AISI 4130/4140 grade. Either aluminum bronze (grade 9D) or 316 stainless steel valves with ceramic gates are commonly used.

All stainless steel (304 or 316) wellheads have been used successfully where the produced CO_2 gas contains more than about 0.04 m^3 (1 bbl) of water per 28 m^3 (1 million ft^3). However, the wellhead temperature must be below 65°C (150°F) to avoid risk of CSC cracking of the austenitic stainless steels by chlorides in the produced water (see Section 1, Item 7.07).

No reports have been found on use of high chromium ferritic stainless steel wellheads for this application. The stabilized 18 Cr-2 Mo ferritic stainless can be machined and welded, is priced in the range of the 18 Cr - 8 Ni austenitic 316 stainless and is not susceptable to CSC.

A complete line of wellhead accessories including gate valves, tubing hangers and adapter flanges are available which are protected internally by corrosion resistant alloy liners applied by the hot isostatic pressing (HIP) process.[7] With increasing availability, more widespread use of internally clad equipment is anticipated.

c. *Gas Gathering Lines*

Wet (water containing) CO_2 gas collected from CO_2 supply wells, or CO_2 rich gas from production facilities, is handled in carbon steel gathering lines which may be either threaded and coupled or welded joint ERW pipe.

Bare steel lines are protected internally with chemical inhibitors. Water soluble inhibitors are preferred by many operators especially where there is sufficient water to accumulate in low segments of the flowlines. When H_2S is also present in CO_2 rich wet gas from production separators, oil soluble/water dispersible inhibitors such as those commonly used for downhole well treating are employed for gathering line protection. Inhibitor treating rates are adjusted to continuously provide low corrosion rates on weight loss test coupons (see Section 1, Chapter 10).

Gas gathering lines can also be protected by internal plastic coatings (such as Baker BTS Plasticap 400 or AMF Tuboscope TK69) or by high pozzolan sulfate resistant cement linings.

Buried lines are coated externally, usually with coal tar enamel, and cathodically protected.

Aluminum bronze (grade 9D) or 410 stainless steel gate valves (such as Cameron FL[7]) are used at temperatures below 120°C (250°F). Higher temperatures are not likely in the gathering systems.

d. *Compressors Stations*

AISI 304L stainless steel has provided good corrosion protection for the compressor station wet piping, and cooling coils.[6]

Compressor piping and interstage coolers constructed of carbon steel have also been protected by continuous upstream injection of oil soluble/water dispersible inhibitors. Inhibitor loss through the water drains has been a problem and multipoint inhibitor injection has been used to insure that all surfaces are constantly filmed. Excess carry-through inhibitor must be removed ahead of the dehydrators to prevent fouling.

The wet CO_2 stream should pass through the tube section of heat exchangers rather than through the shell which is usually constructed of carbon steel. Tubes made of 316L stainless steel have performed satisfactorily at temperatures below about 95°C (200°F). Ferritic stainless alloys such as E-Brite 26-1 (26 Cr-1 Mo, ASTM Grade XM-27, Airco Vacuum Metals) should be suitable for this type application at higher temperatures in the presence of chlorides and H_2S. Tubes internally clad with a high nickel alloy such as Inconel 625 should also perform well.[7]

e. *CO_2 Pipelines and Distribution Lines*

Dehydration is the primary method of corrosion control for pipelines and distribution lines between compressor stations and injection wells. The CO_2 is usually dehydrated between the third and fourth stages of compression by counter current stripping in dehydration towers using triethylene glycol (TEG). Residual water content should be sufficiently low that no liquid water phase forms at the lowest temperature and highest pressure encountered in the compressed gas distribution system (50 ppm maximum at SACROC[6]). Type 304 stainless steel is generally used for dehydrator internals. The 18 Cr - 2 Mo ferritic stainless should also be satisfactory.

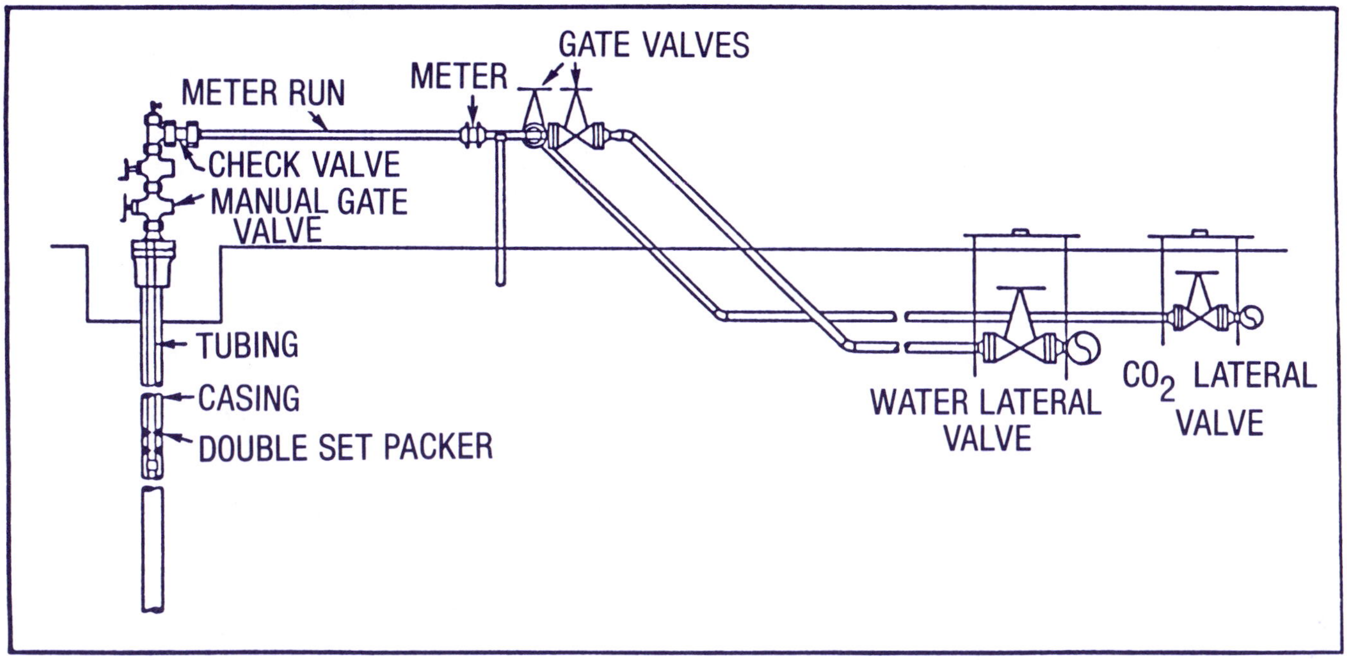

Figure 1.9-1. Typical CO_2/water injection well. (After Newton[6]).

f. CO_2/Water Injection Wellheads

An injection well arrangement for alternated CO_2/water injection (WAG) is illustrated in Figure 1.9-1. In another design, the CO_2 and water injection flowlines are connected separately to the injection wellhead.

Common practice now is to completely isolate the incoming water flowline with a blind flange during the CO_2 injection periods. This has eliminated severe corrosion problems created when high pressure CO_2 seeps past ordinary ball or gate valves and into the shut-off water line. The CO_2 line is similarly blocked during water injection cycles.

In early installations, internally plastic coated meter runs and carbon steel wellheads were tried but corroded rapidly at points where coating damage occurred, particularly at flange faces. Wellhead and valve parts that were coated internally with electroless nickel failed catastrophically due to intense galvanic corrosion at breaks in the nickel coating.[6] Failure of 410 (13% Cr) stainless steel wellheads was also reported. Type 316 stainless steel trees, meter runs and valves were used for replacement with no apparent problems. Ceramic gates in 316 stainless steel valves have given good service.

During the last 3 years, over 500 injection wellheads constructed of 392 aluminum bronze have been installed in a new CO_2 project in the Permian basis. Prior tests indicate these should give good service.

g. Water Gathering and Injection Systems

A variety of materials are used for water distribution lines depending on size and pressure, including fiberglass reinforced polyester or epoxy plastic (FRP) up to about 13.8 MPa (2000 psi) (see Chapter 5, Item 5.02-3).

Cement lined pipe in sizes from 7 to 36 cm (3 to 14 in.) OD has been installed. Cement linings are high density sulfate resistant grade with artificial pozzolans but no sand or inert fillers.[9] The linings are centrifugally cast inside the pipe joints at a coating shop.

API RP10E Specification describes joint welding procedures to be used in the field for cement lined pipe. Alternately, joining by couplings with molded plastic coupling protection inserts or with rubber ring seals may be used. Welding is the most commonly used procedure. Newton reports numerous leaks at field welded joints when asbestos gaskets are used at the joints for protection of the cement linings during the welding operation.[6] Grout (such as X-Pando Special Compound No. 2) is now used instead of the asbestos gaskets in the SACROC project and experience with the grouted joints is reported to be good. Others contend that asbestos gaskets are generally preferable.[9] API RP10E allows use of either gaskets or grout at welded joints.

Water holding tanks are internally coated with glass filled epoxy plastic and gas blanketed to exclude air (see Section 1, Item 8.14).

h. *Downhole Corrosion in Injection Wells*

Some of the most intense corrosion encountered in CO_2 projects (up to 6 mm/yr or 240 mpy) has occured on the exterior of the injection tubing and probably inside the casing also. The tubing strings are set on casing packers and the annuli are filled with deaerated salt water treated with water soluble inhibitors.

During CO_2 injection cycles, the gas has leaked through tubing joints and up past the packers, saturating the salt water packer fluid. Leakage past the packers has been controlled by use of plastic coated double set packers fitted with CO_2 resistant elastomer seals. Use of special PTFE (Teflon) thread lubricant such as Baker Super B has reduced frequency of joint leaks.[10]

Tubing that is internally plastic coated with shop applied thin film (0.18 mm or 6-8 mils) epoxy modified phenolics over phenolic primer (such as Baker BTS PA400 or AMF Tuboscope TK69) has held up satisfactorily. However, corrosion of coupling and pin threads with subsequent leakage has occurred. The internal coupling corrosion problem has been greatly reduced by coating them with polyphenylene sulfide coatings such as Tuboscope Coupling Guard or Baker BTS Ryton.

Some operators are choosing thicker film (0.32 mm or 16 mil) internal tubular coatings to reduce wire line and mechanical damage during well testing and workovers. Examples are the fusion applied epoxy-phenolics such as AMF Tuboscope TK77 or Baker BTS Plasticap 226 and modified high build epoxies such as Baker BTS Plasticap 521.

A program of periodically pulling and inspecting injection tubing has been a factor in increasing service life by preventing the rerunning of defective joints.[6] Inspection procedures include visual examination after cleaning couplings and pins, electromagnetic and ultrasonic wall thickness gauging. With wall thickness reductions of 30% or less, the joints are reworked and recoated as required prior to further use.

In the SACROC operation, some used tubing has been cement lined for reuse as injection tubing. Other operators have reported cement lining of used tubing for use as flowlines. An acid resistant cement (Permian Enterprises Flouroline) has been developed which allows use of cement lined injection tubing in CO_2 injection wells.[9] The lining face is chemically treated to provide resistance not only to CO_2, but also to stimulation acids used to treat limestone formations.

i. *Producing Wells and Flowlines*

CO_2 breakthrough and increased water cut at producing wells can occur within a few months after initial injection into fractured dolomite formations. Corrosion rates typically increase 2 to 4 fold on wellhead test coupons and downhole equipment failure rates increase accordingly in wells with minimal or no inhibitor treatment.

Many early tubing failures were attributed to ringworm corrosion at the upset ends of the joints. In common with previous experience in gas condensate wells producing CO_2 rich gas, the ringworm corrosion problem has been eliminated by full length normalizing heat treatment of tubing after upsetting (see Section 1, Item 1.14).

Ordinary batch treatments down the casing annulus with filming amine corrosion inhibitors are usually continued in rod pumped wells. Overall results have been satisfactory even after the CO_2 content of the produced gas increases to 15% or more where bottom hole temperatures are below about 70°C (160°F). Treating frequencies are increased from perhaps once every 7 to 10 days to 2 or 3 times each week as CO_2 and water production increases.

Rod pumps using Monel 400 for barrels, plungers and all wetted parts have operated without problems.[11]

Rod body failures have been drastically reduced by switching to fiberglass reinforced plastic (FRP) sucker rods and by plasma coating steel rods with 316 stainless steel overcoated with baked on epoxy. Careful design and operation of fiberglass rodstrings is necessary since the standard API guidelines for fatigue life for steel rods do not apply.[12] FRP sucker rod fatigue life predictions and design guidelines are supplied by the manufacturers. Stainless steel coated or corrosion resistant alloy rods are generally used for the lower 1/3 of the rod strings (for weight) when FRP rods are installed.

Increased water production places greater stress on pump rods and often necessitates installation of electric submersible pumps (ESP) to economically handle the larger fluid volumes.

Electric cable failures are a major problem with ESP due to permeation and swelling of the jacket elastomers by CO_2 and methane. Gas resistant jacketing materials encased in metal armor is generally cost effective by providing longer runs between pump changes.

Corrosion of ESP motor and pump housings has caused most failures in SACROC.[6] Squeeze inhibition is not considered practical due to the very high fluid production rates and continuous treatment below the motor assembly via capillary tubing (see Chapter 8, Item 8.12d) would present mechanical problems due to space limitations. Type 304 stainless steel has given good service for motor and pump housings. The 18 Cr - 2 Mo ferritic stainless should also perform well especially in hot (> 100°C or 212°F) wells producing high chloride brines.

Existing bare steel production flowlines are protected by the increased inhibitor treatments down the producing wells, at least during the early life of the CO_2 projects. Replacements are commonly made with fiberglass reinforced epoxy pipe or with internally epoxy coated steel pipe for high pressure service. External protection of buried flowlines is provided by coatings, e.g., coal tar enamel with glass wrap, and cathodic protection.

j. CO_2 Removal Plants

CO_2 is extracted from produced gas for recycling to the injection operations. Both hot potassium carbonate and monoethanolamine (MEA) processes are used at SACROC.[6] Reboiler shells for hot potassium carbonate plants are now constructed with 316L stainless steel and 304L is used for piping replacement. Upper wall sections of stripping columns are internally clad with 304L stainless steel. The ferritic stainless steels containing at least 18% Cr and 1 to 8% Mo should be suitable cladding materials for this service.

Corrosion in MEA plants built of carbon steel with stainless steel trays, overhead piping and valves, can be severe when conditions are such that MEA degradation products (HEED) become concentrated. The corrosion can be minimized with special MEA inhibitors and close control of the sweetening operation (see Chapter 8, Item 8.16).

3. **Alloys for CO_2/H_2S Environments**

Alloys with as little as 13% Cr content (AISI 410/420) have good resistance to corrosion by CO_2 only at CO_2 partial pressures (P_{CO_2}) up to 3.0 MPa (435 psi) and temperatures up to 150°C (300°F).[13] Resistance increases with increasing Cr content and the 18% Cr-8 Ni (304 and 316) stainless steels have been found to be satisfactory at CO_2 concentrations equivalent to P_{CO_2} = 15 MPa (2177 psi) at low temperatures (< 80°C or 176°F) as discussed in the preceding portion of this chapter. However, the 18 to 25% chromium alloys with less than 30% Ni are generally susceptible to sulfide stress cracking (SSC) and stress corrosion cracking (SCC) as discussed in Section 1, Item 7.09. Further, their resistance to pitting corrosion diminishes in the presence of H_2S and chlorides, especially at high temperatures (> 80°C) or with contamination by O_2[14]

The potential for failure of high Cr alloys in the presence of H_2S, together with their relatively high cost, has prompted the search for more economical resistant alloys. Mention was made previously of the use of aluminum bronze alloys, such as alloy 9D and nickel aluminum bronze. These have a proven record of corrosion resistant in sour water injection systems (see Chapter 8, Item 8.14) and they are apparently providing good service in CO_2 EOR operations.

Alloys with more than 90% aluminum (Al) content are used for construction of decks (roofs) on sour oil storage tanks and in gas plant and refinery operations where the metal is exposed to low salinity or condensed water in presence of CO_2, H_2S and O_2 at low temperatures (< 40°C or 104°F). Data recently reported,[15] indicate that certain Al alloys such as Alcoa 6061-T6, X6013-T6, and 5086 may be useful in CO_2 EOR environments including wet field CO_2 and CO_2/H_2S mixtures. Caution should be exercised in exposing Al alloys to produced brines containing Ca and Mg salts in addition to NaCl.

9.02 Steam Flooding and In-Situ Combustion Projects

Both of these are thermal recovery techniques in that the oil bearing reservoirs are heated. Steam flooding is still a viable enhanced recovery method and will probably remain so, especially for heavy oils. Underground (in-situ) combustion projects are declining in number, due in large part to extremely severe corrosion problems in both injection and producing wells.

1. **Corrosion in Steam Floods**

Perhaps the most publicized corrosion problem in steam flooding is that of failure of the water tubes in steam generators. Water side scale deposits cause local overheating and rupture of tubes when the boiler feedwater is not properly softened to essentially zero hardness (see Section 2, Chapter 1). The problem has been largely overcome by using two water softeners in series (one primary and one for polish softening) with a third unit under regeneration.

Even with series softeners, hardness removal is not always complete and at least 1 to 2 mg/L of hardness (calcium, magnesium and iron in solution) often remains. Precipitation of residual hardness as scale in the boiler is prevented by continuously feeding a chelating chemical into the boiler feedwater downstream of the softening units. Chelating chemicals such as sodium EDTA keep the hardness materials in solution. However, an excess of chemical over the amount required to prevent hardness precipitation is very corrosive and has been the cause of rapid failure of boiler tubes by deep, isolated pitting. Good practice requires careful operation of the softening units so that no more than about 10 mg/L of chelating chemical treatment is required.

Fresh (low salinity) water used for steam generation must be treated for complete removal of dissolved oxygen upstream of the softening units. Traces (< 5 mg/L) of H_2S are removed with chlorine (see Section 1, Chapter 6).

Steam from the boilers usually contains CO_2 which can cause corrosion of injection lines and injection well tubing. However, this is generally not a problem because of high alkalinity (pH 10 to 12) of the liquid water coming from the boilers that is injected along with the steam (typically 80% steam and 20% liquid water by weight). Caustic (sodium hydroxide) is added to the boiler feedwater (usually as a 50% water solution) to maintain the high alkalinity in the boiler output. The added caustic alkalinity prevents CO_2 corrosion in the injection system and also helps prevent precipitation of silica scale in the boilers.

One of the major problems in the steam injection wells is casing failure due to compressive stresses created by thermal expansion followed by tensile stressing during cooling. Most failures are located at the couplings.[11] Steel grade P105 is probably the best of the common low alloy steels used for casings. Pre-stressing the casing in tension prior to cementing helps alleviate thermally induced stress cycles. Special expansion joints are often employed in both the tubing and casing strings of injection and producing wells.

2. **Corrosion in Combustion Operations**

Economical solutions have not been developed for many of the serious corrosion problems encountered in combustion operations where some of the reservoir oil is literally burned in place to generate heat and steam.

Air is introduced by means of injection wells and a combustion zone or fire front is created which expands through the reservoir toward the producers. The bank of heat and steam that is generated drives oil and water ahead but enough oil is left behind to support propagation of the fire front. In the burning process, carbon dioxide, carbon monoxide and a variety of corrosive organic acids are formed. These and excess oxygen at high temperatures create an extremely corrosive environment.

Casing joints and well screens made of high chromium and nickel alloys such as Incoloy 825 and Inconel 625 (see Table 1.7-10, Chapter 7) have been tested in both injection and producing wells. General performance was better than lower alloys but even these have failed when exposed to temperatures in excess of 650°C (1200°F) in the presence of the combustion products.

Successful operation of firefloods (from a corrosion standpoint) appears to be primarily a matter of avoiding exposure of well equipment to the excessive heat. Procedures to initiate combustion a safe distance away from injection wells have been worked out but rock geometry often makes such control difficult. In general, substantial destruction of the lower exposed producing well equipment can be expected when the combustion front arrives and passes. However, ordinary carbon steels such as J-55, K-40, etc., have failed prematurely and caused early abandonment. Producing well screens and casing liners made of 304 or 316 austenitic stainless steels have failed by chloride stress cracking at temperatures above 100°C (212°F) but below 210°C (410°F). Balancing alloy cost versus prolonged life (prior to front arrival) for the lower end of producing wells, it appears that 9% Cr - 1 Mo ferritic stainless steel is in general a good choice as a material of construction.

In addition to temperature, oxygen from the injected air is another key factor affecting corrosion rates at the producing wells. Prior to oxygen breakthrough, ordinary filming amine corrosion inhibitors have given adequate protection by the usual batch injection treatment method. The quantity and frequency of application for the inhibitor treatments must be increased, roughly in proportion to the amount of oxygen reaching the producing wells.

Oxidation inhibitors have performed quite satisfactorily for protecting air delivery lines and injection well tubing during air injection cycles. However, these products have failed to control corrosion by injection water containing dissolved oxygen. Best practice is to deaerate the injection water during the water injection cycles.

9.03 Micellar (Surfactant)/Polymer Flooding

The major corrosion problems reported in micellar-polymer flooding have been in the injection system. The micellar surfactants generally consist of substantially oil soluble sulfonates mixed with various alcohols and water. A slug or bank of the surfactant mixture is injected and followed by water containing polymers such as polyacrylamides or xanthans. The polymers reduce mobility of the driving water to retard forward fingering and to promote piston-like displacement of the leading surfactant bank.

Concentrated surfactant solutions have a strong detergent effect on surfaces of metal tanks and lines and tend to remove millscale and metal oxides, thus exposing the surfaces to general corrosion by dissolved oxygen. Exposure times prior to and during injection are usually limited and damage to the field equipment is of no great economic importance. Contamination of the surfactant solutions by dislodged rust and surface soil can be harmful however, and the solutions are usually filtered to remove suspended debris while they are being injected.

Injection of chase water containing polymers can continue for several months or years and common practice is to deaerate the water to prevent rusting and consequent contamination by iron oxides. Dissolved oxygen also degrades the polymer chemicals. Because metal ions such as cobalt affect polymer properties, non-catalyzed oxygen scavenger chemicals are used.

Sulfate reducing bacteria (see Section 2, Chapter 3) often thrive in the deaerated polymer treated water and the generated H_2S can, in time, cause significant corrosion damage. However, a more serious problem is plugging of the injection wells by mixtures of iron sulfide corrosion products and incompletely dissolved polymer. A variety of biocide chemicals have been used to retard bacterial growth and iron sulfide generation. If the water is kept totally air-free, biocides such as glutaraldehyde, thiocarbamates and quaternary ammonium chlorides have proven effective. Oxidizing biocides such as chlorine cannot be used because they degrade the polymers.

REFERENCES

(1) Ostroff, A. G.: *Introduction to Oilfield Water Technology*, (NACE), 1979.

(2) Dewaard, C., and Milliams, D. E.: "Carbonic Acid Corrosion of Steel," *Corrosion* (NACE), vol 1, no 5, May, 1975.

(3) Burke, P. A., and Hausler, R. H.: "Assessment of CO_2 Corrosion in the Cotton Valley Limestone Trend," *Materials Performance* (NACE), August, 1985.

(4) Newton, L. E., Jr., and McClay, R. A.: "Corrosion and Operational Problems, CO_2 Project, SACROC Unit," SPE Paper No. 6391, March, 1977.

(5) Gill, T. E.: "Ten Years of Handling CO_2 for SACROC Unit," SPE Paper No. 11162, September, 1982.

(6) Newton, L. E., Jr.: "SACROC CO_2 Project—Corrosion Problems and Solutions," *Materials Performance* (NACE), May, 1984.

(7) *Cameron Products for Enhanced Oilfield Recovery*, Cameron Iron Works, P. O. Box 1212, Houston, TX 77251-1212.

(8) Martin, R. L., "Corrosion Control by Inhibitors in Producing Oil Wells in Carbon Dioxide Flooded Fields," *Advances in CO_2 Corrosion* (NACE), vol 2, p 250, 1985.

(9) Permian Enterprises, Inc., P. O. Box 2745, Midland, TX 79702.

(10) Baskin, G. M., and Spriggs, D. M., "Wellbore Operations in CO_2 Flooding—A Case History," *Advances in CO_2 Corrosion* (NACE), vol 2, p 258, 1985.

(11) Craig, B. D.: *Practical Oil-Field Metallurgy*, PennWell Publishing Company, Tulsa, OK, 1984.

(12) Hallden, D. F.: "Enhancement of Fiberglass Sucker Rod Design is Offered," *Oil and Gas Journal*, September 30, 1985, p 58.

(13) Ikeda, A., Mukai, S., and Ueda, M., "Corrosion Behavior of 9 to 25% Cr Steels in Wet CO_2 Environments," *Advances in CO_2 Corrosion* (NACE), vol 2, p 47, 1985.

(14) Masamura, K., et al., "Pitting Corrosion Resistance of High Alloy OCTG in CO_2 Environment as Affected by Chlorides and Sulfides," *Advances in CO_2 Corrosion* (NACE), vol 2, p 80, 1985.

(15) Hafeez, A., and Wanderer, E. T., "Aluminum Alloys for Use in CO_2 Enhanced Oil Recovery (EOR) Systems," *Advances in CO_2 Corrosion* (NACE), vol 2, p 113, 1985.

10 Detecting and Monitoring Corrosion Activity

10.01 Early Detection

Evidence that corrosion is occurring at a serious rate is often discovered initially by visual examination of damaged equipment.The item may be broken, cracked or pitted and usually has residues of corrosion reaction products on the surface. Ideally, the attack should not have been allowed to progress that far. Consequently one of the most frequently asked questions is, "How can I detect and measure corrosion activity before a failure occurs?" About as much effort has been made to find easy answers to this question as has been applied to development of corrosion prevention techniques.

The difficulties of early detection stem largely from two factors:

1. Non-uniform rate of attack at different locations in production systems, and
2. Insolubility of corrosion reaction products.

Corrosion detection and assessment of intensity would be much simpler if both these situations did not exist. All that would be required is monitoring of dissolved metal concentration in the water phase and calculation of penetration rate per unit area of metal surface exposed.

In practice, corrosion monitoring is usually restricted to one or more of the following procedures:

1. Measure relative corrosiveness of system fluids with sensors and assume equipment corrosion rates are proportional.
2. Measure actual corrosion penetration on exposed equipment via calipers, ultrasonic thickness measurements, or radiography.
3. Analyze produced fluids for dissolved iron and compute normalized overall corrosion rate.

10.02 Devices for Monitoring Fluid Corrosiveness

Corrosion sensors, in general, are most useful for indicating: (1) differences in corrosion rates at various points in surface systems (e.g., up and downstream of tanks or pumps), (2) differences at specific locations over a length of time, and (3) changes in corrosion rate as a result of inhibitor treatment. The values they measure are related to real system corrosion rates but should not be interpreted as true representations. Data obtained with the testing devices are more properly considered as measurements of the fluid corrosivity at the point of exposure.

Table 1.10-1 describes some popular types of corrosion sensors or monitors used for corrosion rate testing. Corrosion sensors such as test coupons and electronic meter probes are most frequently mounted through tees in flowlines for easy retrieval. Various downhole sensors, including wire-line retrievable coupons and radioactive tubing subs, have been used. However, most operators monitor subsurface corrosion by tests at the wellhead.

10.03 Characteristics and Limitations of Corrosion Rate Monitors

1. **Weight-loss coupons**

   Test coupons such as those illustrated in Figure 1.10-1 provide results within one to three months. When running repeat tests with coupons at a specific location, exposure times should be held reasonably constant for reliable comparison between tests. The reason for this is that initial corrosion rates on the clean coupons are high but tend to diminish with time as illustrated in Figure 1.10-2. Coupons exposed for two or three months will generally indicate lower corro-

Table 1.10-1
Devices for Measuring Fluid Corrosivity

Sensor Type	*Exemplary Source*	*Functional Nature and Application*
Coupon Flat Rod Ring	Inhibitor Suppliers, Coupon Suppliers* or User Fabricated	Exhibits weight loss and pitting
Nipple or Spool	User Fabricated	Same as coupons but is part of piping system
Electrical Resistance Probe	Rohrback Instruments, Santa Fe Springs, CA (formerly Magna) or Petrolite Instruments, Houston, TX	Direct reading of corrosion rate from electrical resistance increase as sensor size decreases
Polarization Resistance Probe	Rohrback Instruments or Petrolite Instruments	Direct reading of corrosion rate from electrode potential (E) and applied current (I) relationship which has the units of R
Hydrogen Probe or Patch	Fincher Engineering Houston, TX, or Petrolite Instruments	Detects rate of H_2 penetration through probe or pipe wall
Galvanic Probe	Fincher Engineering or Thomas Kelley Co., Tulsa, OK	Detects changes in corrosivity as a function of current flow between dissimilar metal electrodes

*Metal Samples Company
Route 1, Box 152
Munford, Alabama, U.S.A., 36268

sion rates than coupons exposed only one month, even when actual corrosivity of the water being tested remains constant.

Coupons should always be positioned so that they are exposed to the corrosive water phase. Figure 1.10-3 shows a coupon installation in a pipeline water trap on an oil or gas pipeline. Coupons exposed to only the oil or gas phase will not reflect corrosiveness of water flowing along the bottom of the pipeline.

Notice that in Figure 1.10-3, a coupon lubricator is used to insert and extract the coupons while the pipeline is under pressure. Coupon lubricators are commercially available from companies such as Fincher Engineering, Houston, Texas, and Rohrback Cosasco Systems, Santa Fe Springs, California, or Reading, Berkshire, England. The lubricator is screwed into a full opening gate valve through which the coupon is inserted by means of the holding rod. The valve is left open while the coupon is in place and the lubricator assembly holds the full line pressure (up to 2.1 MPa or 300 psig). For extraction, the coupon is pulled up through the valve into the lubricator housing, the valve is closed and the whole lubricator assembly with the coupon inside is unscrewed for removal. The holding rod is made of stainless steel or Monel, and the coupon is electrically isolated from it with a plastic chuck to prevent galvanic corrosion of the coupon while in use.

Test coupons are cleaned and weighed before and after exposure in the field. In some cases actual average and maximum pit depth is measured, but most usually the weight loss is converted to an average penetration (corrosion) rate such as mils penetration per year or micrometers (μm) per year. Calculations of various corrosion rate expressions from steel coupon data are shown in Table 1.10-2.

Chemical companies usually supply test coupons and process them as part of their service. The Metal Samples Company, Inc., Route 1, Box 152, Munford, Alabama 36268, U.S.A., sells a variety of test coupons and holders including a pressure retrieval system (coupon lubricator) and drill-pipe corrosion rings. Details on corrosion testing with coupons, including flat, rod, and ring types are given in NACE Standard RP-07-75.[1]

2. **Test Nipples (Spools)**

Reliable data that closely indicates actual system corrosion can be obtained with test nipples that are used as part of a flowline. Two types are shown in Figure 1.10-4. These should be exposed three to nine months or even longer. After long-term exposure to total flow conditions, the nipples are split and examined for pitting and scale accumulation. If the nipples are prop-

Table 1.10-2
Corrosion Rate Expressions

$$mpy = \text{mils* penetration/year} = \frac{W \times 2831}{A \times T} \quad \text{(for steel)}$$

where: W = specimen weight loss, grams
A = specimen area, square inches
T = exposure time, days

*mil = 0.001 inch

mdd = milligrams/square decimeter**/day
= mpy × 5.47 (for steel)

**decimeter = 0.1 meter = 3.937 inches
square decimeter = 100 cm^2 = 15.5 in^2

mm/yr = millimeters/year

$$= \frac{mpy}{40}$$

μm/yr (or anum) = micrometers/year
= mpy × 25.4

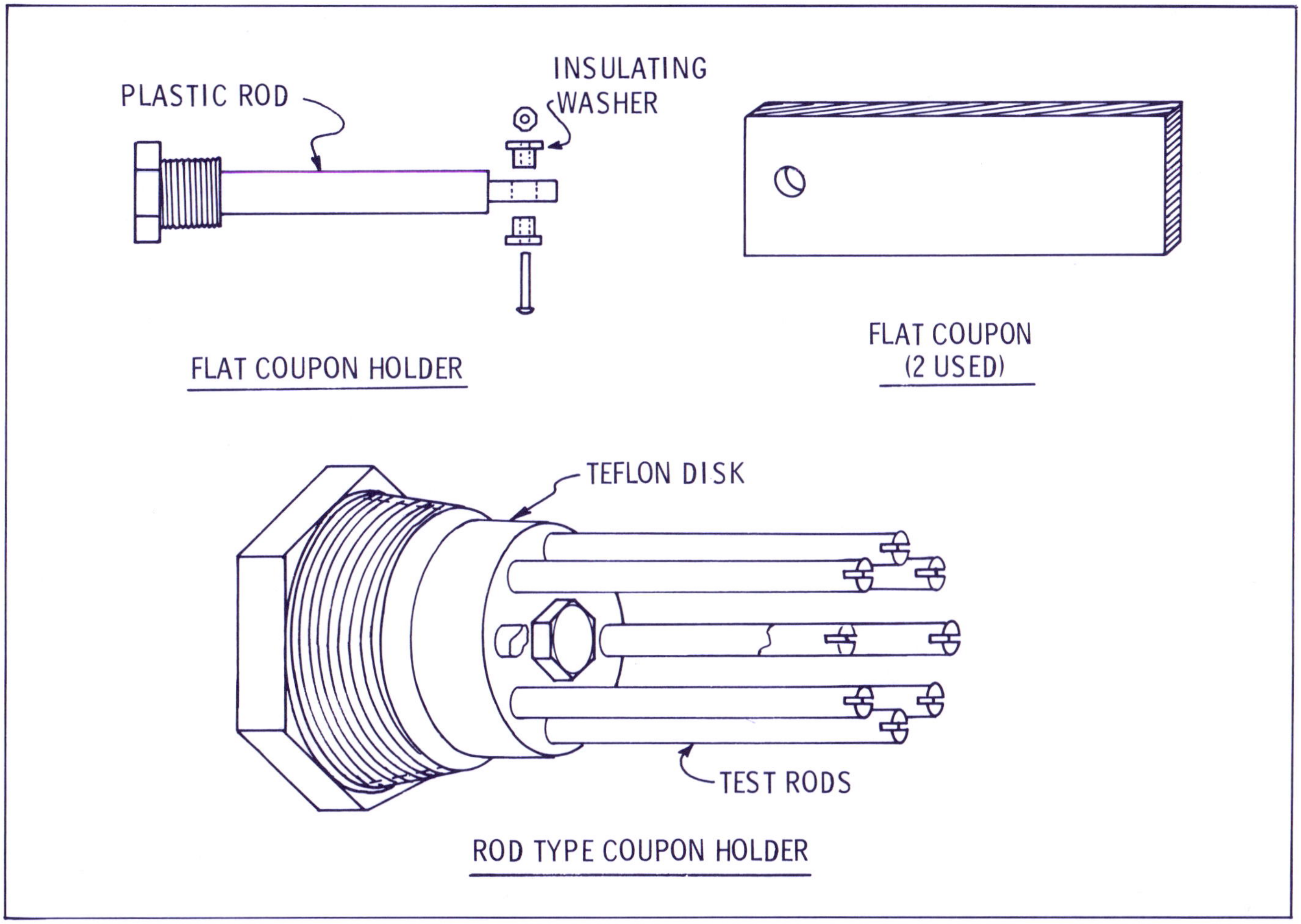

Figure 1.10-1. Two types of corrosion test coupons and holders.

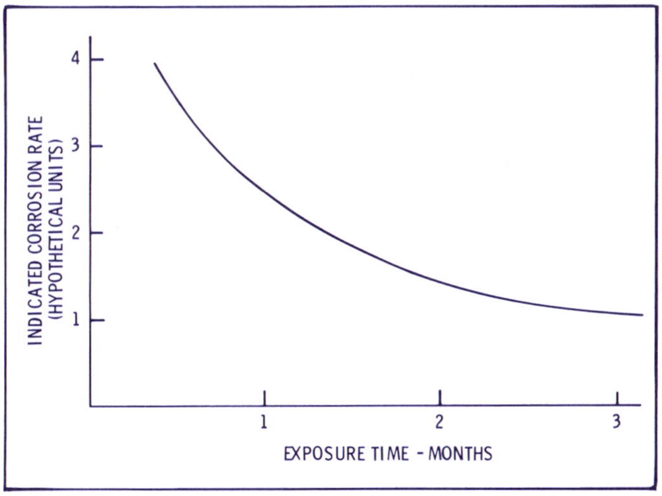

Figure 1.10-2. Illustration of test coupon data variance with time.

erly prepared and weighed initially, their long-term weight loss provides the best available measure of true system corrosion rate in pipes or flowlines.

3. **Corrosion Meters**

Two types of direct-reading electronic corrosion rate meters are used. Each type employs a probe sensor that is extended through a tee or nipple into a flowline, pipeline or vessel wall. An electrical lead connects between the probe and the meter, which may be portable or permanently mounted for central station readout. Some central station installations in plants or offshore platforms use several permanently mounted probes at various locations. Periodic readings are taken and printed systematically by means of the meter instrument and an electronic switching device that contacts each probe in turn on a regular basis.

Figure 1.10-5 illustrates features of a probe used with *electrical resistance* corrosion meters such as the

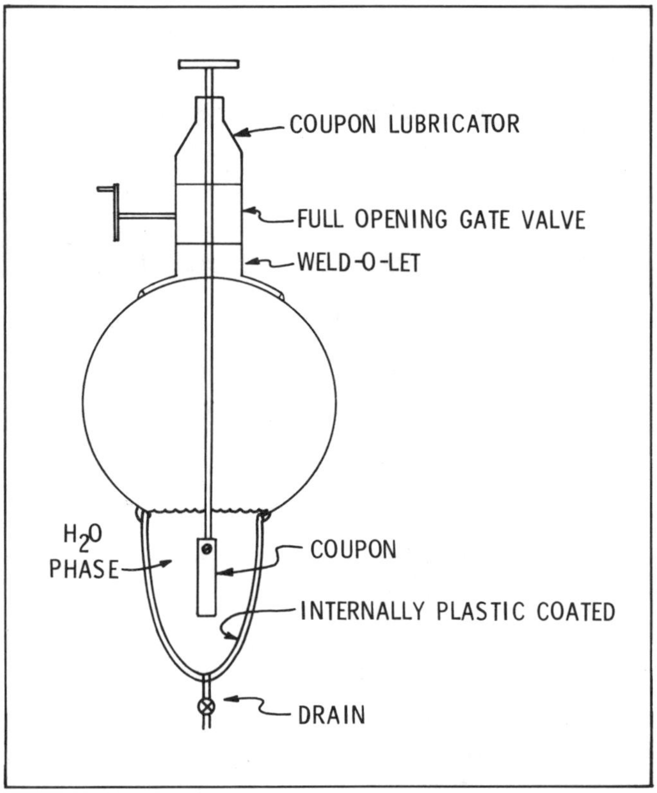

Figure 1.10-3. Flat coupons installed in water trap welded to bottom of a pipe line.

Magna Corrosometer™ sold by Rohrback Instruments. Petrolite Instruments also sells electric resistance type corrosion meters. A variety of probe sizes and shapes are available.

The sensor element exposed to corrosive fluids is a small diameter wire, tubing or a thin metal strip. As small amounts of metal are lost from the sensor element by corrosion, its resistance to electrical current flow increases, and the change is translated to corrosion rate (mpy) by the electronic meter. Since the device measures actual corrosion or metal loss on the sensor, it is not necessary for the probe to be immersed in a continuous electrically conductive water phase. Readings can be made in oil or gas lines for example as long as there is enough water present for corrosion of the sensor element to occur. Stable readings can usually be obtained within a few hours after the sensor probe is installed and changes in corrosion rate can be rapidly sensed for up to several months until the sensing element is corroded through. A flush mounted resistance probe for measuring corrosion rate in thin layers of water at the bottom of gas or oil lines is available from Corrocean A/S (Trondheim, Norway, N-7001).

Essentially *instantaneous* readings of corrosion rate can be obtained with the Petrolite (Tretolite) PAIR™ and Rorhback Corrater™ corrosion meters. These do not measure corrosion of metal directly as do test coupons or resistance probes. Instead, an electro-

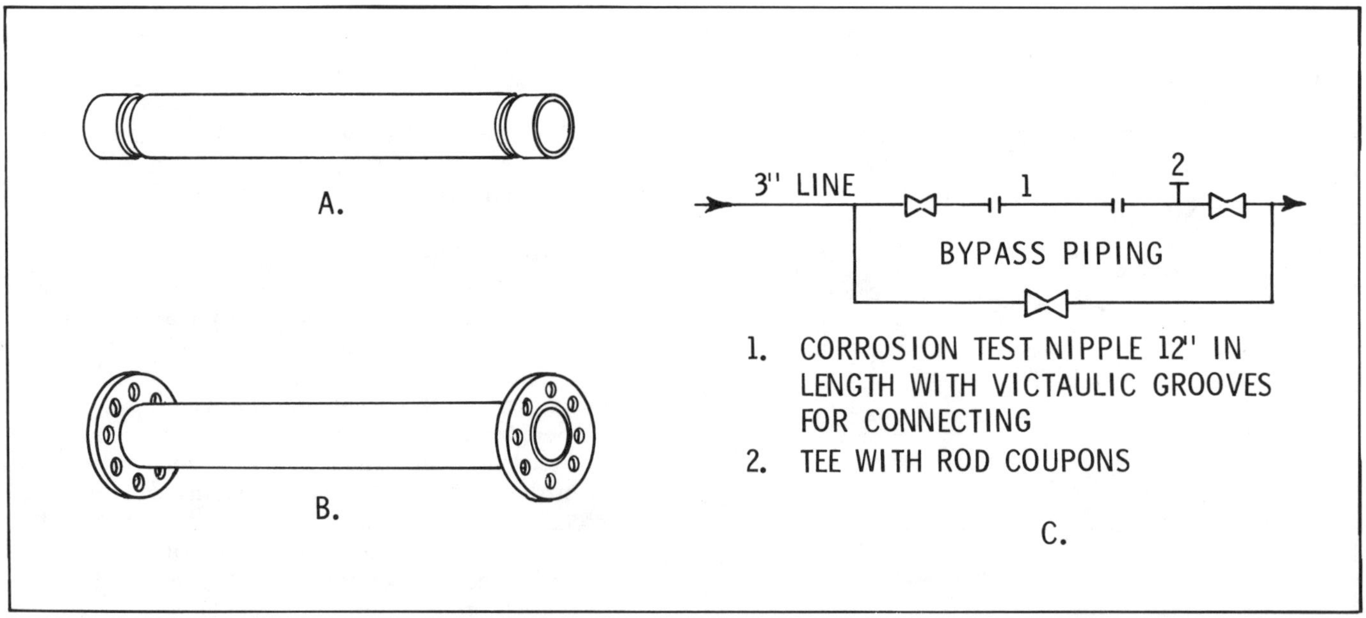

Figure 1.10-4. Corrosion test nipples. A. Nipple with grooves for Victaulic couplings. B. Spool with flanges. C. Bypass piping arrangement with nipple and test coupon combination.

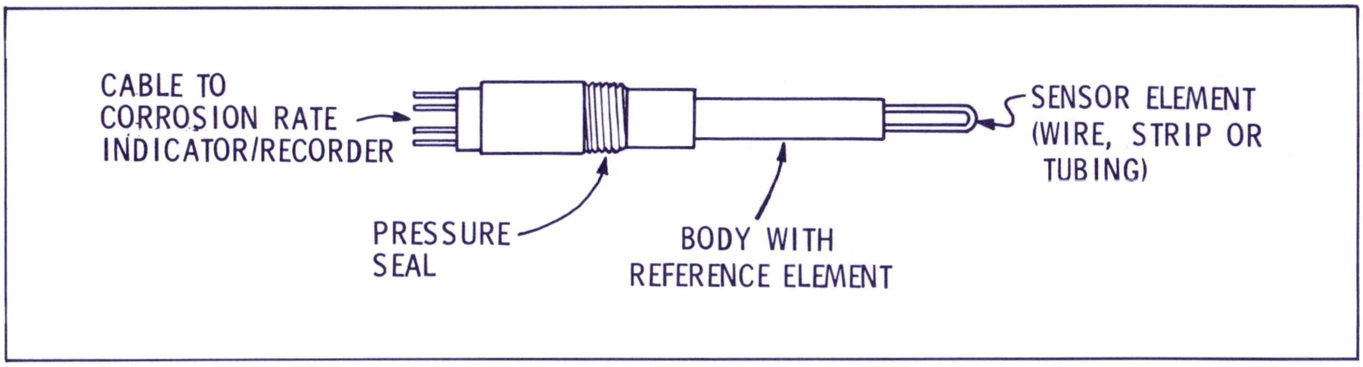

Figure 1.10-5. Electrical resistance probe.

chemical technique called linear polarization or polarization resistance is used (see Appendix Item 1A.10). The test probes (Figure 1.10-6) utilize either two (Corrater) or three (PAIR) metal electrodes that are immersed in a continuous electrolyte (salt water). Petrolite Instruments also offers a flush-mounted probe for use with their PAIR™ corrosion meter (see photo, Figure 1.10-7). The three electrodes of the flush-mounted probe are in the form of concentric rings with insulating plastic surrounding the inner two. The probe is mounted through the wall on the bottom side of oil or gas flowlines or pipelines so that the flat face of the probe is flush with the inner pipe surface. In this position, the flush-mounted probe senses even thin layers of water moving along the bottom of the pipe and eliminates the need for water traps (Figure

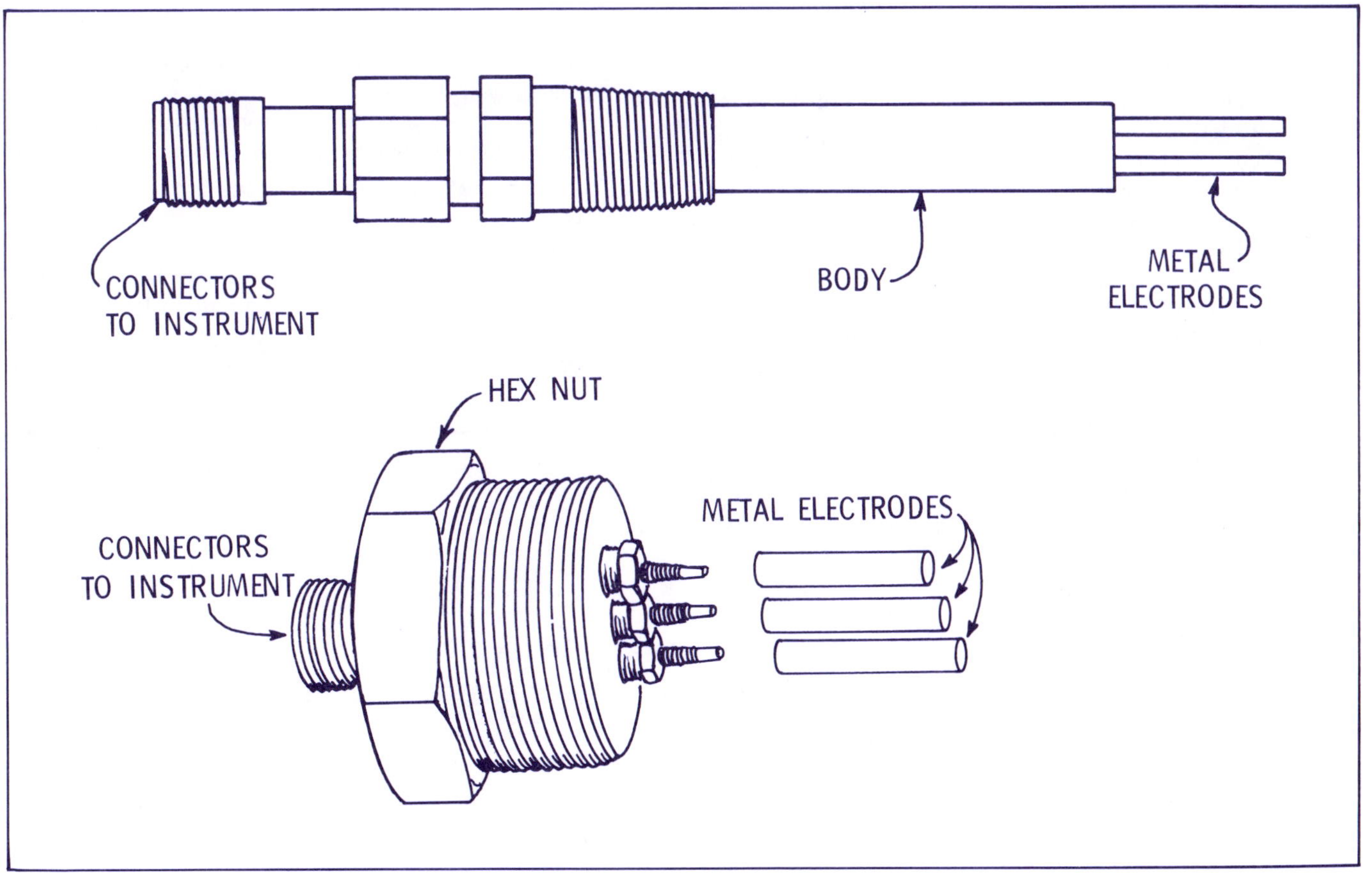

Figure 1.10-6. Illustration of two and three element polarization resistance probes.

Figure 1.10-7. Flush mounted polarization resistance probe. (Photo courtesy of Petrolite Instruments, Inc., Houston, Texas).

1.10-3). Scouring by solids in high velocity streams helps prevent fouling of the electrodes and improves sensitivity of the flush mounted corrosion rate sensor.

In both the two- and three-element polarization resistance probes, the electrical potential of one of the electrodes is monitored with respect to one of the other electrodes as a small electrical current is applied. The amount of current required to shift the potential of the working electrode by a fixed amount (usually 10 to 20 millivolts) is proportional to corrosion intensity. The electronic meter translates the amount of current applied to a readout number indicating corrosion rate as mpy. The corrosion rate reading is accurate for the corrosion taking place on the probe elements at the instant the reading is taken. However, accumulations of corrosion products, mineral scale or heavy hydrocarbons reduces sensitivity and probes left in the system for long periods may fail to respond as conditions change in the system being monitored.

Many users believe that electrical resistance-type probes are less affected by fouling than the polarization resistance-type and the former appears to be gaining in popularity. It should be noted, however, that both types are subject to the same basic limitations as corrosion test coupons. They can only read corrosion rate on the probe sensor, and this rate is, at best, an indicator of average overall corrosion in the system being tested. When either type of corrosion meter probes is left installed for long periods, they should be cleaned and inspected on a regular basis and back-up weight-loss test coupons should be exposed simultaneously in at least one location in the system.

4. **Hydrogen Probes and Patches**

Hydrogen probe or patch sensors are designed to measure corrosion by sensing the amount of hydrogen that migrates through the wall of a pipe or tank. Some of the atomic hydrogen, H^O, generated at cathode sites on metal being corroded by substantially *oxygen-free* water containing H_2S, penetrates into and through low hardness steel (see Section 1, Item 1.17 and Item 7.06). The H^O that passes completely through the steel wall changes to molecular H_2 gas and is swept away. Theoretically the amount of hydrogen passing through the wall is proportional to the corrosion rate.

The Petrolite Corporation supplies a device called the Hydrogen Patch Probe™, which is mounted on the outside wall of a vessel or pipe. Atomic hydrogen penetrating the wall is detected by electrochemical reduction and an electronic readout instrument indicates relative corrosion rate. Hydrogen pressure probes (Figure 1.10-8) operate on the same basic principle

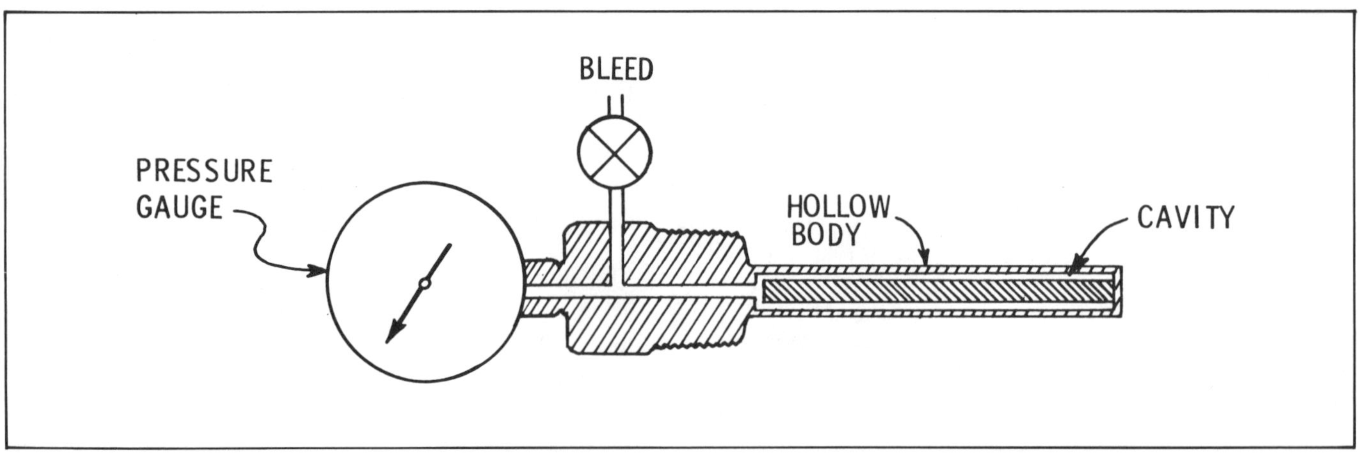

Figure 1.10-8. Hydrogen pressure probe.

except pressure of trapped molecular hydrogen is measured with a gauge. The pressure probe is mounted through a tee or nipple into a pipe or tank. As the thin wall of the probe body corrodes, H_2 is trapped inside and builds pressure. A rate of pressure increase greater than about 7 kPa (1 psig) per day is considered a signal of significant corrosion.

Hydrogen probes or patches are generally not reliable for quantitative indication of corrosion rate but may be used to detect very rapid corrosion in air-free sour systems. They do not function in aerated fluids.

5. **Galvanic Probes**

These are simply two dissimilar metal rods, usually steel and brass, mounted on but insulated from a threaded pipe plug. When the two rods are immersed in aerated salt water and electrically connected through a milliammeter, the dissimilar metal couple generates a small electrical current that varies in strength with the amount of oxygen present. The activity of the dissimilar metal corrosion cell and thus the amount of current flow is largely determined by the concentration of oxygen, which depolarizes the brass cathode and keeps the galvanic cell action going. Therefore, the galvanic probes together with a small recording milliammeter are useful to signal an influx of oxygen into water streams, and they are used for that purpose downstream of deaerators in waterflood and drilling operations.

In air-free systems such as deaerated drilling muds, the current output of the galvanic probe couple may surge when H_2S contamination occurs. However, the current output in any field situation is not a reliable measure of corrosion rate.

10.04 Iron Content Analysis of Produced Water and the COPRA Correlation

Iron content determinations (iron counts) are the most widely used method for monitoring downhole corrosion rates in gas and gas condensate wells that produce little or no H_2S, viz., sweet production. Carbon dioxide is the primary corrosive agent. Organic acids are frequently present and tend to keep iron in solution for at least a few minutes (prior to air oxidation) after a water sample is taken from the wellhead or flowline. Dissolved iron content analysis can be run directly on a freshly caught water sample using a simple colorimetric test kit such as the Hach Ferro-ver (Hach Company, P. O. Box 389, Loveland, CO 80539, U.S.A.). Alternately, each water sample is acidified with several drops of concentrated hydrochloric acid (pH 1 to 2) to hold iron in solution and groups of samples are submitted to a laboratory for iron determinations.

The concentration of dissolved iron tends to increase with increasing corrosion rate but iron count values per se are generally *not* reliable as indicators of corrosion intensity, especially at pits. To compensate for variations in water production between wells, some operators determine average iron concentration (mg/L or ppm) and compute total iron loss rate for each well as pounds or kilograms per day. This method has greatest value for indicating changes in iron loss rate over time or before and after initiation of inhibitor treatments.

Some sand formations containing clays, such as the chlorites, produce water with a natural iron content which must be accounted for in corrosion studies. The background iron concentration is usually constant in relation to the volume of formation water in the total produced water at the wellhead (formation plus water of condensation).

Wells that produce formation water can be assumed to have corrosion iron loss over the full length of the tubing string and iron count data can be used to calculate a uniform overall corrosion rates as follows:[2,3]

(Eq. 1.10-1) $$C.R. = \frac{C_{Fe} \cdot Q_{H_2O}}{0.082\ L\ d}$$

where: C.R. = uniform corrosion rate (mpy)
C_{Fe} = iron count in produced water (ppm or mg/L)
Q_{H_2O} = total water production (bbl/day)
L = length of tubing (ft)
d = inside diameter of tubing (in.)

When caliper survey or tubing inspection data indicate tubular corrosion is primarily concentrated between certain depths in the well, the length of affected areas can be substituted for total length (L) in the Equation 1.10-1. Localization of corrosion in the upper portion of the tubing string may be caused by water condensation due to cooling. If this zone can be determined or calculated, its length can be used for L (see Figure 1.8-3).

It has been established that corrosion rates in flowing gas wells are a function of gas and water production rates and are strongly influenced by the water composition. To account for both gas and water, Gatske and Hausler[2] developed the COPRA correlation (Corrosion Rate-Production Rate) to predict actual well fluid corrosivity prior to equipment failure and for evaluation of inhibitor treatments.[4] For the COPRA correlation, the uniform corrosion rate (C.R.) computed from iron counts (Equation 1.10-1) is normalized with respect to water production as follows:

(Eq. 1.10-2) $$K_G = \frac{C.R.}{Q^1_{H_2O}}$$

where: K_G = normalized or relative average corrosion rate
$Q^1_{H_2O}$ = bbl H_2O/MMcf*

(*MMcf = million cubic feet of gas).

The calculated K_G factor is then plotted as a function of gas production rate as illustrated in Figure 1.10-9. The data points represent K_G at specific gas rates and a series of connected points produces a line curve the slope of which indicates corrosivity of a group of wells irregardless of differences in water and gas production rates. Figure 1.10-9 illustrates reduction in K_G/MMcf/day for a range of gas production rates as a result of treating the wells with corrosion inhibitors.

A COPRA type correlation can be made using test coupon corrosion rate (C.R.) in lieu of iron count data when there is reason to believe the coupon data are representative.

10.05 After-the-Fact Corrosion Detection

Metal loss that has already taken place on equipment can be measured by several means. Three types of tools for in-place inspection are:

1. Downhole tubing or casing calipers,
   a. Mechanical feelers (measures internal diameter and pit depth)
   b. Electromagnetic induction types (measures total reduction in wall thickness)
   c. Combination of a and b tools;
2. Ultrasonic wall thickness testers;
3. Radiography,
   a. X-Ray,
   b. Gamma ray.

Downhole inspection with various type calipers is expensive but justifiable when compared with the high cost of tubing and casing failures. Details of caliper tools and data interpretation are available from companies such as the J. C. Kinley Co., Houston, Texas, and worldwide affiliates (Otis Engineering Co., Ltd., Flopetrol Johnston, Expro (N/S), Ltd., and Stravanger Oilfield Services).

1. **Tubing Inspection**

   Tubing caliper surveys are commonly run in gas, condensate and oil wells where iron count or wellhead coupon test data indicate downhole corrosion may be occuring at a serious rate. Direct contact or feeler type calipers are most often used for internal

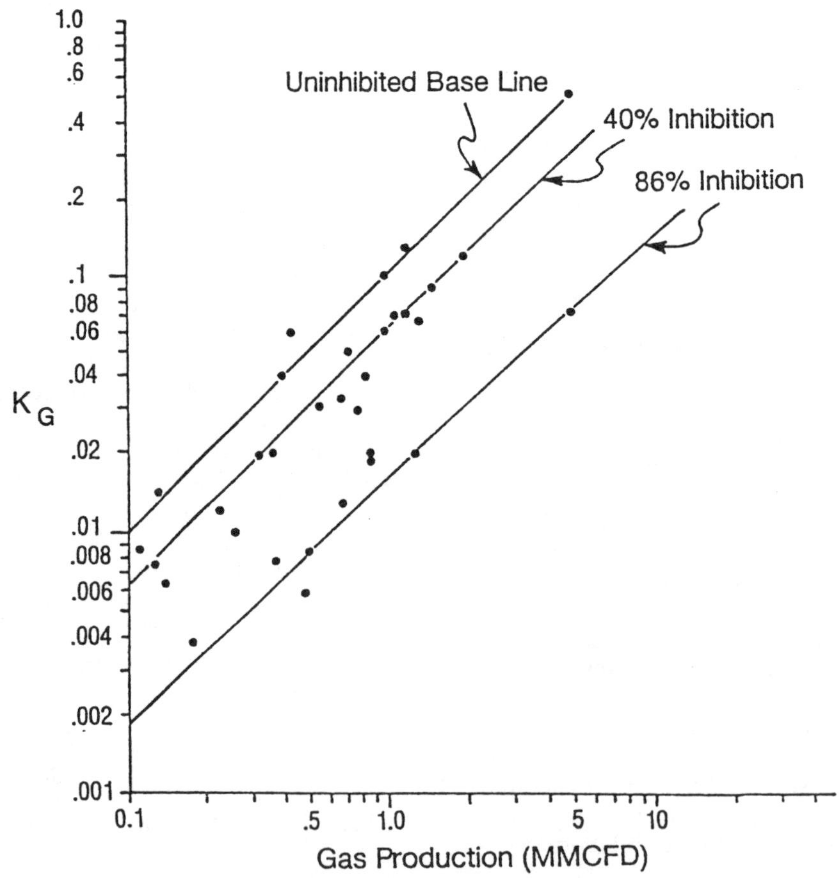

Figure 1.10-9. COPRA Correlation for a group of gas wells before and after initiation of a corrosion inhibition program. (After Burke and Hausler[3]).

tubing inspection. The contact feelers generally indicate location of severely corroded areas but may miss isolated pits. For a single pit as small as 3 mm (0.125 in.) in diameter, the probability of detecting it with one caliper run up the tubing is about 15%.[5] The probability increases to 80% if the survey is rerun ten times in the same tubing. Repeat surveys are normally run from bottom to top while the tool is in the hole.

Maximum pit penetration may be obscured by deposits but the probability of extreme pit depths or total wall penetration can be computed using extreme value statistics calculations as discussed by Eldridge[6] and Pilkington.[7] With Eldridge's method, pit depth rank charts are prepared using statistical theory of extreme values in which the number of maximum pit depths measured (percent of wall penetration) in each joint is considered. A graphical plotting procedure is used to predict the probabilities of deeper pitting or total penetration in the string. A simple step-wise procedure for the analysis is available from Kinley.[5] Pilkington developed pit distribution charts to supplement the pit depth rank charts to illustrate location of significant corrosion damage throughout the tubing string. Hausler and Burke applied the extreme value statistical analysis method to caliper data from a group of wells with various water and gas production rates.[8] Their goal was to establish a value of K_G (normalized corrosion rate) for a COPRA type correlation (see Item 10.04).

2. **Casing Calipers**

Electromagnetic casing corrosion detectors measure reduction in casing wall thickness by generating a magnetic flux that passes out through the casing wall from one end of the tool and is received by an internal sensor. Returning magnetic flux characteristics (density and phase) are influenced by metal thickness and are not affected by presence of liquids or surface deposits. The wall thickness measuring tool cannot distinguish internal and external metal loss and may not detect perforations smaller than 2.5 cm (1 in.). An electromagnetic device is available for measuring internal casing diameter. Combination tools simultaneously measures wall thickness and internal diameter to detect and differentiate both internal and external corrosion damage.

3. **Ultrasonic Thickness Gauges for Surface Testing**

Many producing companies train selected personnel in nondestructive testing techniques, especially use of ultrasonic thickness gauges. Modern gauges such as the Model DM3 manufactured by Krautkramer Branson (P. O. Box 350, Lewiston, PA 17044, U.S.A.) are small, easy to calibrate and use and have direct digital readout of metal thickness.

Periodic ultrasonic wall thickness measurements, at selected inspection points, are especially recommended for vessels and surface piping locations where failure is intolerable. On offshore production platforms, tube turns and similar locations where erosion by solids is likely should be checked at least annually.

Pipe or tank wall thickness is determined with an ultrasonic instrument by measuring the time required for a high-frequency sound wave to pass through the metal. The sound wave is generated by a piezoelectric transducer that is pressed against the metal surface. The hand-held transducer (search unit) receives high-frequency electrical impulses from a power source in the main instrument and converts the electrical energy to sound vibrations. A sound vibration or wave passes at a known uniform speed through the metal and is detected by a receiving transducer which converts the sound back to an electrical impulse. Since the speed of sound travel is constant for the metal being tested, the time interval between the input and received electrical pulses is a function of metal thickness. After wall thickness is initially measured at a certain spot on the pipe or tank wall, repeat tests made several months later at the same spot will reveal the amount of thinning or metal loss that has occurred at that location.

Sound waves generated by the transducer do not propogate efficiently through air and the transducer face must make solid contact with the metal being tested. Flat face transducers are used on tank walls and large diameter pipe. Curved face transducer search units are available for use on various sizes of small pipe. General practice for external thickness tests is to select a spot on the metal surface, smooth that area with a hand file or grinder and then cover the smoothed area with a coupling fluid such as oil or glycerine. The couplant fluid aids transmission of ultrasonic sound between the search unit and the metal. Repeat tests for detecting thickness changes are always made at the same spot for greatest accuracy.

Three basic methods of time interval ultrasonic testing are used. The pulse echo technique utilizes a single transducer to emit a pulsed beam of ultrasonic sound through the couplant fluid into the external surface of the metal. At the opposite (internal) side of the metal wall, the sound signal is reflected or echoed back to the sending transducer. The time interval between the initial pulse and the return echo is measured by the calibrated test instrument to determine thickness. Some older instruments have a cathode ray tube (CRT) that displays spikes of the input and return signal, and the distance between the peaks is measured manually with a transparent scale that reads thickness units. Newer instruments measure the time interval electronically and display a digital readout of thickness. Compact models are available with all components in a single case.

Another type of search unit operates on a similar principle, but the unit head contains two transducers set at an angle to each other. One transmits and the other receives.

In the through transmission method, separate transducers are placed on each side of the metal to send and receive the ultrasonic beam. Since both sides must be accessible, this method is not used for routine thickness tests in the field.

4. **Radiography**

Radiographic imaging is achieved by passing x-rays or gamma rays through valve bodies, elbows or other items. The radiograph is recorded by film and reveals flaws such as internal surface pits, wall laminations or cracks. Radiographs can provide positive visual evidence of defects and corrosion damage.

10.06 Visual Inspection and Analysis

No test instrument or inspection tool has ever been developed that can benefit a corrosion control program as much as thorough on-site inspection by interested personnel. Direct examination and a few simple tests on failed equipment will often reveal the basic cause of the problem.

A practical outline for field investigation of failures is tabulated in Table 1.10-3. The listed steps can be used as a guide for preparation of a data acquisition form. Pre-prepared forms, with blank spaces for observations and comments, are quite helpful for use by field crews.

Table 1.10-3
Scheme for Field Inspection of Equipment Failures

1. Record specific identification of failed item and position of use.
2. Describe the gross nature of the failure, i.e.,
   - Longitudinal rupture
   - Transverse fracture
   - Perforation
   - Wall thinning—internal and external
   - Internal and external pitting
3. Examine overall and local deposits.
   - Color
   - Texture
   - Thickness and distribution
4. Test solubility and reaction of deposits in dilute acid as discussed in item 10.07.
5. Clean a number of local areas and pits by wire brushing and scraping. Visually examine configuration and distribution of pits, cracks and thinned areas.
6. Photograph the overall failure and take close-ups of local areas before and after cleaning.

Table 1.10-4
Characteristic Features of Some Corrosion-Related Failures

Appearance	*Probable Contributing Factors*
Small conical pits with steep sides and smooth edges. Pits filled with black deposit.	Hydrogen sulfide attack (H_2S may be natural or generated by bacteria) Figure 1.10-12
As above plus transverse cracks.	Hydrogen sulfide attack with tensile stress (stress corrosion fatigue)
Transverse fracture with little or no pitting but with black deposit.	Sulfide stress cracking Excessive metal hardness
Round bottom connecting pits with sharp sides. Grey deposit but pit bottoms are bright.	Carbon dioxide attack (Section 1, Chapter 8, Figure 1.8-3)
General thinning with sharp feathery or weblike residual metal. Little or no deposits.	Mineral acid corrosion
Rust deposits. Shallow, widespread pitting or deep pits under rust nodules.	Oxygen corrosion
Single, isolated pits in a row on one side.	Electrolytic corrosion due to current discharge
Worn or abraded areas with numerous small pits.	Erosion by solids or metal rubbing in presence of H_2S, CO_2 or O_2

Table 1.10-4 is a summary list of observations that can be made to help identify the most probable causes of equipment failure. For example, the appearance of a common type of sucker rod failure is shown schematically in Figure 1.10-10. Figure 1.10-11 is a photograph of the broken end of a sucker rod that failed by corrosion fatigue after six months use in a sour-producing well.

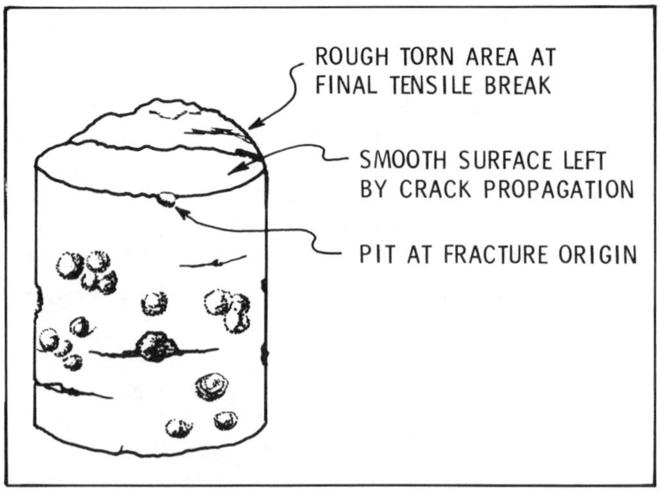

Figure 1.10-10. Anatomy of a sucker rod break due to stress corrosion fatigue.

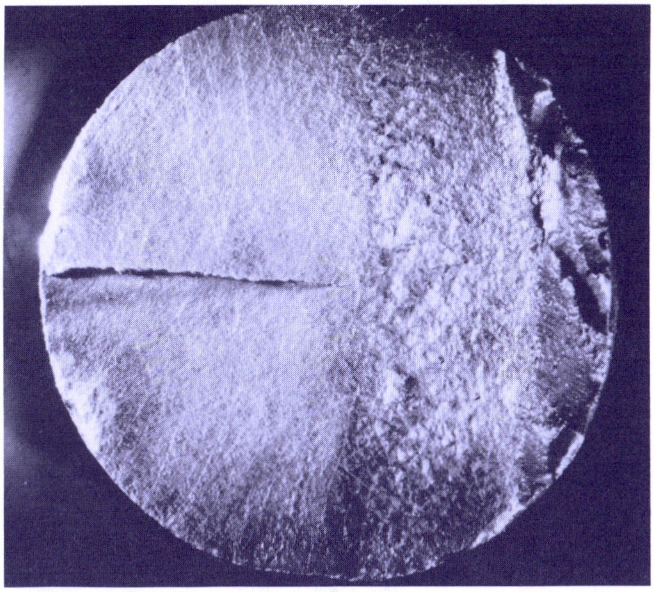

Figure 1.10-11. Cross section of a sucker rod that failed by corrosion fatigue.

Note pitting along the left hand edge, the smooth metal face where the brittle fracture occurred and the final (non-brittle) separation on the right hand side. The final, non-brittle tensile break probably happened rapidly after the effective diameter of the rod was reduced (by initial slow cracking) to the point that it could no longer sustain the tensile load.

Similar fracture characteristics, viz., a smooth transverse break with torn metal peaks on one side, are often observed on collars and tubing that have failed by sulfide stress cracking (see Section 1, Item 7.05).

Table 1.10-5
Acid Solutions for Deposit Testing

Solution 1	
Water (potable or tap water)	400 ml
Isopropyl alcohol	150 ml
Concentrated HCl acid[1] (clear, pure acid required)	450 ml

[1]Add acid *last*, slowly with stirring

If available, add five to ten milliliters of a liquid detergent or wetting agent to final solution.

Solution 2

To 100 ml of Solution 1, add approximately 5 gm (about 1 level teaspoon) of sodium arsenite (Na_2HAsO_3) powder and stir to dissolve.

Caution: Protect eyes and skin when preparing acid solutions. Do not inhale or ingest the sodium arsenite powder. Wash skin thoroughly with soap and water if contacted by sodium arsenite.

10.07 Tests for Identifying Corrosion Products

Step 4 of Table 1.10-3 calls for testing surface deposits in a dilute acid solution. Recipes for two test acids are listed in Table 1.10-5.

Acid Solution 1 can be used for testing the general nature of heavy deposits scraped from corroded metal surfaces. Drop approximately one cm^3 of the deposit into approximately 25 ml of Solution 1 held in a small glass container or test tube. Observe rate of solution and effervescence (gas evolution). If slow gas formation is apparent, carefully sniff the vapors for the characteristic rotten egg odor of H_2S. Chemical test kits for detecting H_2S are commercially available.[9]

The presence of both evolved H_2S *and* a black color of the original sample indicate the deposit contains iron sulfide (FeS) corrosion product. Detection of FeS cor-

Figure 1.10-12. Sulfide pitting in a water injection flowline. H_2S created by sulfate reducing bacteria.

rosion product is informative, especially on well equipment and in water injection systems where bacteria may be active.

When the deposit fizzes and evolves gas rapidly, it probably contains calcium carbonate ($CaCO_3$) scale. Combinations of $CaCO_3$ and FeS occur frequently.

Iron carbonate (Fe_2CO_3), also called siderite, has a grey appearance and emits CO_2 gas when dissolved in acid. The CO_2 evolution can sometimes be observed by simply placing a few drops of Solution 1 directly on the deposit.

Rust or iron oxide is usually reddish brown and dissolves slowly in acid without effervesence. Millscale (Fe_3O_4) dissolves in acid slowly or not at all. When dry, millscale can be attracted by a magnet.

Acid Solution 2 is prepared specifically for detecting iron sulfide directly on metal surfaces. This test is suggested when inspecting equipment that has failed by brittle fracturing. Fracture failures may result from several causes but one of the most common in oil production operations is sulfide stress cracking (see Section 1, Chapter 7). In this situation, a thin, black deposit is usually present but pitting or other signs of corrosion may not be apparent. The test for iron sulfide consists of simply placing a few drops of Solution 2 directly on the metal near the facture. Appearance of a yellow precipitate (suspension) in the acid indicates iron sulfide corrosion product is present (disregard clear yellow color with no precipitate).

10.08 Record Keeping and Economic Analyses

Many engineers and operations personnel detest preparation of detailed reports. However, these records are a vital part of the corrosion monitoring process. Starting with data and observations from the original failure analysis reports (Items 10.06 and 10.07), failure trends and problem locations can be systematically identified. Similar processing and cross correlation of corrosion rate test information provides a warning of potential trouble. The importance of these efforts cannot be overemphasized.

Burdensome data accumulation and processing can be made easier by simple report forms, ledger-type data entry sheets and a preplanned scheme for periodic summarization. Many companies utilize computer programs for the purpose.

Ultimately, an economic appraisal must be made to clarify the following questions:

1. Is any corrosion control measure justified?
2. What is the economic advantage of treatment over equipment replacement, considering the expected life of the operation?
3. How do alternative control measures compare economically, considering the probability of success, present worth and annual cost?
4. How can management be convinced that preventive expenditures are justified when some unknown parameters exist?

The general approach is to use some type of economic analysis technique that provides a value for pay-out period or return on investment. Unfortunately, the technical uncertainties of corrosion prediction and control, complicated by economic variables such as future worth, produce a formidable task. Calculated risks are involved in most managerial decisions on corrosion control expenditures. Minter emphasized this fact and illustrated statistical methods used to determine the most economical corrosion control method compatible with over-all operations in three large offshore fields.[10]

A systematic procedure for making economic appraisal is presented in NACE Standard RP-02-72.[11]

REFERENCES

(1) NACE Standard RP-07-75, "Preparation and Installation of Corrosion Coupons and Interpretation of Test Data in Oil Production Practice," NACE, 1440 South Creek Drive, Houston, Texas 77084, U.S.A. (latest revision).

(2) Gatzke, L. K., and Hausler, R. H., "A Novel Correlation of Tubing Corrosion Rates in Deep, Hot Gas Wells with Water and Gas Production Rates," *Advances in CO_2 Corrosion* (NACE), vol 1, 1984, p 87.

(3) Chambers, J. B., and Hillegeist, R. B., *Petroleum Technology*, July, 1975, p 701.

(4) Burke, P. A., and Hausler, R. H., "Assessment of CO_2 Corrosion in the Cotton Valley Limestone Trend," *Advances in CO_2 Corrosion* (NACE), vol 2, 1985, p 133.

(5) J. C. Kinley Company, 5815 Royalton, Houston, Texas 77081, U.S.A.

(6) Eldridge, G. G., "Analysis of Corrosion Pits by Extreme Value Statistics and Its Application to Oil Well Tubing Caliper Surveys," *Corrosion* (NACE), vol 1, no 1, January, 1957, P 51t-60t.

(7) Pilkington, P. E., "Simplified Presentation of Oil and Gas Well Caliper Surveys," *Materials Performance* (NACE), vol 1, no 5, May, 1962, p 42.

(8) Hausler, R. H., and Burke, P. A., "Corrosion Monitoring in Sweet Production: Field Experiences with Iron Counts, Coupons, and Calipers," *Advances in CO_2 Corrosion*, (NACE), vol 2, 1985, p 161.

(9) Hach Chemical Co., Loveland, Colorado 80539, U.S.A.

(10) Minter, J. F., "Calculated Risk as a Tool in Corrosion Economics," *Materials Protection*, (NACE), March, 1965, p 35-42.

(11) NACE Standard RP-02-72, "Direct Calculation of Economic Appraisals of Corrosion Control Methods: NACE 1972," NACE, 1440 South Creek Drive, Houston, Texas 77084, U.S.A.

SECTION 2

Water Technology

GENERAL INTRODUCTION TO SECTION 2

The first chapter teaches in plain language the rudiments of water chemistry necessary to enable the reader to interpret and use water analysis reports. That information is applied in Chapter 2 to the problem of mineral scale deposition in production operations.

Chapter 3, called Practical Microbiology, is designed to develop basic familiarity with the most important groups of microorganisms, how to judge bacteria count data, and how to cope with bacterial infections in water injection systems.

Chapter 4 deals with water quality for injection purposes. This is a controversial subject and the primary goal is to provide a working concept of how and why the intake capacity of injection wells often declines and to offer a procedure for rating relative injection quality of various waters. Removal of suspended matter for quality improvement is discussed in Chapter 5 of this section with emphasis on remedies for oil carryover.

Corrosion in water injection systems was treated previously in Section 1, Chapter 8, Items 8.14 and 8.15.

1 Water Properties and Components

1.01 Properties of the Water Substance

The usual chemical formula for water is H_2O. All this tells us is that for each *atom* of oxygen in the *molecule*, there are invariably two atoms of hydrogen. In other words, the water substance is formed by a combination of the *elements*, oxygen and hydrogen, in a fixed ratio. Water is, therefore, a discrete *compound*; being composed of at least two elements chemically combined in definite proportions by weight.

The molecular structure of water can be conveniently depicted as shown in Figure 2.1-1. The unique physical and chemical characteristics of *pure* water are inherently related to the structure and behavior of the individual water *molecules*.

Unlike many other compounds, water is extremely polar; viz., opposite sides of the molecule tend to behave somewhat like the poles of a magnet. This behavior results from both hydrogen atoms being positioned on one side of the molecule. The side with the hydrogen atoms has a small positive charge while the opposite oxygen side is more negatively charged. Because of the charge distribution, the water molecule spins with a measurable eccentricity called the dipole moment. The dipole moment of water is 1.84 Debye units compared with 0 for organic solvents such as benzene or carbon disulfide. Water is a much more powerful solvent for minerals and crystalline compounds in general than are the less polar organic solvents.

Another distinctive characteristic of water molecules is their powerful attraction for each other. Referring to Figure 2.1-1, the atoms of hydrogen are attached to an oxygen atom by *valence* bonds represented by solid lines. Valence defines the degree of combining power of each type of element. Oxygen, having a valence of 2, requires 2 atoms of hydrogen to form the water molecule because hydrogen has a valence of only 1. However, each hydrogen atom is also attracted to oxygen atoms in nearby water molecules and form a hydrogen bond as indicated by the dashed line in Figure 2.1-1. Hydrogen bonds are much weaker than valence bonds.

Depending on temperature, water may exist as single H_2O molecules (in water vapor) or as a chain of up to 8 molecules linked together in polymeric fashion by the hydrogen bonding forces. This is why water remains a liquid at ordinary temperature. Consider, for comparison, H_2S (hydrogen sulfide) that ordinarily is a gas even though each molecule of H_2S is approximately twice as heavy as H_2O. In H_2S there is no hydrogen bonding as in water.

Hydrogen bonding effects also explain why ice (solid state water) floats. In the ice state, water molecules orient in groups and form distinct crystals with some open spaces in the crystal lattice. On melting, the spaced array collapses and the cold liquid is more dense than the slightly cooler solid. Were it not for this unique density inversion, the world's oceans might freeze from the bottom up.

Although most water molecules tend to cling together by hydrogen bonding as just discussed, a few actually separate as follows:

$$\text{(a) } \underset{\text{water}}{HOH} \rightarrow \underset{\text{hydrogen ion}}{H^+} + \underset{\text{hydroxyl ion}}{OH^-}$$

This separation process is called *ionization*. In pure water with nothing dissolved in it, only about one molecule in ten million undergoes ionization. How the concentration of hydrogen ions and hydroxyl ions are influenced by other materials dissolved in water and how this affects water properties will be discussed later.

Table 2.1-1 lists numerical values of some important physical properties of pure water.

1.02 Water Constituents

Water is such a powerful solvent that it dissolves at least a small amount of most inorganic matter and also

Table 2.1-1
Some Physical Properties of Water

Temp °C	Density g/ml	Specific Volume ml/g	Surface Tension dynes/cm[1]	Vapor Pressure mmHg	Viscosity Centipoises
0	0.9982	1.00013	—	—	Solid
4	1.0000	1.00000	75.06	6.101	1.567
30	0.9957	1.00440	71.18	31.824	0.801 (1.000 at 20.2°C)
100	0.9584	1.04340	58.85	760.000	0.284

[1]Vs. Air

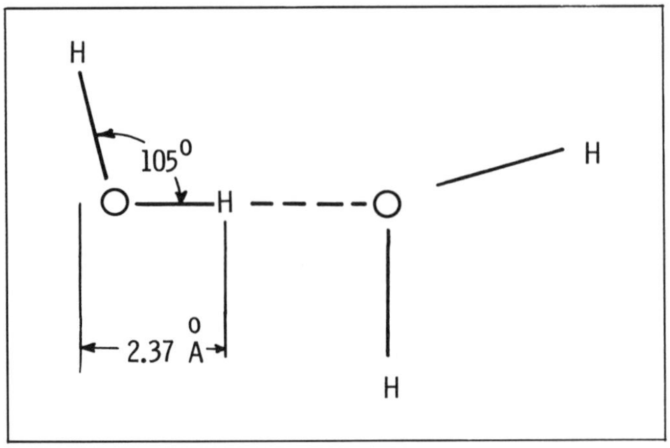

Figure 2.1-1. Structure of the water molecule.

many organic compounds. For the most part, it is these materials, either in solution or suspended, that cause water to exhibit unwanted properties.

In oil production operations, the major problems caused by or promoted by water are:

1. equipment corrosion,
2. injection well plugging,
3. scale deposition,
4. microbiological fouling.

As a result of its solvent power, liquid water does not exist as a pure substance in nature. Thus, what is broadly referred to as water, is really a range of aqueous solutions of various components and concentrations. The most common contaminants in water are dissolved minerals, gases and suspended insoluble matter such as clays and organic matter. In Table 2.1-2, waters are classified or named according to their sources, uses or gross properties.

1.03 Behavior of Dissolved Substances

Readers not familiar with common natural elements and their chemical symbols should refer to Item 1.05 and the first two columns of Table 2.1-3. The list shows the names of the substances that are commonly dissolved in water and their corresponding chemical symbols as shown in water analysis reports. The table will be discussed in more detail later.

Although it is occasionally done, it is incorrect to consider most substances existing in solution as combined compounds. A water analysis indicating the concentrations of dissolved salts in *molecular* form is only conjecture and is based on calculations of hypothetical combinations. Water analyses usually list *ion* concentrations.

Table 2.1-2
Classifications and General Characteristics of Water

	Typical TDS[1] Range—mg/L	Dissolved Gases[2]			Other Common Constituents
		O_2	H_2S	CO_2	
1. Salinity Classes					
a. Fresh	<2,000				
b. Brackish	2,000–10,000				
c. Brine	>10,000				
d. Hard	<2,000				$CaCO_3$ hardness >100 mg/L
e. Soft	<2,000				$CaCO_3$ hardness <60 mg/L
2. Source Classes					
a. Produced (Formation)	5,000–150,000		P	P	oil, suspended solids
b. Seawater	35,000–40,000	P			silt, slimes
c. Surface (lakes, rivers)	2,000–5,000	P			silt, clays
d. Well water	2,000–10,000	V	V	P	
3. Use Classes					
a. Injection	Variable				Dependent on source:
b. Cooling	200–2,000	P			bacteria, slimes
c. Process waste	5,000–20,000	P	V		chemicals, solids and
d. Municipal waste	200–3,000	P	V		organics

[1]TDS = total dissolved solids
[2]P = commonly present
mg/L = milligrams/liter
V = variable but possible

Table 2.1-3
Common Dissolved Constituents of Water

	Chemical Symbol[1]	Atomic or Radicle Weight	Equivalent Weight
1. *Cations*			
Sodium	Na^{+}	23.0	23.0
Potassium	K^{+}	39.1	39.1
Calcium	Ca^{++}	40.1	20.0
Magnesium	Mg^{++}	24.3	12.2
Iron	Fe^{++}	55.8	27.9
Manganese	Mn^{++}	54.9	27.5
Barium	Ba^{++}	137.4	68.7
Strontium	Sr^{++}	87.6	43.8
Aluminum	Al^{+++}	27.0	9.0
2. *Anions*			
Chloride	Cl^{-}	35.5	35.5
Sulfate	$SO_4^{=}$	96.0	48.0
Carbonate	$CO_3^{=}$	60.0	30.0
Bicarbonate	HCO_3^{-}	61.0	61.0
Hydroxide	OH^{-}	17.0	17.0
3. *Gases*[2]			
Oxygen	O_2		
Hydrogen Sulfide	H_2S		
Carbon Dioxide	CO_2		

(1) + and − valences shown are those usually valid in water solutions.
(2) Water analyses show concentrations of dissolved gases in their molecular form as indicated.

Inorganic dissolved salts can best be considered as existing in solution as positively (+) charged *cations* and negatively (−) charged *anions*. For example, when a crystal of sodium chloride (table salt) dissolves in water, the crystalline structure disintegrates and ions start to form, as follows:

$$\underset{\text{sodium chloride molecule}}{NaCl} \leftrightharpoons \underset{\text{sodium ion (cation)}}{Na^{+}} + \underset{\text{chloride ion (anion)}}{Cl^{-}}$$

The molecule of NaCl is stable and electronically neutral. As it dissolves, the sodium *atom* loses an *electron* to the chlorine *atom*; thus, both become *ions* because of their charges. It is these charges that distinguish ions from atoms.

Presence of charged ions allows the water to become electrically conductive. Salt (NaCl) in solution also increases water viscosity, density and surface tension. From this, it may be seen how the properties of water are altered by substances dissolved in it.

When water dissolves NaCl, the negative oxygen atoms of water molecules are attracted to the positive sodium ion while the positive (hydrogen) side of the water molecules attach to the negative chloride. This attraction is called *hydration* and is sufficient to overcome forces that tend to reform the salt crystal. Ultimately the solution becomes saturated with NaCl; that is, the *solubility limit* is reached. About 330 grams (g) of NaCl will dissolve in a liter (L) of water at 20°C (64°F). If the saturated solution is cooled or concentrated by evaporation, sodium and chloride ions start to recombine and visible salt crystals reform. This behavior is indicated by the double arrows in the previous equation depicting ionization of NaCl.

Salt deposits occasionally form in gas wells when formation water (saturated with NaCl) enters the wellbore and is cooled and evaporated by evolution of gas. Deposits are removed with fresh water.

Another example of salt behavior is that of calcium sulfate, $CaSO_4$. Many producing formations contain anhydrite crystals or unhydrated $CaSO_4$. Formation water dissolves some amount of anhydrite depending on temperature, pressure and concentrations of other salts such as NaCl. Solubility varies from about 1.5 to 6.0 g/L.

Both pressure and sodium chloride increase the amount of anhydrite that dissolves. When pressure declines as the formation water moves into a rock fracture or the wellbore, the concentration of $CaSO_4$ (calculated as such) exceeds the solubility limit (at a lower pressure) and solid phase $CaSO_4$ scale is formed.

The reactions may be depicted as follows:

Solution under pressure:

$$\text{(a)}\quad \underset{\text{anhydrite}}{CaSO_4} \rightarrow \underset{\text{calcium ion (cation)}}{Ca^{++}} + \underset{\text{sulfate ion (anion)}}{SO_4^{=}}$$

Precipitation at reduced pressure:

$$\text{(b)}\quad Ca^{++} + SO_4^{=} + 2H_2O \rightarrow \underset{\text{gypsum or calcium sulfate dihydrate}}{CaSO_4 \cdot 2H_2O}$$

At this point, we should also note that the calcium ion, Ca^{++}, has two positive valence charges whereas sodium, Na^{+}, has only one. The sulfate ion, $SO_4^{=}$, has a negative valence of two; thus, one part of $SO_4^{=}$ can combine with one part of Ca^{++} and form the neutral $CaSO_4$ salt. Complex ions such as $SO_4^{=}$, which are composed of two or more atoms of different elements, are called *radicles*. The radicle complex is so stable that even solution in water does not separate the constituent atoms, and the whole radicle takes on a charge to become an ion. One part of $SO_4^{=}$ requires two parts of Na^{+} to form highly soluble sodium sulfate salt, Na_2SO_4.

Going back to the previous reaction equation (b), which represents precipitation of calcium sulfate at reduced pressure, note that water, H_2O, is involved. The dihydrate or gypsum form (as shown) predominates at temperatures below about 120°C (250°F) and is the precipitate form of calcium sulfate scale most frequently found in producing wells. Anhydrite can precipitate at higher temperatures, for example, in boilers.

Calcium carbonate, $CaCO_3$, is another salt that causes troublesome deposits or scales in producing operations. The chemistry of $CaCO_3$ solution and precipitation is quite interesting and oil production personnel should be familiar with it.

Solid phase $CaCO_3$ exists as such in formation rocks along with many other calcium containing (*calcareous*) minerals. Most of these have low solubility in neutral water but $CaCO_3$ dissolves easily in water that contains carbon dioxide, CO_2.

For practical purposes, the water solubility of $CaCO_3$ is directly related to the concentration or partial pressure* of CO_2 in the gas phase in equilibrium with the water. Solution and precipitation of $CaCO_3$ may be illustrated by the following equations:

(*c*). *Solution of $CaCO_3$ by CO_2 at elevated pressure*

$$CaCO_3 + CO_2 + H_2O \rightarrow \underset{\text{calcium bicarbonate}}{Ca(HCO_3)_2}$$

(*d*). *Ionization of soluble calcium bicarbonate in water*

$$Ca(HCO_3)_2 \rightarrow \underset{\text{calcium ion}}{Ca^{++}} + \underset{\text{bicarbonate ion}}{2HCO_3^-}$$

(*e*). *Dissociation of bicarbonate and release of CO_2 at reduced pressure*

$$2HCO_3^- \rightarrow \underset{\text{carbonate ion}}{CO_3^=} + CO_2 \uparrow + H_2O$$

(*f*). *Precipitation of insoluble calcium carbonate*

$$\underset{\substack{\text{calcium ion}\\\text{(equation b)}}}{Ca^{++}} + \underset{\substack{\text{carbonate ion}\\\text{(equation c)}}}{CO_3^=} \rightarrow CaCO_3 \downarrow$$

In these equations two additional complex ions or radicles, HCO_3^- and $CO_3^=$ are introduced. Most natural waters contain bicarbonate, HCO_3^-, in proportion to the amount of CO_2 present. Because CO_2 is a gas, it quickly escapes from water at reduced pressure and the water analysis usually does not show CO_2 as such.

The HCO_3^- (equation d) persists for some time in the water after pressure is reduced, but part of it gradually decomposes according to equation e. Both heat and reduced pressure force equation e to the right, forming $CO_3^=$ that then reacts with Ca^{++} to reform $CaCO_3$.

While the solubility of $Ca(HCO_3)_2$ is quite high, the solubility of $CaCO_3$ is less than 0.05 g/L in CO_2 free water. $CaCO_3$ scale deposits accumulate in oilfield equipment such as heater treaters, injection wells under vacuum, and many other locations. Scale problems and their prediction will be dealt with in more detail in Chapter 2 of this section.

*Partial pressure = mol fraction of CO_2 in gas × total pressure (see Section 1, Item 8.11).

1.04 Hydrolysis Reactions

An important chemical reaction involving water (as H_2O) is the *hydrolysis* of certain gases and salts. Hydrolysis is a chemical reaction in which a compound reacts with water to form an acid, a base or both. Such reactions change the acidity or alkalinity of the solution and thus its tendency to precipitate scale, corrode metals, and alter other chemicals.

A substance that dissolves in water to generate hydroxyl ions (OH^-) or carbonate ions ($CO_3^=$) is referred to as a *base* and the solution is said to be basic or alkaline. Bases neutralize acids. Examples of bases are:

1. Ammonia, NH_3 (ammonium hydroxide is NH_4OH),
2. Sodium hydroxide, NaOH (caustic soda),
3. Sodium carbonate, Na_2CO_3 (soda ash),
4. Sodium bicarbonate, $NaHCO_3$ (baking soda).

Acids are substances that increase the hydrogen ion (H^+ concentration when dissolved in water and have the ability to neutralize bases. Some common acids are:

1. Hydrochloric, HCl (muriatic acid),
2. Sulfuric, H_2SO_4,
3. Acetic, CH_3COOH,
4. Carbonic, H_2CO_3 (CO_2 in H_2O).

Bases and acids react to form salts that may be neutral, alkaline or acidic (when dissolved). The neutral salt, NaCl, is formed by reaction of a strong acid, HCl, with a strong base, NaOH.

Alkaline salts, formed by reaction of a strong base with a weak acid, dissolve in water to produce basic or alkaline solutions. An example is sodium carbonate (soda ash) formed by reaction of the NaOH, a strong base, with H_2CO_3, a weak acid.

Acid salts hydrolyze in water to reform, in effect, the strong acid and weak base that generated them initially. Following is a sample reaction:

$$\underset{\text{iron chloride}}{2\,FeCl_3} + 6\,H_2O \rightleftarrows \underset{\substack{\text{iron hydroxide}\\\text{(a weak base)}}}{2\,Fe(OH)_3} + \underset{\substack{\text{hydrochloric acid}\\\text{(a strong acid)}}}{6\,HCl}$$

A water solution of iron chloride is acidic.

1.05 Water Constituents Found by Analysis

Water analyses (or mineral analyses as they are sometimes called) show concentrations of the major constituents that existed in a sample of the water when it was analyzed in the laboratory. A series of analytical tests are conducted to determine the concentration of dissolved ions that are usually present. A typical routine analysis will find more than 99 wt.% of all substances present in solution.

Table 2.1-3 lists the soluble constituents that are most often shown by water analysis reports. Notice that they are separated into three groups:

1. Positively charged cations,
2. Negatively charged anions,
3. Gases.

All the cations are metal elements and all the anions are non-metal elements or groups of non-metal elements (radicles). These substances exist in solution as charged ions as discussed under Item 1.03.

In addition to the soluble matter and gases dissolved in the water sample, many analysis reports also show the total amount of *insoluble* materials that were present. These are usually small particles that were suspended in the water when the sample was taken. Detailed analysis of the suspended solids in water samples is usually not performed but their general nature may be indicated in the report, i.e., calcium carbonate scale, iron sulfide, clay, sand, etc. Table 2.1-4 lists the types of insoluble materials most often found in oilfield waters.

Table 2.1-4
Suspended and Organic Matter Found in Water

Insoluble Inorganics	*Organics*
Calcium Carbonates and Sulfates	Bacteria, Molds and Yeasts
	Algae, Protozoa, Diatoms
Clays	Decaying Vegetation
Quartz or Sand	Tannins—Wood Extracts
Iron, Manganese, Aluminum Oxides	Soaps and Polymers
	Oil
Iron Sulfides	
Barium Sulfate	
Feldspar	
Silica[1]	

[1]Silica may exist either suspended or dissolved (as SiO_2).

1.06 Some Properties of the Metal Cations and Their Salts

Each of the cations listed in Table 2.1-3 can, under proper conditions, combine with any of the anions to form a salt. When the combination occurs, the total number of the positive (+) valence charges of the cations must exactly equal the total of the negative (−) anion valences. For example, one monovalent Na^+ can combine with one monovalent Cl^- to form NaCl salt. Two Na^+ ions are required for one $SO_4^=$ ion to form Na_2SO_4. Only one divalent Ca^{++} is required for one divalent $SO_4^=$ to form the $CaSO_4$ salt. Two Al^{+++} ions combine with three $SO_4^=$ ions to form $Al_2(SO_4)_3$ and so on.

Each of the more than 45 salts possible from the Table 2.1-3 list possess their own physical and chemical properties, and detailed knowledge of all of them is not necessary in petroleum operations. However, the following general observations will be helpful:

1. All the common salts of sodium (Na) and potassium (K) are very water soluble. Their chloride and sulfate salts are neutral but the bicarbonates, carbonates and hydroxides are alkaline.
2. Chloride salts of Ca, Mg, Ba, and Sr are water soluble but solubilities of their sulfate salts vary as follows:

   $$BaSO_4 < SrSO_4 < CaSO_4 < MgSO_4$$
   Lowest *Highest*

   The carbonate and hydroxide salts of Ca, Mg, Ba, and Sr all have low water solubility with $Mg(OH)_2$ being least soluble in neutral non-acidic water.
3. The chlorides and sulfates of iron, manganese and aluminum are water soluble and their solutions are acidic. Their other common inorganic salts (carbonates, hydroxides, sulfides, etc.) have low solubility in water but are soluble in acids.

1.07 Water Properties Determined by Analysis

The presence of dissolved or suspended matter in various amounts causes water to exhibit certain qualities or characteristics that are measured as the following properties:

1. **Acidity**

   Capacity to neutralize bases or alkaline materials. Usually caused by carbon dioxide, organic acids, mineral acids or hydrolysis of acid salts.
2. **Alkalinity** (basicity)

   Capacity to neutralize acids. Usually caused by bicarbonate (HCO_3^-), carbonate ($CO_3^=$), and hydroxyl (OH^-) ions.
3. **pH**

   A measure of the hydrogen ion activity.

Acidity, Alkalinity and pH are interrelated properties and are important with respect to the ability of a water to cause corrosion, scale and other problems. Alkalinity is routinely determined quantitatively in the laboratory but is normally *not* expressed as such in water analysis reports. The water analyst converts alkalinity to HCO_3^-, $CO_3^=$ or OH^- and the concentration of those ions are shown in the water analysis. Acidity is also a quantitative determination but is usually a special test.

The expression, pH, is derived from the German word *potenz* (power) and *H*, the symbol of hydrogen. The nu-

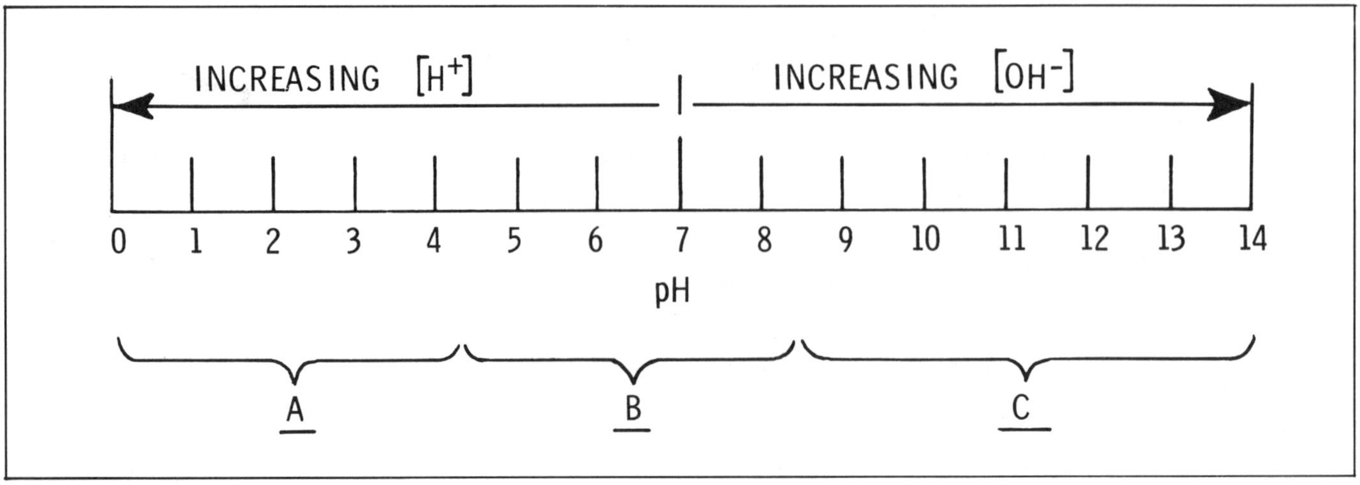

Figure 2.1-2. Scale of pH values and the three zones of alkalinity.

merical value of pH is an indicator of the *intensity* of the acidic or basic properties of water (not the quantitative *capacity* to neutralize). Figure 2.1-2 depicts a scale of pH values as used on pH meters.

In pure water at pH 7, both the hydrogen ion concentration, (H^+), and the hydroxyl ion concentration, (OH^-) (see equation a, Item 1.01) are equal and the solution is said to be *neutral*. Acidic solutions have pH values of less than 7. The stronger the acid, the lower the pH number. Basic solutions have pH values greater than 7 with alkalinity increasing as the pH grows larger (and vice versa).

Indicated on Figure 2.1-2 are three zones of pH: A, B, and C. Zone A is from pH 4.2 down and alkalinity, as such, is actually zero. High concentrations of CO_2 can force the pH of mineral free water down into this range. Water solutions of organic acids such as formic or acetic, sometimes used for well acidizing, produce pH values of 1 to 2, whereas strong acid solutions, such as 10% HCl, are below a pH of 1. Spent HCl solutions from well stimulation treatment typically show pH values of 2 to 3.5. A water sample with a pH of less than 4 should be considered as contaminated since natural waters are never this acidic. All low pH waters are corrosive to steel.

Most natural waters fall into the pH range represented by zone B of Figure 2.1-2. Bicarbonate, HCO_3^-, alkalinity increases over this pH range (4.2 to 8.3) and free CO_2 acidity declines. Being volatile, CO_2 escapes readily from water as discussed under Item 1.03, and the water pH increases. This effect has two very important practical ramifications, viz.:

a. Corrosiveness tends to decrease and carbonate scaling tendency tends to increase as CO_2 is lost and alkalinity increases.
b. Unless special sampling procedures are used to prevent loss of CO_2, the bicarbonate concentration (alkalinity) and pH shown by the water analysis *do not* accurately reflect the original water composition.

Zone C of Figure 2.1-2 is the range of highest alkalinity. Between about 8.3 to slightly above pH 12, both HCO_3^- and $CO_3^=$ ions exist in solution. Some natural waters have a pH in the lower end of this range. However, $CO_3^=$ alkalinity most often results from decomposition of HCO_3^- as depicted by equation e under Item 1.03. Appearance of $CO_3^=$, with attendant precipitation of $CaCO_3$ scale, occurs when seawater is deaerated with gas or by vacuum (see Section 1, Chapter 6). A similar situation results when water with high bicarbonate alkalinity is used in open cooling systems. Waters with high bicarbonate alkalinity lose CO_2 and tend to form $CaCO_3$ scale when heated in heater treaters or after entering injection wells under vacuum.

Above about pH 12, hydroxide alkalinity appears due to presence of free hydroxyl, OH^-, ions. Such high alkalinity is reached in boiler blowdown water and when caustic is added to drilling fluids, for example. When OH^- is shown in a water analysis, the HCO_3^- will be zero because the two ions react.

Buffer capacity is another alkalinity related water property. Solutions that resist change in pH when acids or bases are added are called buffer solutions. High bicarbonate concentration can cause a buffering effect. The HCO_3^- reacts with either a strong acid or a strong base (and thus impedes pH change) as indicated:

$$\text{(a) } HCO_3^- + \underset{\text{acid}}{H^+} \rightarrow H_2O + CO_2\uparrow$$

$$\text{(b) } HCO_3^- + \underset{\text{base}}{OH^-} \rightarrow CO_3^= + H_2O$$

Due to reaction (b), bicarbonate ions and hydroxide alkalinity do not co-exist in dilute solutions.

4. **Salinity**

   This term is used to classify the dissolved mineral content of waters. Chloride salinity indicates the total concentration of chloride, Cl^-, present. The salinity value is high in strong salt waters or brines. Sodium chloride, NaCl, salinity is similar except the chloride content determined by analysis is expressed as NaCl.

5. **Total Dissolved Solids (TDS)**

   The sum of all dissolved ions (cations plus anions) shown by the water analysis. TDS by evaporation is determined by carefully heating and evaporating a small volume (usually 50 to 100 ml) of water and weighing the dried salt residue. Results are shown as milligrams per liter (mg/L) or parts per million (ppm). The TDS by evaporation number is usually lower than the TDS by summation of ions determined by analysis due to loss by evaporation.

6. **Density**

   Weight per unit volume expressed as grams per liter (g/L), pounds per gallon (lb/gal), kilograms per cubic meter (kg/m^2), etc.

7. **Specific Gravity (Sp. Gr.)**

   The *ratio* of the density of the water being tested to the density of pure water (no dissolved salts) at a stated temperature. Being a ratio, sp. gr. has no units such as g/L or lb/gal. Specific gravity is usually measured with a hydrometer, which is a slender sealed glass instrument that is weighted on the bottom. The hydrometer floats at various depths in a volume of water being tested and the value of sp. gr. is read from a calibrated stem at the surface of the water. Hydrometers are usually calibrated at 4°C where the density of pure water is exactly 1.000 (see Table 2.1-1). If the temperature of the water being tested is higher than 4°C, the measured temperature is recorded and reported, i.e.,

   $$\text{Sp. Gr. at } \frac{20°\text{C}}{4°\text{C}}, \frac{25°\text{C}}{4°\text{C}}, \text{ etc.}$$

   Dissolved salts increase water density and, therefore, specific gravity. Because the amount of increase is different with different salts, the density or specific gravity values cannot be used directly as a measure of TDS. However, the following equation provides an approximation that is sufficiently accurate for *estimating* total dissolved solids (TDS) in oilfield waters having salinities of 10 to 150 g/L (10,000 to 150,000 mg/L or ppm):

   $$\text{TDS as g/L} = (\text{Sp. Gr.} - 1) \times 1380$$

   or

   $$\text{Sp. Gr.} = 1 + \frac{\text{TDS (g/L)}}{1380}$$

   (Average error = ±6%.
   Computed from 1000 water analyses)

8. **Total Suspended Solids (TSS)**

   The weight of insoluble, suspended matter filtered from a measured volume of water, preferably expressed as milligrams per liter (mg/L). Plastic membrane (Millipore) test filters, retaining solids larger than 0.45 micrometers (microns or μm), are commonly used to determine TSS (see Chapter 4).

9. **Turbidity**

   An optical property relating to light adsorption and scattering in water. This empirical measurement provides a number whose magnitude is affected by the amount and kind of suspended matter. The actual concentration of suspended matter is not indicated and there is generally no reliable correlation between turbidity readings and the weight of suspended solids. However, the turbidity test is easy to perform with commercial turbidimeters and a *change* in turbidity usually signals a change in kind or amount of suspended solids.

   An accurate and easy to use turbidimeter is shown schematically in Figure 2.1-3. A unit of this type used downstream of floodwater filters can rapidly sense and signal solids leakage through the filter beds, for example.

10. **Biochemical Oxygen Demand (BOD)**

    The quantity of dissolved oxygen required during stablization of decomposable organic matter by aerobic biochemical action.

11. **Chemical Oxygen Demand (COD)**

    The quantity of oxygen consumed during chemical oxidation of organic matter under controlled conditions. The value is used for monitoring plant wastes and streams but does not directly indicate the quantity of carbonaceous matter nor the amount of organics that are biologically oxidizable.

1.08 Units of Concentration

Water analysts use several terms or units to express concentrations of dissolved and suspended substances in the water analysis reports. Oil production personnel should understand the meaning of the units and how to use them for comparison and calculations.

Table 2.1-5 lists the seven common units of concentration, their abbreviations and a brief description of each. More detailed explanations of the units are given in the following paragraphs. The first three, milligrams per liter, parts per million and milliequivalents per liter should be committed to memory.

(1) and (2) The preferred unit for indicating the weight of each constituent per unit volume of water is *milli-*

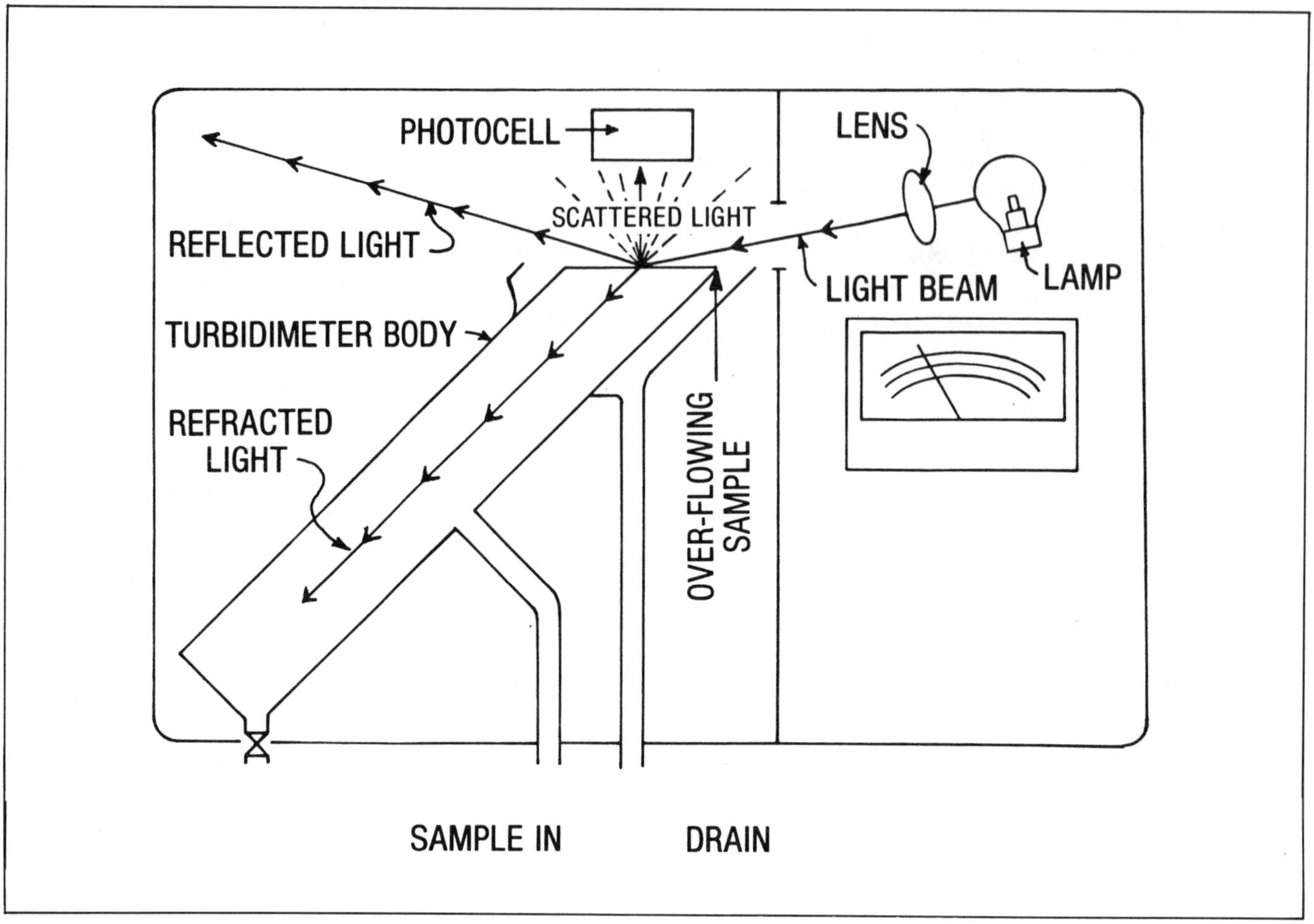

Figure 2.1-3. Flow diagram of a surface scatter turbidimeter. (Courtesy Hach Company).

Table 2.1-5
Concentration Units for Water Constituents

Unit	*Abbreviation*	*Description*
(1) Milligrams per liter	mg/L	Number of milligrams of substance in each liter of water
(2) Parts per million	ppm	Approximately equal to mg/L $\frac{\text{ppm}}{\text{sp. gr.}} = \text{mg/L of water}$
(3) Milliequivalents per liter	meq/L	$\frac{\text{mg/L}}{\text{equivalent weight}}$
(4) Equivalents per million	e.p.m.	$\frac{\text{ppm}}{\text{equivalent weight}}$
(5) Grains per gallon	gpg	mg/L × 17.1
(6) Calcium Carbonate Equivalent (of any ion)	mg/L as $CaCO_3$	$\left(\begin{array}{c}\text{mg/L of ion from}\\ \text{water analysis}\end{array}\right) \times \left(\frac{50}{\begin{array}{c}\text{equivalent weight of ion}\\ \text{from Table 2.1-3}\end{array}}\right)$
(7) Weight Percent	wt %	g/100 milliliters (ml) or $\frac{\text{mg/L}}{10{,}000}$

grams per liter, mg/L. However, *parts per million*, ppm, is still used, especially to indicate concentration of dissolved O_2 or H_2S. For many purposes, mg/L and ppm can be used interchangeably. In low salinity waters having a sp. gr. of close to 1.000, the error is small since the water density is 1 kilogram (or 1 million milligrams) per liter. Therefore, 1 milligram in 1 kilogram is 1 part per million by weight in low salinity water. Precise correlation of ppm and mg/L can be computed for any water by taking sp. gr. into account (see Table 2.1-5).

(3) One very important unit of concentration that is not readily understood is *milliequivalents per liter* (meq/L). It is actually quite simple.

Referring to Table 2.1-3, column 3, note that each cation and anion has a particular atomic or radicle weight assigned to it. For the single element ions, viz., all the cations and the chloride ion, the *atomic weight* values are the *individual element* weights.

Element weights are based on a standard that states that carbon-12 has a weight of exactly 12. Weights of other elements are compared with the standard, i.e., the atomic weight of potassium is 39.1 which is

$$\frac{39.1}{12}$$

or 3.258 times heavier than carbon-12.

The radicles (sulfate, bicarbonate, carbonate and hydroxide) are assigned *radicle weights* that are the sum of the atomic weights of their constituent elements. For example, the standard atomic weight of oxygen, O, is slightly less than 16 and the standard atomic weight of hydrogen, H, is slightly more than 1. Therefore, the radicle weight of the hydroxide (or hydroxyl) ion, OH^-, is 17 as shown in Table 2.1-3.

Equivalent weights shown in column 4 of Table 2.1-3, are the atomic or radicle weights divided by the number of element or radicle valences (see Item 1.01). For example, sodium has a valence of 1 (indicated by the single + by the symbol) and its atomic and equivalent weights are the same. When the ion valence is 2 or more, the equivalent weight is 1/2, 1/3 or other fraction of the atomic or radicle weight as can be seen by examining the numbers in Table 2.1-3.

A *milliequivalent* of any ion or salt is the equivalent weight divided by 1000.

From Table 2.1-5, it may be seen that milliequivalents per liter, meq/L, of any ion is simply the milligrams per liter concentration (shown by the water analysis) divided by the equivalent weight of that ion (from Table 2.1-3). For example, 1000 mg/L of Ca^{++} is equal to:

$$\frac{1000}{20} \text{ or } 50 \text{ meq/L of } Ca^{++}$$

Likewise, 1000 mg/L of $SO_4^{=}$ is equal to:

$$\frac{1000}{48} \text{ or } 20.8 \text{ meq/L of } SO_4^{=}$$

The number of milliequivalents per liter (meq/L) for each ion shown in a water analysis, is usually listed alongside the column showing the milligrams per liter (mg/L). The reason for listing meq/L is that this unit of concentration is useful in making certain chemical calculations such as scale prediction. For example, assume that a water analysis shows 1000 mg/L each of Ca^{++} and $SO_4^{=}$ and you want to know how much calcium sulfate ($CaSO_4$) is in the water. The meq/L of each of these ions is 50 for Ca^{++} and 20.8 for $SO_4^{=}$ as was calculated previously.

Even though the concentrations of Ca^{++} and $SO_4^{=}$ expressed as mg/L are equal, we see that the concentration of $SO_4^{=}$ as meq/L is less than the meq/L of Ca^{++}. In this case, the maximum concentration, as meq/L, for $CaSO_4$ is limited by and cannot be greater than the meq/L of $SO_4^{=}$. Therefore, the concentration of $CaSO_4$ in this particular water is taken as 20.8 meq/L.

When the meq/L of Ca^{++} in a water analysis is less than that of $SO_4^{=}$, the meq/L of $CaSO_4$ is equal to the meq/L of Ca^{++}.

To translate the 20.8 meq/L of $CaSO_4$ in the previous example back to mg/L or ppm, the equivalent weight of $CaSO_4$ must be known. This is easily determined by referring back to Table 2.1-3 for the equivalent weights of Ca^{++} and $SO_4^{=}$ and adding the two numbers, viz.:

$$\underset{\text{(equivalent wt. of } Ca^{++})}{20} + \underset{\text{(equivalent wt. of } SO_4^{=})}{48} = \underset{\text{(equivalent wt. of } CaSO_4)}{68}$$

$$\text{From Table 2.1-5, meq/L} = \frac{\text{mg/L}}{\text{equivalent weight}}$$

or meq/L × equivalent weight = mg/L

and 20.8 × 68 = 1414 mg/L $CaSO_4$

Meq/L will be used for scale prediction calculations in Chapter 2.

(4) and (5) Equivalents per million and grains per gallon are rarely used in modern water analysis reports but are listed for reference if they are ever encountered.

(6) Calcium carbonate ($CaCO_3$) equivalent (unit 6 in Table 2.1-5) is still used in water softening calculations and is the standard unit of expression for *alkalinity* and *hardness*.

a. Hardness is a water property that was originally a measure of the amount of soap required to produce foam or suds in cleaning operations such as washing dishes or bathing in the home. Referring back to Table 2.1-2, notice that low salinity fresh waters (such as waters used in homes or as cooling and boiler feedwater in industrial plants) are classed as hard or soft depending on their hardness expressed as $CaCO_3$.

   The total hardness value of a water, as $CaCO_3$, is computed from a water analysis by converting the mg/L of Ca^{++}, Mg^{++} and Fe^{++} to $CaCO_3$ equivalent by the calculation procedure shown in Table 2.1-5 and taking the sum:

$$\text{Total hardness as } CaCO_3 = (Ca^{++} \text{ as } CaCO_3) + (Mg^{++} \text{ as } CaCO_3) + (Fe^{++} \text{ as } CaCO_3)$$

b. As discussed under Item 1.07, the alkalinity of a water results from the presence of HCO_3^-, $CO_3^=$ or OH^- ions. In most natural waters, only HCO_3^- ions or HCO_3^- and a small amount of $CO_3^=$ are present. Total alkalinity as $CaCO_3$ is computed by converting the alkalinity ions to their $CaCO_3$ equivalent and adding the $CaCO_3$ values. If both HCO_3^- and $CO_3^=$ are present, total alkalinity is the sum of HCO_3^- (as $CaCO_3$) plus $CO_3^=$ (as $CaCO_3$). If $CO_3^=$ and OH^- are present (rare), total alkalinity is the sum of $CO_3^=$ (as $CaCO_3$) plus OH^- (as $CaCO_3$).

Examples of total hardness and total alkalinity calculations are given in Item 1.09.

(7) Weight percent is used to indicate high concentrations such as the amount of various salts added to water to increase density.

1.09 Examples of Water Analysis Calculations and Interpretations

Table 2.1-6 shows tabulated analytical results (water analyses) from four water samples, as analyzed by commercial testing laboratories. The data will be used to illustrate how water analyses are employed for certain calculations and correlations.

1. Sample 1 water was obtained from a shallow (depth: 35 m or 115 ft) well completed in an alluvium sand. Intended use of the water was for the cooling system and as boiler feed in a gas processing plant. Oxygen content and pH of the water were measured at the wellhead during sample collection.

   Sample 1 was labeled as fresh water, but from inspection of the water analysis results, the well water is more properly classed as a brackish water (see Table 2.1-2). The analysis also shows that the major dissolved salt constituent is NaCl.

   The water contains no hydroxide (OH^-) or carbonate ($CO_3^=$) and therefore, alkalinity is due entirely to the 165 mg/L of bicarbonate (HCO_3^-). Alkalinity expressed as $CaCO_3$ equivalent was computed for this water as follows:

$$\text{Total alkalinity as } CaCO_3 = 165 \times \frac{50}{61}$$

$$\text{or} \quad 135 \text{ mg/L}.$$

   For this calculation, the concentration of bicarbonate, HCO_3^-, was taken from the water analysis and the equivalent weight of HCO_3^- was taken from Table 2.1-3.

Table 2.1-6
Examples of Water Analyses

	SAMPLE 1		*SAMPLE 2*		*SAMPLE 3*		*SAMPLE 4*	
	Fresh Water		*Seawater*		*Formation (Connate) Water*		*Produced Water*	
Ion	*mg/L*	*meq/L*	*mg/L*	*meq/L*	*mg/L*	*meq/L*	*mg/L*	*meq/L*
Na^+	1364	59.3	11,144	484.5	31,602	1374.0	28,543	1241.0
K^+	8	0.2						
Ca^{++}	101	5.1	464	23.2	18,662	933.1	14,010	700.5
Mg^{++}	28	2.3	1,350	111.0	2,838	232.6	2,470	202.5
Fe^{++}	3	0.1						
Ba^{++}	0							
Sr^{++}	0							
Cl^-	2265	63.8	19,900	562.1	89,499	2,521.1	75,500	2126.8
$SO_4^=$	24	0.5	2,600	54.1	410	8.5	432	9.0
HCO_3^-	165	2.7	149	2.5	617	10.1	502	8.2
$CO_3^=$	0		0		0		0	
OH^-	0		0		0		0	
TDS	3955		35,607		143,559		121,457	
Sp. Gr.	1.003		1.026		1.104		1.088	
pH	7.9		7.8		6.2		7.4	
pHs @ 50°C	6.9							
O_2	2.1		3.8					
H_2S	0				139		190	

Hardness was computed to determine water softening capacity required if the water were used as boiler feed, as follows:

$$\text{Total hardness as } CaCO_3 = Ca^{++} + Mg^{++} + Fe^{++}$$

expressed as $CaCO_3$

$$\text{or, for Sample 1: } \left(101 \times \frac{50}{20}\right) + \left(28 \times \frac{50}{12.2}\right)$$

$$+ \left(3 \times \frac{50}{27.9}\right) = 373 \text{ mg/L as } CaCO_3$$

Notice in the analysis of Sample 1 that both the actual pH value and a value for pH_s is shown. As mentioned, actual pH was measured at the wellhead when the sample was taken. The pH_s value was computed after the anlysis was completed. pH_s is the theoretical pH that the water would have if it were exactly saturated with $CaCO_3$. Calculation of pH_s for predicting $CaCO_3$ scaling tendency from water analyses will be discussed in detail in Chapter 2 of this section.

For Sample 1, actual pH was higher than pH_s at 50°C (122°F) and this indicated that $CaCO_3$ scale precipitation would probably occur if this water were used in the gas plant as planned.

Based on the water analysis and the calculations just described, the company in charge of water management for the new plant concluded that the well water was unsuitable for its intended use for the following reasons:

a. *Excessive Hardness*

The projected cost of softening the water (removing hardness ions) was prohibitive since hardness would have to be reduced from 373 mg/L as $CaCO_3$ to near zero for boiler feed use.

b. *Scale Precipitation*

Deposition of calcium carbonate scale in lines and heat exchangers was a virtual certainty even if the raw water were used "once through." Scale control would be especially difficult if the water were concentrated (in the cooling towers) by a factor of 3 as originally planned.

c. *Corrosion and Bacterial Problems*

Presence of dissolved oxygen and iron in the water at the supply wellhead indicated corrosion of the well pump and tubing would be serious and bacterial fouling was likely.

An alternate supply of water from a surface source was obtained and the water well was abandoned.

2. Samples 2, 3 and 4 are water analyses from a waterflood operation. The seawater (Sample 2) was injected into a producing reservoir containing connate (formation) water of the composition listed under Sample 3. Sample 4 was a produced water sample from an oil well.

The seawater used for flooding contained dissolved oxygen and was therefore considered corrosive. Designers of the waterflood equipment specified installation of gas stripping towers to remove dissolved oxygen to a level of about 0.2 mg/L and use of a sulfite oxygen scavenger to reduce residual O_2 to less than 0.05 mg/L for corrosion prevention (see Section 1, Chapter 6).

Calculations, performed by the water analyst from the water analysis alone, indicated that during gas stripping (for O_2 removal) the seawater alkalinity would increase and the pH would rise from the original 7.8 to 8.3 or 8.4 due to generation of $CO_3^{=}$ by decomposition of HCO_3^{-} (see Item 1.03). It was later determined, however, that the available stripping gas (O_2 free) contained 5 mol % CO_2. The pH of the seawater, in equilibrium at 0.14 MPa (20.3 psig) with the gas containing 5 mol % CO_2, was found to be 6.5 and consequently, the $CaCO_3$ scaling tendency was actually negative. If vacuum deaeration or stripping with a CO_2 free gas had been employed, treatment of the deaerated seawater with a calcium carbonate scale inhibitor chemical would have been required.

The mineral content of Sample 2 as shown is typical of most seawaters, but some have slightly higher total salinities. Notice in the analysis that values are not listed for certain constituents such as potassium. Omission of a concentration value for a particular ion usually indicates the concentration of that constituent was not determined. Zero value for concentration indicates none was present. Notice also that in all the analyses the sums of the meq/L values of the cations (+) and of the anions (−) are exactly equal. When Samples 2, 3 and 4 were analyzed, the only cations actually determined in the laboratory were calcium and magnesium. The meq/L of both Ca^{++} and Mg^{++} were substracted from the total meq/L of all the anions and the difference was assigned by the water analyst to sodium Na^{+}. From the calculated meq/L for Na^{+}, the mg/L for that constituent was calculated as follows:

$$Na^{+} \text{ as mg/L} = \text{meq/L} \times 23$$

(equivalent weight of Na^{+})

The computed values for mg/L of Na^{+} were reported as shown in the analysis reports. This is an accepted practice and is satisfactory for analysis of most oil field waters.

Sample 4 represents water from a producing well after breakthrough of the injected seawater. Comparison of the mineral contents as shown by the analyses indicates the concentrations of ions in Sample 4 falls

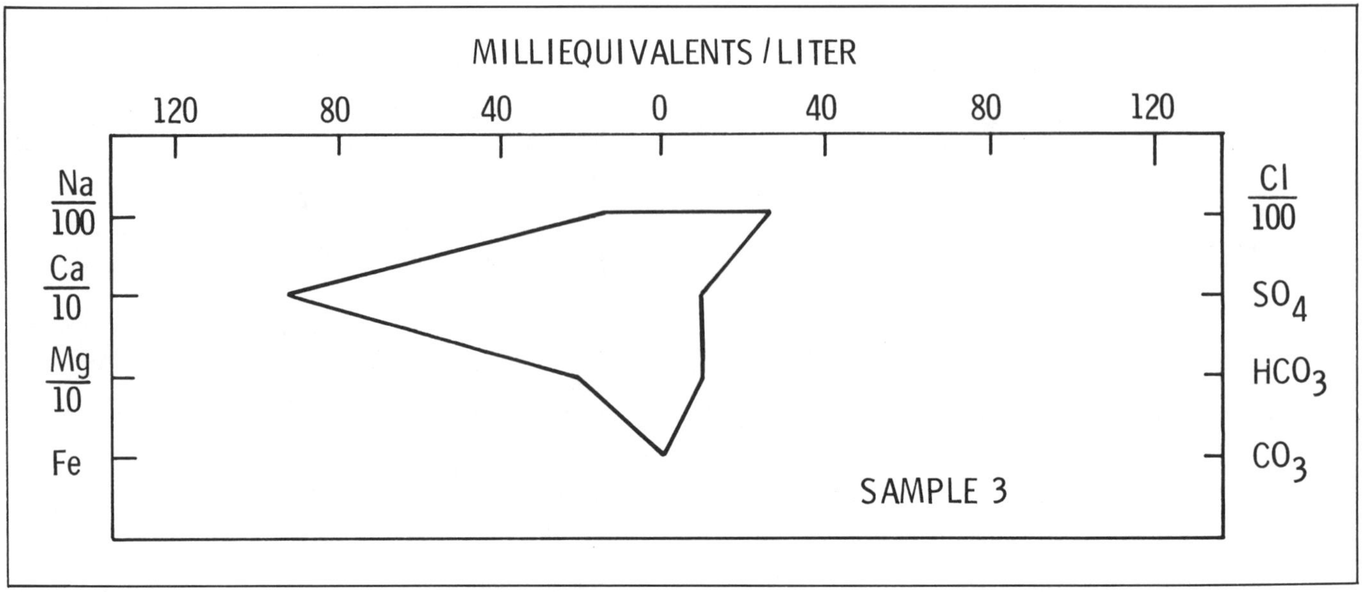

Figure 2.1-4. Example of a water analysis pattern (Stiff Method).

between that of Samples 2 and 3. The produced water appeared to be a simple mixture of injected and connate water, and it was desirable to determine the volume ratio of the two. Since chloride salts of all cations in the water are soluble (see Item 1.06), the volume ratio of injected seawater and connate water required to produce the composition of the produced water was computed from the chloride content of all three waters as follows:

$$\text{Vol \% connate water in produced water} = \frac{(Cl_{prod} - Cl_{inj}) \times 100}{(Cl_{con} - Cl_{inj})}$$

where Cl_{prod} = mg/L Cl^- in produced water
Cl_{con} = mg/L Cl^- in connate water
Cl_{inj} = mg/L Cl^- in injection water

Calculation:

Vol % connate water
in produced water

$$= \frac{(75{,}000 - 19{,}900) \times 100}{(89{,}400 - 19{,}900)} = 80\%$$

$$\text{Vol \% injected in produced water} = 100 - 80 = 20\%$$

1.10 Graphical Plots of Water Analysis Data

In the examples just discussed, Sample 4 (produced water) appeared to be a mixture of injection and connate waters. This impression resulted from visually comparing concentrations of individual ions in the three water analysis reports. To make such comparisons easier, several procedures have been developed to convert the digital analysis data to graphical plots that form distinctive geometric patterns for each separate water. Figure 2.1-4 illustrates one type of water analysis pattern for Sample 3 (the connate water). In this example, the meq/L of each ion in the Sample 3 analysis report, were divided by 1, 10, or 100 and the results were plotted as shown in Figure 2.1-4. Notice that all cation values are plotted on the left hand side of the graph and all the anion values are plotted on the right hand side. Where the water analysis shows zero or no concentration value for ions such as Fe^{++} and $CO_3^{=}$, these are plotted as zero in order to complete the plot pattern. The top horizontal scale can be varied as necessary to cover the required range of ion concentrations. However, the scale must be the same for all waters to be compared. Alternately, many water analysis laboratories use mg/L for the concentrations of ions instead of meq/L and the top scale is logarithmic, e.g., 10, 100, 1000, 10000 mg/L.

If the water analysis data for Samples 2 and 4 were plotted exactly as shown for Sample 3 in Figure 2.1-4, the three patterns could be layed side by side and visually compared. Since the ion concentrations for the three waters are all different, the three patterns would be different and one might be able to see that Sample 4 is probably a mixture of Samples 2 and 3. It would certainly be apparent that the produced water (Sample 4) is not 100% connate or 100% injected water.

For comparing water analyses where there is reason to believe that one water is a mixture of two or more different waters, use of a graphical plotting method as shown in Figure 2.1-5 is recommended. For this ex-

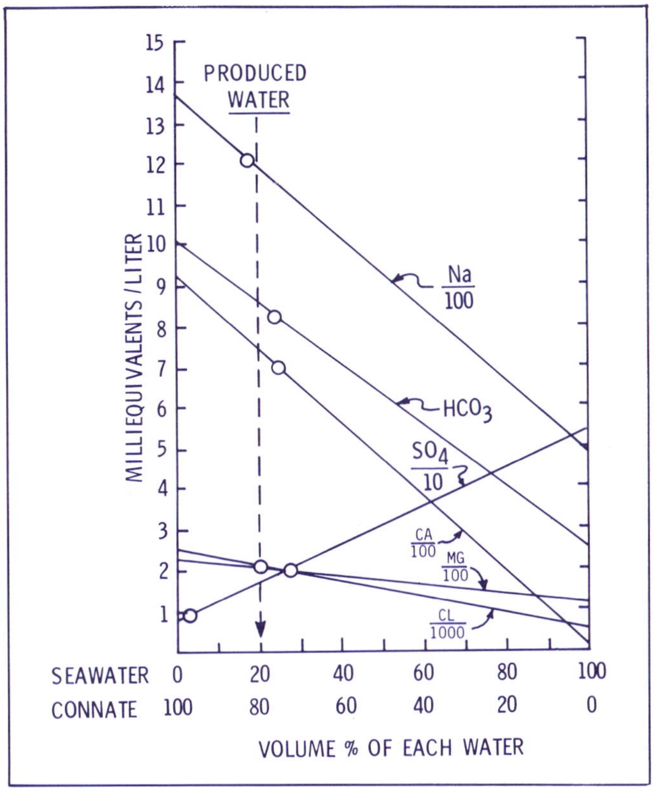

Figure 2.1-5. Graphical plot of water analyses.

ample, an appropriate scale of meq/L values was chosen as shown on the left hand side of Figure 2.1-5. The same concentration values are used for the right hand vertical scale but the numbers are not shown on the graph. The bottom scale of the graph was arranged to indicate zero to 100% seawater in one direction and 0 to 100% connate water in the opposite direction. The points along the left vertical scale were plotted to represent the meq/L of each ion in the connate water analysis (Sample 3). The meq/L concentration values from the water analysis were first divided by 1, 10, or 100 to make them fit conveniently on the chosen scale as shown. The same divisors were used for plotting the meq/L of each ion in the seawater on the right vertical scale. Tie lines connecting the concentrations of each ion on the two vertical scales represent all possible mixtures of the seawater and connate water. After plotting the concentration of each ion shown by the analysis of the produced water on the appropriate tie lines and drawing a best fit ordinate, the volume ratio of the two original waters in the mixed produced water was readily seen.

Results of the graphical procedure agree with the proportions of seawater and connate computed from their chloride contents alone (Item 1.08b). Notice, however, that the $SO_4^{=}$ concentration in the produced water was considerably lower than expected. This observation indicated a possible $CaSO_4$ scale problem and deposits of $CaSO_4$ were found in the producing well pump.

This method of plotting water analysis data has also been useful for other purposes. In one case, for example, the ion distribution in produced water from a particular producing well did not conform to any combination of injected and original connate water. The data points for the produced water plot on a graph similar to Figure 2.1-5, were widely scattered, and it was concluded that an unknown foreign water was probably present. Further investigation revealed that a hole had been corroded through the casing of that producing well, and water from an upper zone was leaking into the well. The casing leak was located and repaired.

2 Mineral Scales and Water Compatibility

2.01 Introduction

The name "scale" is broadly applied to any hard deposit formed on equipment in the presence of water. By this definition scale deposits in oil producing operations may be insoluble corrosion products (such as iron sulfide, iron carbonate or iron oxides) or *mineral scales* that are precipitated directly from water. This discussion will deal primarily with the latter.

It should be kept in mind that corrosion products, bacterial matter, heavy oil and formation fines often codeposit along with the precipitating mineral scales. The mixed composition of a scale shown by a composite chemical analysis can be misleading, especially if the scale deposit is layered. When taking samples of deposits for laboratory analysis to determine their source, multiple portions of the total deposit thickness should be selected or if feasible a representative section of the scaled equipment should be submitted.

2.02 Mineral Scales and Problems of Predicting Their Occurrence

The mineral scales most frequently encountered in oil and gas production operations are listed in Table 2.2-1. Typical ranges of their water solubilities are also listed for comparison.

Calcium carbonate scale ($CaCO_3$) is probably the most common type, but $CaSO_4$ (gypsum) is frequently encountered. The occurrence of $BaSO_4$ is less widespread than the calcium scales but it is a serious problem in many locations such as the Gulf Coast region of the U.S.A. Both $BaSO_4$ and $SrSO_4$ appeared with increased frequency after seawater flooding was initiated in relatively new fields in the North Sea and Gulf of Suez. Scaling by mixtures of $CaSO_4$ and $SrSO_4$ has been recognized as a potential problem in portions of the huge Ghawar Field in Saudi Arabia where seawater is being injected.[1] Limited solubility is the underlying reason why these substances form scale deposits that reduce well productivity and clog perforations, pumps, valves and flowlines.

The solubility or amount of scaling material that can be held in water solution is influenced by temperature, pressure, and concentration of other ions and dissolved gases. Precise calculation of scale solubility in any specific water under various conditions is a formidable task and complex computer programs have been developed for the purpose. Simpler methods can be used for practical estimation and will be discussed later.

When both solubility and concentration are known for a specific water composition, a value called the scaling tendency or scale index (SI) can be computed:

$$SI = \text{Concentration} - \text{Equilibrium Solubility}.$$

Positive SI numbers indicate scale can precipitate because the concentration of scale material exceeds its solubility. Negative SI numbers result when solubility is higher than concentration. The calculated SI is not quantitative for $CaCO_3$, but most predictive calculations for $CaSO_4$ indicate the amount of total scale that *may* form. None of the commonly used SI calculation methods provide information on the *rate* of precipitation (scaling kinetics) nor whether an adherent deposit will result.

To complicate matters even more, the scale prediction calculations must utilize data from water analyses that often are not truly representative. For example, a water sample collected at a wellhead may have already experienced partial loss of barium, strontium, calcium, sulfate and carbonate ions by scale precipitation downhole. Further loss may occur in the sample bottle prior to analysis.

2.03 Solubility, Supersaturation and Deposition

In their natural state most waters such as seawater, ground waters or formation (connate) waters are at equi-

Table 2.2-1
Common Oilfield Scales

Composition	*Typical Water Solubility (Mg/L)*[1]
Calcium Sulfate—$CaSO_4$[2]	1500–6000
Strontium Sulfate—$SrSO_4$	300–600
Calcium Carbonate—$CaCO_3$[3]	10–80
Barium Sulfate—$BaSO_4$	3–60

[1]At atmospheric temperature and pressure
[2]Precipitates as:
$CaSO_4 \cdot 2H_2O$ (dihydrate or gypsum)
$CaSO_4 \cdot \frac{1}{2}H_2O$ (hemihydrate)
$CaSO_4$ (anhydrite)
[3]$CaCO_3$ solubility increases in proportion to concentration of CO_2 dissolved in the water

librium with respect to their dissolved mineral constituents. Equilibrium solubility of minerals simply means that the rate of solution of the specific mineral with which the water is in contact, is exactly balanced with the rate of reprecipitation of that mineral. For example, a formation water in contact for a long period of time with $CaSO_4$ crystals (anhydrite) and $CaCO_3$ (limestone) in the rock will contain some constant concentration of Ca^{++}, SO_4^{++} and HCO_3^- ions. In addition, there will be certain levels of concentration of Na^+, Mg^+ and Cl^- in solution and generally lower amounts of K^+, Fe^{++}, Ba^{++}, Sr^{++}, SiO_2 and trace metals.

Dissolved gases such as CO_2, H_2S and light hydrocarbons are usually present in ground and formation waters. Dissolved air (principally O_2 and N_2) is the major gas constituent of surface waters.

The concentration of each dissolved constituent is controlled by (1) availability for solution, (2) inherent solubility, (3) presence of other ions and dissolved gases and (4) physical conditions (primarily temperature and pressure but water motion can be important). In other words, solubility is not a fixed value and can vary considerably even for low solubility substances such as those listed in Table 2.2-1.

When a water is saturated with a scale forming material under equilibrium conditions, a change in conditions can either reduce or increase solubility and trigger precipitation or cause more of the material to dissolve. Precipitation occurs when concentration exceeds solubility under the new conditions.

Precipitation to relieve excess concentration is normally not instantaneous and the water solution can persist in an oversaturated or *supersaturated* state for a considerable period of time. This is particularly true for the sulfate scales (calcium, barium and strontium sulfates). Supersaturation represents the quantity of scale forming material that is in excess of solubility and is therefore available for precipitation.

Even after precipitation starts, only part of the new scale crystals adhere to the rock or metal surfaces and build solid deposits.

The mechanics of scale deposition are quite different from that of precipitation. As logic would predict, scale deposits tend to build on rough surfaces particularly where some pre-existing scale is present.

Fluid velocity is also an important variable in the creation of deposits, but the effect is not simple. During the process of precipitation and deposition, the new scale crystals must lose both their layer of hydrating water and their surface electrical charge. Fluid motion promotes both of these effects by sweeping the starting crystals to surfaces where dehydration, charge dissipation and compaction can more readily occur. High velocity also tends to carry away the least adherent particles leaving only the most dense and hard to remove residue.

Velocity is involved in yet another deposit promoting action, viz., the orifice effect. All mineral scales and barium sulfate in particular are prone to deposit downstream of constrictions such as casing perforations, chokes and pump passageways. In many cases, the only significant deposits in wells are at these locations. This can be attributed in part to reduction in solubility as a result of pressure drop. All the sulfate scales including barium sulfate are somewhat more soluble at high pressure, and a reduction in pressure is one of the physical conditions that can create supersaturation. An effect of possibly greater practical importance, however, is the rapid relief of supersaturation caused by the sudden pressure drop at an orifice. This effect can provide ample scaling material to create a deposit in a limited zone at the downstream side of the constriction.

Definite prediction of the amount and location of scale deposits is practically impossible. However, knowledge of the opportunity for scale to form is essential for advance planning in water handling operations such as waterfloods and cooling systems.

2.04 Importance of Scale Prediction Calculations

The inability to predict exactly where and to what extent scale deposits will form might lead one to question whether scaling tendency calculations are worthwhile. One useful application was described in Section 2, Item 1.09a. In that example, calculation of the $CaCO_3$ scaling tendency for the proposed plant cooling water (Sample 1, Table 2.1-6) predicted a potentially serious scale problem and was instrumental in the decision to avoid use of the water.

Scale prediction calculations are especially important in water injection operations. When two or more waters are to be mixed prior to injection, the operator must know in advance if scale solids that can plug injection wells are likely to form as a result of mixing. It often happens that mixtures of incompatible waters are scale forming even when the individual waters are not. This situation can also develop when a single water that is incompatible with the in-place formation water is used for flooding. For example, the seawater (Sample 2) and formation water (Sample 3) listed in Table 2.1-6 and discussed in Section 2, Item 1.09b, individually have little or no tendency to precipitate $CaSO_4$ scale. However, mixtures of the two waters are oversaturated with $CaSO_4$, and gypsum ($CaSO_4 \cdot 2H_2O$) scale has formed in and around the producing wells after seawater breakthrough. The problem and consequences of water incompatibility will be discussed in more detail later in this chapter.

Methods for performing the scale prediction calculations based on water composition will now be described. Emphasis will be given to practical guidelines on conditions that promote scaling and to calculation procedures that field personnel can use without detailed knowledge of the physical chemistry involved.

2.05 Prediction of Calcium Carbonate Scaling

The basic chemistry of $CaCO_3$ precipitation was outlined in Chapter 1, Item 1.03 and prediction was introduced in Item 1.09.

The Langelier equation[2] is most often quoted as the foundation of calculations for predicting the occurence of $CaCO_3$ scale. The value of pH_s (pH a water would have if it were exactly saturated with $CaCO_3$) is given by:

$$pH_s = (pK_2' - pK_s') - pCa^{++} - pAlk$$

where pK_2' and pK_s' are empirical constants, pCa^{++} is the negative logarithm of the calcium ion concentration in mols per liter, and pAlk is the negative logarithm of total alkalinity expressed in equivalents per liter.

Because high total dissolved solids (TDS > 5000 mg/L) influences the validity of the Langelier Saturation Index, Stiff and Davis[3] extended and simplified the empirical calculations for application to oilfield brines. By their method:

$$pH_s = pCa + pAlk + K$$

where K is a constant, the value of which depends on TDS and temperature.

To account for differences in the influence of various dissolved ions on the value of K, all major ions shown by the water analysis are first converted to a common basis called *ionic strength*. Table 2.2-2 provides factors for converting individual ion concentrations (from the water analysis) to ionic strengths. The values of individual ionic strengths are then added to obtain the total ionic strength for that water. Total ionic strength (sum of ionic strengths of individual ions) is then applied to the chart of Figure 2.2-1 to obtain a value of K at various temperatures. Values for pCa and pAlk are obtained from Figure 2.2-2 using mg/L of Ca^{++} for pCa and the *sum* of the mg/L $HCO_3^=$ and $CO_3^=$ for pAlk.

Table 2.2-2
Factors for Converting Ion Concentrations to Individual Ionic Strengths

Ion	*Mg/L*[1]	*Meq/L*[1]
Na^+	2.2×10^{-5}	5×10^{-4}
Ca^{++}	5.0×10^{-5}	1×10^{-3}
Mg^{++}	8.2×10^{-5}	1×10^{-3}
Cl^-	1.4×10^{-5}	5×10^{-4}
$SO_4^=$	2.1×10^{-5}	1×10^{-3}
HCO_3^-	0.8×10^{-5}	5×10^{-4}

[1]Concentration values taken from a water analysis are multiplied by the listed factors to obtain ionic strength.

Values obtained as described are substituted in the Langelier equation to obtain a scale index, viz.,

$$SI = pH_{(actual)} - pH_s \text{ or}$$

$$pH - (pCa + pAlk + K)$$

The SI number derived may be negative (when $pH_s >$ pH), zero or some positive number. A negative SI value indicates the water, as represented by the water analysis, will not precipitate $CaCO_3$ at the temperature of the calculation. A positive SI number indicates the water is oversaturated with respect to $CaCO_3$ and scale precipitation is possible. An example calculation is given later.

Based on correlations with actual field scale problems, serious scale deposition usually occurs when the calculated SI value is +1.0 or greater.

Reliability of the SI calculation for $CaCO_3$ scale prediction is improved when both pH and alkalinity are determined immediately after the water sample is collected in the field. In rare cases, the water sample is collected in a pressure container (to avoid loss of dissolved gases) and transported to a laboratory for analysis. As a practical matter the scale prediction calculations must usually rely on analysis of aged samples.

The pH value shown by the water analysis report can be as much as 1.0 or more pH units higher than that of the water as it existed in equilibrium with produced gas in the field system. The pH change results from loss of dissolved gas and conversion of HCO_3^- ions to $CO_3^=$ ions, which increases alkalinity (see Chapter 1, Item 1.03). Both pressure reduction and heating promotes CO_2 loss and alkalinity increase.

Experience has shown that $CaCO_3$ scale is most likely to occur in oilfield water systems (at reduced pressure) when the produced gas (originally present in the water) contains more than 2 mol % CO_2. When both gas and water composition are known. it is possible to back calculate original water pH under field conditions. Computer programs are available for the purpose.

When a computer program is not available or use of a more sophisticated calculation method is not feasible, the following simple method developed by the author can be employed to estimate approximate system pH:

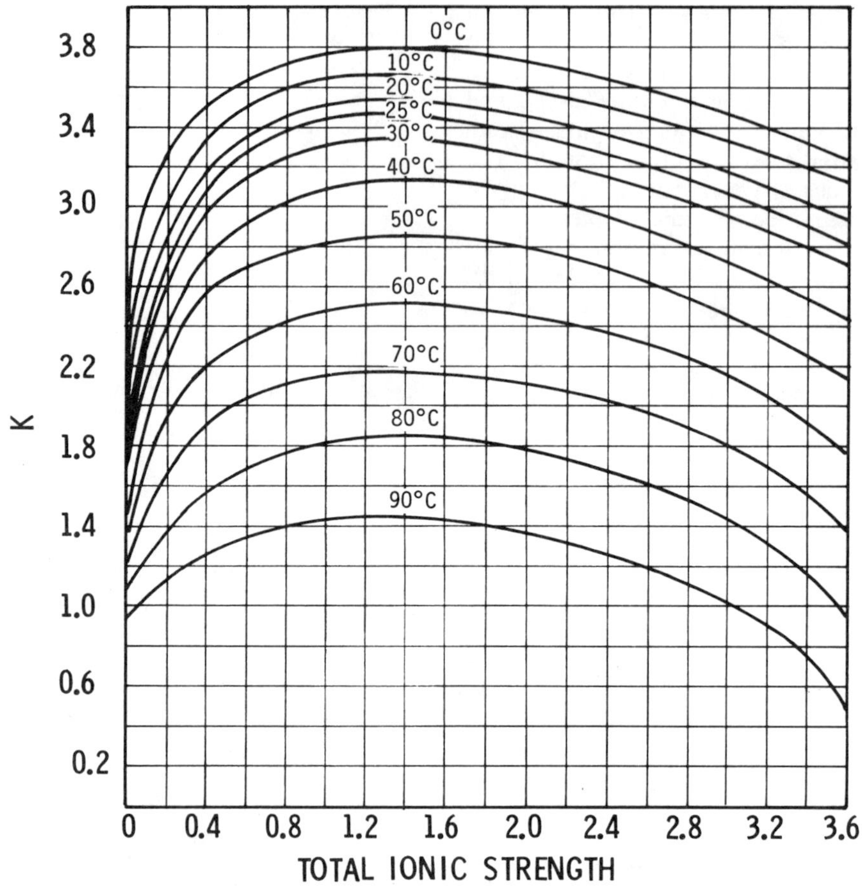

Figure 2.2-1. Values of K at various ionic strengths. (From Stiff and Davis).

Step 1. Calculate ratio R:

$$R = \frac{\text{mg/L } HCO_3^- \times 0.82}{M_{CO_2} \times S_f}$$

where:

M_{CO_2} = mol fraction CO_2 in the gas. (from gas analysis)

S_f = CO_2 solubility factor at various temperatures and total pressure, from Figure 2.2-3.

Step 2. Read equilibrium water pH from Figure 2.2-4 for the value of R calculated by Step 1.

Use of the calculated pH instead of the water analysis pH in the SI calculation often improves reliability of $CaCO_3$ scale prediction.

Example calculations of SI for $CaCO_3$ and system equilibrium pH are given in Appendix 3A, Item 3A.1. A method for calculating pH_s direct from a water analysis using a hand calculator is also shown in Appendix 3A.

2.06 Sulfate Scales—General

Precipitation of calcium, barium and strontium sulfate scales is usually caused by one or more of the following conditions:

1. Mixing of incompatible waters (high sulfate waters such as seawater mixed with brines containing high concentrations of calcium, barium or strontium).
2. Reduction in pressure.
3. Evaporation and concentration of brines by gas evolution in wellbores or by heating in surface equipment.
4. Temperature change.

As an example of Condition 1, concentrations of the sulfate scales in excess of solubility result when a water with > 10 meq/L sulfate content mixes with a different

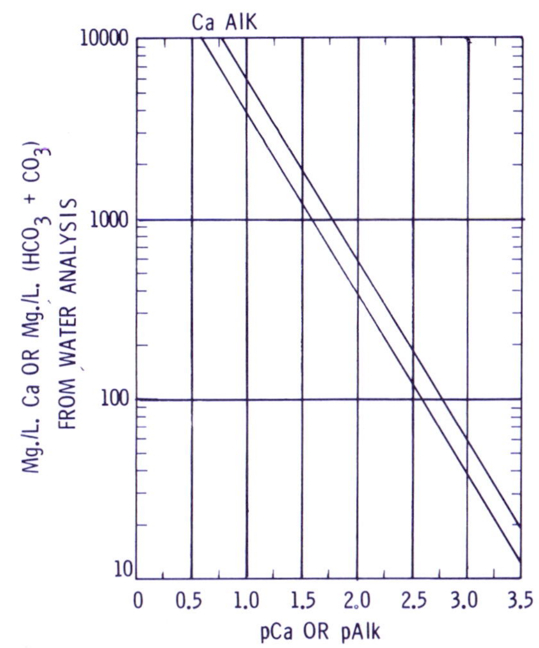

Figure 2.2-2. Chart for converting concentrations of Ca^{++} and (HCO_3^- + $CO_3^=$) to pCa and pAlk. (From Langelier).

water containing high (> 900 meq/L) concentrations of calcium or strontium, or essentially any amount of barium. The excess of concentration over solubility (supersaturation) precipitates as scale to re-establish solution equilibrium.

The water solubility of $CaSO_4$ (anhydrite) at 100°C (212°F) decreases from about 2100 mg/L at 1000 bars pressure (14,500 psi) to about 850 mg/L at 100 bars (1450 psi).[4] Solubilities of $BaSO_4$ and $SrSO_4$ also decrease with reduction in pressure. The pressure effect can shift solubility of the sulfate salts sufficiently to cause a saturated water solution to become supersaturated. This situation can occur when a formation water reaches a wellbore, for example. However, it is unlikely that pressure drop alone can account for the heavy scaling that often occurs in wells producing from low pressure (> 35 MPa or 5000 psig) reservoirs. As discussed in Item 2.03, the pressure drop at orifices and similar constrictions can promote rapid relief of supersaturation and consequent scale deposition. In most cases of severe scaling, supersaturation is the result of previous mixing of incompatible waters and pressure reduction probably contributes most by accelerating precipitation.

The solubility of calcium sulfate *decreases* with increasing temperatures over certain ranges. Calcium sulfate (gypsum) solubility in pure water at 40°C (104°F) is approximately $1^1/_2$ times greater than the solubility at 100°C (212°F). Large amounts of $CaSO_4$ scale often form

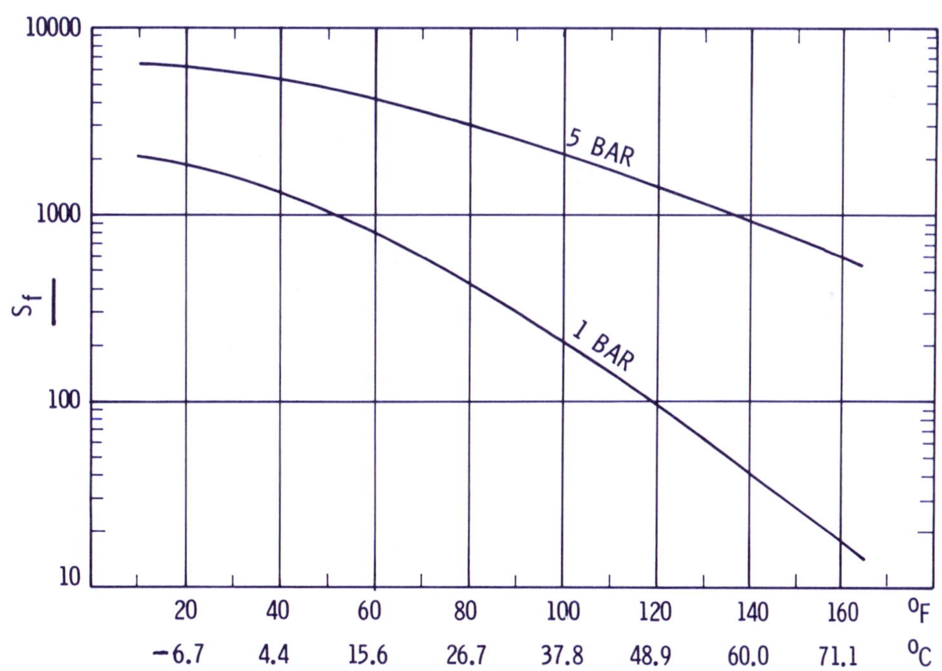

Figure 2.2-3. Values of S_f at various temperatures and system pressures (BAR = 1 ATMOSPHERE or 14.5 PSIG).

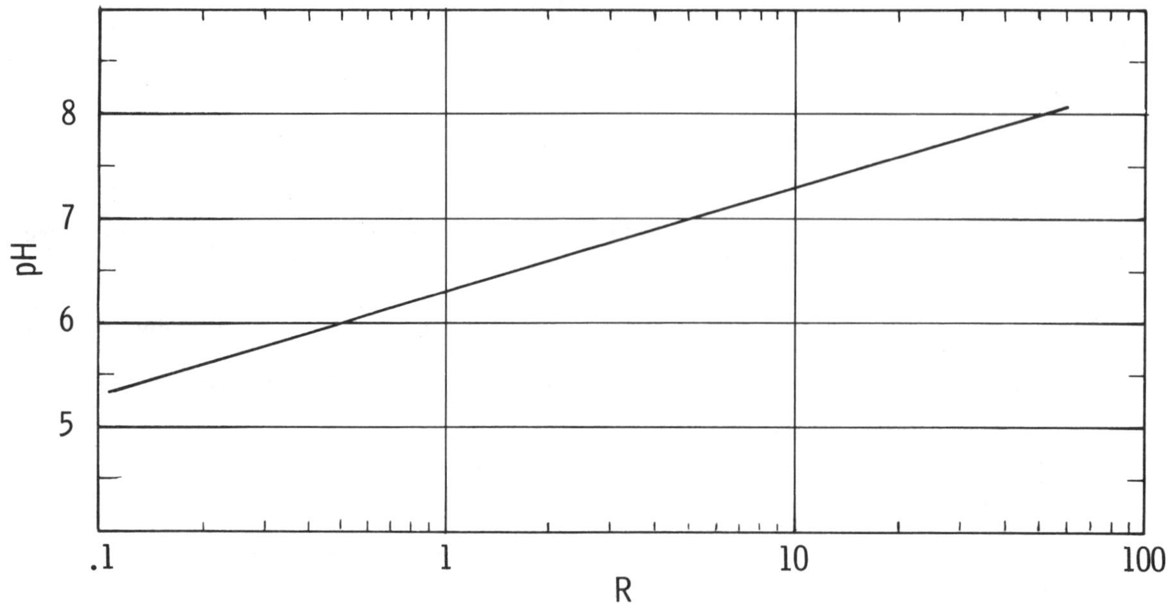

Figure 2.2-4. Approximate pH vs R Values.

in heated emulsion treating vessels especially when production from different formations are mixed. In most severe cases, the high rates of scaling are the combined result of mixing incompatible waters and the temperature effect on $CaSO_4$ solubility.

In contrast to calcium sulfate, the solubilities of barium and strontium sulfates are highest at high temperatures and decline on cooling. The consequence is that formation waters saturated at reservoir temperature with these salts can form scale as the produced water cools in the well or surface equipment. However, supersaturation created by mixing of different waters is much more likely to cause heavy scaling than is temperature reduction alone.

2.07 Calculation of Sulfate Scale Solubilities

Methods for computing calcium sulfate solubility at various temperatures have been developed by Skillman, McDonald and Stiff,[5] Metler and Ostroff,[6] Carlberg and Matthews[7] and Vetter and Phillips.[8] The Skillman, McDonald and Stiff publication provides information for a computor program.*

Numerous publications provide some data on characteristics and solubilities of barium and strontium sulfates. Some of the most notable are the papers by Templeton,[9] Davis and Collins,[10] Jacques and Bourland,[11] and Weintritt and Cowan.[12] Ostroff[13] provides an equation for calculating barium sulfate solubility in waters that contain predominately sodium and chloride ions and little magnesium or calcium ions.

A comprehensive computor program for calculating water solubilities of all the alkaline earth sulfates and many other limited solubility materials at various temperatures in the presence of other ions is commercially available. The program is called ECES.**

2.08 Practical Estimation of Calcium Sulfate Solubility Directly from Water Analysis

The calculation procedure described here was developed by the author and is designed primarily for use by field personnel who may need to compute calcium sulfate scaling tendency only occasionally or for a limited number of waters or water mixtures at a time.

The empirical method has given satisfactory results for a variety of waters ranging from undersaturated to saturated and oversaturated with calcium sulfate.

*Copies available from the American Petroleum Institute, 211 North Ervay, Dallas, Texas 75201-3688.

**Available from OLI Systems Inc., 52 South Street, Morristown, New Jersey 07960.

Step 1. Examine the water analysis. If dissolved ion concentrations are listed only as mg/L (or ppm), convert to meq/L by the method and examples discussed in Section 2, Chapter 1, Item 1.08.

The concentration (meq/L) of $CaSO_4$ in the water is numerically equal to the *lowest* meq/L value of either Ca^{++} or $SO_4^{=}$ shown by the analysis. Record

$[CaSO_4]$ = meq/L of $CaSO_4$ present.

Step 2. Compute excess common ion, ECI, by subtracting the lowest meq/L of either Ca^{++} or $SO_4^{=}$ from the other. Record

ECI = common ion in excess of $[CaSO_4]$.

Step 3. Add meq/L of Na^+ and Mg^{++} and record

NM = sum of Na^+ and Mg^{++} as meq/L.

Step 4. Compute

$$R = \frac{NM}{ECI}$$

Step 5. Compute

$$K_s = [CaSO_4] - R$$

Interpretations:

K_s = or < 6, water is undersaturated with $CaSO_4$ at 60°C (140°F). Use solubility calculation Step 6.

K_s = or > 9, water is oversaturated with $CaSO_4$ at 60°C. Use solubility calculation Step 7.

Note: K_s values in the range of 6 to 9 are near saturation. When $R < K_s$, use calculation Step 7. When $R > K_s$, $[CaSO_4]$ is approximately equal to solubility at 60°C and S.I. for $CaSO_4$ is near zero.

Step 6. For undersaturated waters ($K_s \leqq 6$), compute

$$S_{60} = [CaSO_4]\left(\frac{936}{NM} + \frac{NM}{2808}\right)$$

where S_{60} is approximate solubility of $CaSO_4$ as meq/L in the water at 60°C and atmospheric pressure.

To obtain a value for solubility at other temperatures, multiply S_{60} times the appropriate factor from Table 2.2-3.

Table 2.2-3

Temperature		
°C	°F	$CaSO_4$ Solubility Factor at 1 Bar*
30	86	1.30
40	104	1.15
50	122	1.05
60	140	1.00
70	158	0.95
80	176	0.90
90	194	0.82
100	212	0.75
110	230	0.67
120	248	0.55
130	266	0.47

*$CaSO_4$ solubility at 60°C increases approximately 2.3% for each 7 MPa (1000 psi) increase in total pressure above atmospheric.

Step 7. For saturated or oversaturated waters ($K_s \geqq 6$), compute

$$M = \frac{R}{[CaSO_4]}$$

When $M > 0.12$, compute $S_{60} = 1.20\ K_s$

When $M \leq 0.12$, compute $S_{60} = 0.68\ K_s$

To obtain a value for solubility at other temperatures, multiply S_{60} times the appropriate factor from Table 2.2-3.

Step 8. Compute

$$SI = [CaSO_4] - \text{Solubility}$$

When SI is a negative number (Solubility > $[CaSO_4]$), the numerical value of SI is the approximate meq/L of $CaSO_4$ the water can dissolve at the temperature of the calculations.

When SI is a positive number ($[CaSO_4]$ > Solubility), the numerical value of SI is the approximate meq/L of $CaSO_4$ scale the water can precipitate at the temperature of the calculation.

meq/L $CaSO_4 \times 68$ = mg/L or g/m^3
meq/L $CaSO_4 \times 23.8$ = lb/1000 bbl

To convert $CaSO_4$ to $CaSO_4 . 2H_2O$ (gypsum) multiply concentration (any units) times 1.68.

Notes: ECI reduces $CaSO_4$ solubility
NM increases $CaSO_4$ solubility
Waters undersaturated at 60°C may be oversaturated at higher temperatures.

Example calculations using this procedure and water analyses listed in Table 2.1-6, are given in Appendix 3A, Item 3A.2.

2.09 Barium and Strontium Sulfate Scales

Most barium sulfate scale deposits also contain some strontium sulfate.[14] Until recently, the appearance of strontium in oilfield scales has been primarily in the presence of barium sulfate.[11,12] Scale deposits that are principally strontium sulfate are now known to occur in the North Sea, Gulf of Suez, Saudi Arabia and several other locations around the world.

As indicated in Table 2.2-1, strontium sulfate is much more soluble than barium sulfate but is less soluble than calcium sulfate. Sodium chloride at concentrations up to about 8 wt % (80,000 mg/L) increases solubility of all the sulfate scales. Presence of calcium and magnesium in brines that contain chloride as the major anion, apparently does not decrease solubility of barium and strontium sulfate.[10]

Excess common ion (ECI) decreases barium and strontium sulfate solubility. As similarly defined under Item 2.08 for $CaSO_4$, excess common ion is the excess of either Ba^{++} or Sr^{++} ions over the $SO_4^{=}$ or the excess of $SO_4^{=}$ over the Ba^{++} or Sr^{++}, when the concentration of each is expressed as meq/L.

Figure 2.2-5 illustrates the level of strontium and barium sulfate solubilities in oilfield brines. The information is useful for detecting potential scaling problems especially in mixed waters. The solubility curves were derived by averaging a series of computer generated solubility values in actual and hypothetical brines with salinities (total dissolved solids, TDS) from 12,000 to greater than 200,000 mg/L. In all cases, the sum of the Ca^{++} and Mg^{++} ion concentrations was less than 20% of the total cations and Cl^- was the major anion. The results agree reasonably well with published data.[10,11,14]

The dashed curves in Figure 2.2-5 for $SrSO_4$ illustrate the influence of excess common ion, ECI. The designations of ECI = 2x or 20x means the meq/L of excess common ion (either Sr^{++} or $SO_4^{=}$) is 2 times or 20 times greater than the concentration of $SrSO_4$ expressed as meq/L.

For comparison with solubility data such as that of Figure 2.2-5, the concentration of $SrSO_4$ is taken di-

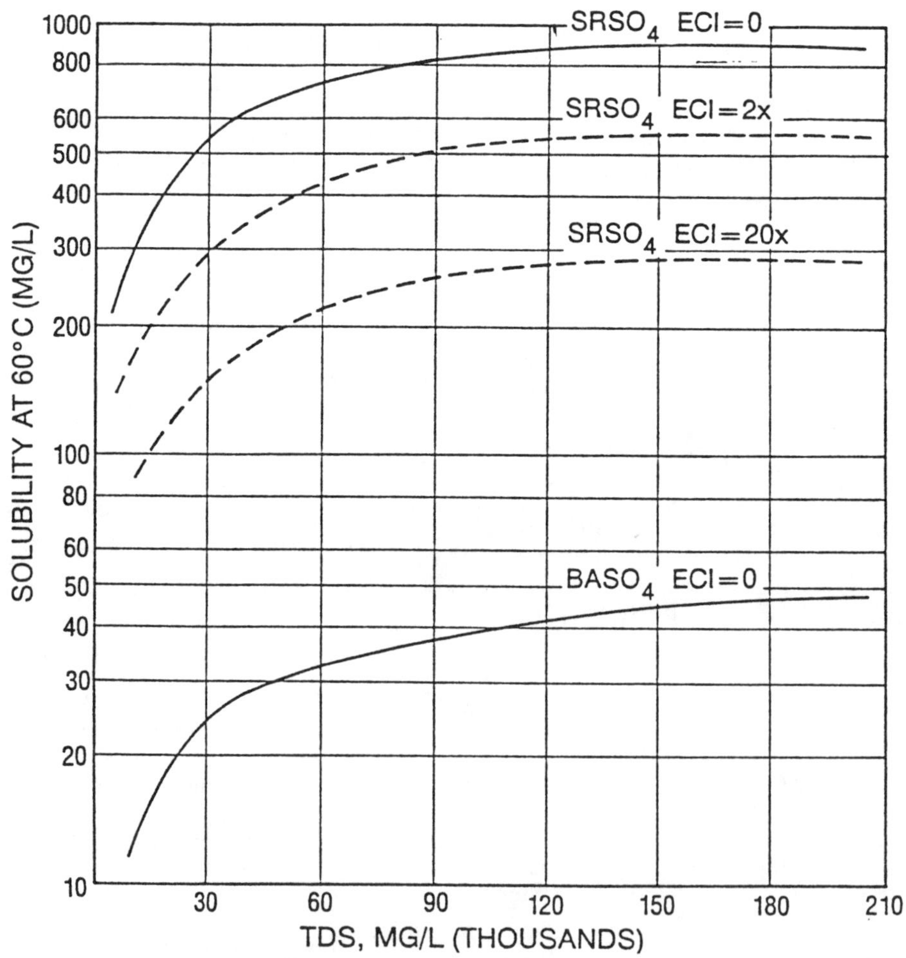

Figure 2.2-5. Approximate solubilities of strontium sulfate and barium sulfate in oilfield brines at 60°C.

rectly from the water analysis as numerically equal to the lowest value (as meq/L) of either Sr^{++} or $SO_4^{=}$.

Large excesses of common ion (usually $SO_4^{=}$) over $SrSO_4$ concentration, decrease solubility of $SrSO_4$ well below that indicated in Figure 2.2-5. Concentrations of excess $SO_4^{=}$ in the order of 95 meq/L reduces solubility of $SrSO_4$ to about 0.2 meq/L in brines containing 50,000 mg/L TDS. $SrSO_4$ concentration expressed as meq/L is converted to mg/L by multiplying times the equivalent weight, 91.6.

The concentration of $BaSO_4$ in a water or water mixture is determined from the water analysis in the same way as that described for $SrSO_4$. Because the solubility of $BaSO_4$ is so low, indications of possible $BaSO_4$ precipitation are most often found by computing the hypothetical analysis of a mixture of two or more waters in lieu of an actual analysis.

The final hypothetical composition of a mixture of waters is easily determined from the water analysis of each component water. Multiply the concentration of each ion listed in the individual analysis times the volume percent, expressed as a decimal fraction, of that water in the final blend. The sum of the fractional concentrations of each ion is the total concentration of that ion in the mixture. An example calculation for determining composition of a mixture of two waters is given in Appendix 3A, Item 3A.2(b).

The lower curve in Figure 2.2-5 indicates approximate solubility of $BaSO_4$ in oilfield brines where ECI is approximately zero. Again, excess common ion (usually $SO_4^{=}$) reduces $BaSO_4$ solubility. When excess $SO_4^{=}$ (over the concentration of $BaSO_4$, as meq/L) is 10 meq/L or higher, the solubility of $BaSO_4$ is less than 0.05 meq/L (6 mg/L) in brines with TDS of less than 100,000 mg/L. The meq/L of $BaSO_4$ multiplied times equivalent weight (116.7) is the concentration expressed as mg/L.

When calculation of a hypothetical analysis of a water mixture indicates concentration of $BaSO_4$ will be greater than solubility, precipitation is apt to occur until concentration approaches solubility.

Precipitation of 1 meq/L of Ba^{++} must also precipitate 1 meq/L of $SO_4^{=}$. Loss of $SO_4^{=}$ by precipitation as $BaSO_4$ can be estimated to determine the amount of $SO_4^{=}$ remaining for precipitation of the more soluble $SrSO_4$.

2.10 Water Compatibility for Mixing

The term *compatibility* as used here relates to the chemical reactions resulting when two or more waters are mixed. If the waters are compatible, no new solids are formed on mixing. Incompatible waters interact to form new solids and the total suspended solids (TSS) of the mixture is greater than the sum of the fractional TSS of the individual waters prior to mixing.

Any of a variety of reactions can create new solids. For example, insoluble iron sulfide or iron oxide particles may form when a water containing dissolved iron is mixed with other waters containing dissolved H_2S or O_2.

Any amount of new solids adds to the suspended solids load already present and reduces the quality of the mixed water for subsurface injection. In general, the most serious problems of incompatibility (largest amounts of new solids) result when a water with high sulfate ($SO_4^{=}$) content is mixed with one or more waters with high calcium content. The example calculations given in Appendix 3A, Item 3A.2 (b), shows that a mixture of 20 vol % of seawater and 80 vol % of the example connate water is capable of precipitating 6 meq/L (408 mg/L) of $CaSO_4$ (or 685 mg/L of gypsum) at 60° (140°F). This fact alone shows that the two waters are incompatible and should not be mixed prior to injection. This is true even if no adherent scale deposits form and the new scale solids simply stay in suspension in the mixed water stream.

2.11 Determination of Compatibility

Mineral scaling tendency calculations from the water analyses are helpful for predetermination of compatibility prior to water mixing. The water compositions should also be examined for presence of iron, hydrogen sulfide and oxygen that may form iron precipitates. Minor amounts of other precipitatable constituents such as manganese may also be present.

In addition to scaling calculations, compatibility studies often include laboratory mixing tests to physically measure the quantities of scale solids formed. Samples of the waters to be mixed are individually prefiltered to remove existing solids. Measured amounts of each filtered water are blended in a suitable (usually glass) container to provide the expected range of proportions, i.e., 25/75, 50/50 and 75/25 parts by volume. The mixtures are then held (closed) for 24 to 48 hours at expected temperature. Heavy precipitates are visible but it is good practice to refilter the mixed waters and weigh the amounts of new solids. The mixing tests can be carried out under air free nitrogen to prevent oxidation and precipitation of iron. This is not necessary if iron is absent and mineral scales are the primary precipitates.

2.12 Incompatibility and Production Well Scaling

Incompatible waters generally should not be allowed to mix prior to injection because of the possibility of injection well plugging. However, injection waters that are incompatible with the in-place connate water are frequently used for flooding.

Interwell plugging due to water incompatibility has not proven to be a serious problem. The mixing zone within the formation is limited and any solids that may form are distributed through a large volume of rock porosity because of delayed precipitation and water movement.

Bulk mixing during miscible displacement of connate

water by injection water does not occur except in fractures or rock voids. However, as the flood front approaches a producing well, the opportunity for bulk mixing increases exponentially. In many cases, injection water fingers to the producing well through fractures or high permeability zones and mixes with connate water close to or in the wellbore. Scale deposited in close proximity to the wellbore is detrimental to well productivity.

When water incompatibility is known to exist in a waterflood operation, the usual practice is to periodically squeeze treat the producing wells with scale inhibitor (see Item 2.14 below). Treatment of the injection water for scale control at the producing wells is futile due to adsorption loss of the inhibitor chemicals within the formation.

Even when scaling tendency calculations and water mixing tests predict that scale desposition is likely after injection water breakthrough, scale inhibition treatment of all producing wells is often not necessary. In some fields where floodwater incompatibility exists, only 50 to 60% of the producers have suffered enough scale accumulation to justify scale removal and inhibition treatments.

Some operators install perforated metal test coupons at the producing wellheads to detect early signs of scale accumulation. This type of testing is not always reliable, especially during early stages of scaling.

A diligent program of monitoring the production from each well is helpful in deciding which producers are good candidates for scale treatment. Breakthrough and rate of volume increase of the incompatible injection water can usually be detected by the chloride ion concentration test (see Chapter 1, Item 1.09b). The producers with the slowest rate of increase in the volume percent of injection water are most apt to suffer scale buildup. Another guideline indicator of detrimental scaling is that the average oil production over a period of several days or a few weeks declines at a faster rate than the increase in total water production. When total water production and volume percent of injection water increases at a comparatively rapid rate, the expense of scale control treatments may not be justified.

2.13 Scale Removal

Calcium carbonate scale and most corrosion products can be dissolved and removed with 15% HCl acid. Oily deposits are often pretreated with a paraffin solvent before applying acid. Wetting agents and alcohols in the acid increase solution rate of oily scale deposits. Iron chelating agents such as HEDTA or citric acid are added when iron corrosion products are present. These additives help prevent reprecipitation of dissolved iron from the spent acid. Acid volumes required for scale removal in and around wellbores are generally 10 to 30% of that required for well stimulation jobs.

Calcium sulfate or gypsum deposits are not removed effectively by acid treatment. These are removed by a two stage treatment consisting of presoaking the deposit with a converter solution and then acidizing with HCl acid. Converter solutions contain caustics (NaOH or KOH), soda ash or ammonium bicarbonate and convert $CaSO_4$ to $CaCO_3$ and $Ca(OH)_2$ which are acid soluble. Organic acids such as hydroxyacetic (glycolic acid) are also used as converters, usually in combination with caustics. Some gypsum solvents are available that do not require acid, but a follow-up acid wash is usually beneficial.

Barium and strontium sulfate scales must be removed mechanically since they are not amenable to removal by acid or other chemicals.

2.14 Scale Inhibition

The three most widely used classes of scale inhibitor chemicals are:

1. Amino-phosphonates,
2. Phosphate esters of amino-alcohols,
3. Sodium polyacrylate polymers.

The chemicals are usually sold as water solutions (20 to 30% active). Alcohols are generally added to lower the freezing point and to hold the inhibitor chemical in solution.

Inhibition of producing wells to retard scale accumulation is most often done by formation squeeze injection of a water solution of the inhibitor. Following are some general guidelines for such treatments:

1. Prior to injection of the inhibitor solution, allow the well to produce long enough to flush out any spent acid, converter solution or scale solvent that may have been used previously to remove existing scale deposits. The volume of produced water required for flushing is at least three times the volume of scale removal chemical and overflush that was previously injected. Injection of inhibitor too soon after a scale removal treatment can result in severe plugging and loss of production.
2. The scale inhibitor chemical should be prediluted with water (in which it is soluble) to a final concentration of 1 to 2 vol %. Higher concentrations generally do not lengthen squeeze life and more chemical is wasted during early returns.
3. Avoid use of phosphate ester base inhibitors for squeezing formations hotter than about 80°C (176°F) when long squeeze life is desired.
4. The *diluted* inhibitor slug volume should be sufficient to provide 0.2 m^3 (55gal) of *original* inhibitor chemical for each 30 m (100 ft) of formation thickness. Alternately, the *diluted* slug volume may be calculated as 1% of the daily water production, with a minimum of 12 m^3 (100 bbl).
5. Inject at a slow rate and overflush with a volume of

oil or water sufficient to displace the inhibitor slug into the formation, plus 10 vol % excess.

6. Precede and follow the scale inhibitor slug with a 0.5–1.0 m^3 (4 to 8 bbl) pad of oil containing a preselected demulsifier chemical.
7. When treating water sensitive formations containing clays, add 2 wt % potassium chloride (KCl) or other anti-swelling clay stabilizer that is known to be effective prior to injecting the inhibitor solution.

Surface treatments for scale inhibition are usually done by continuous, direct chemical feed. If necessary to avoid precipitation in strong brines, the inhibitor chemical may be prediluted with low salinity water and the feed rate increased proportionately. The feed rate should be as nearly continuous as possible. Final inhibitor treating rate (usually 2 to 10 mg/L) is adjusted by visually observing scale accumulation in the equipment or on test coupons. Frequent analysis to monitor the concentrations of scale forming ions in the water throughout the system provides data for a continuing material balance. However, the analytical results must be analyzed statistically to reliably detect small changes due to scale precipitation.

REFERENCES

(1) Lindlof, J. D., and Stoffer, K. G., "A Case Study of Seawater Injection Incompatibility": *Journal Petroleum Technology*, July, 1983, p 1256-1262.

(2) Langelier, W. G., "The Analytical Control of Anti-Corrosion Water Treatment": *Journal American Water Works Association*, vol 28, 1936, p 1500.

(3) Stiff, H. A., and Davis, L. E., "A Method for Predicting the Tendency of Oilfield Waters to Deposit Calcium Carbonate": Petroleum Trans., AIME, vol 195, p 213, 1952.

(4) Dickson, F. W., Blount, C. W., and Tunnell, G., "Use of Hydrothermal Solution Equipment to Determine the Solubility of Anhydrite in Water from 100 to 275C and from 1 Bar to 1000 Bars Pressure": *American Journal of Science*, vol 261, p 61, 1963.

(5) Skillman, H. L., McDonald, J. P., and Stiff, H. A., "A Simple, Accurate, Fast Method for Calculating Calcium Sulfate Solubility in Oil Field Brines": API Paper No. 906-14-1, 1969.

(6) Metler, A. V., and Ostroff, A. G., "The Proximate Calculation of the Solubility of Gypsum in Natural Brines from 28 to 70°C": *Envir. Sci. and Tech.*, vol 1, p 185, 1967.

(7) Carlberg, B. L., and Matthews, R. R., "Solubility of Calcium Sulfate in Brine": SPE Paper No. 4353, 1973.

(8) Vetter, O. J. G., and Phillips, R. C., "Prediction of Deposition of Calcium Sulfate Scale Under Downhole Conditions": Journal Petroleum Technology, vol 22, p 1299, 1970.

(9) Templeton, C. C., "Solubility of Barium Sulfate in Sodium Chloride Solutions from 25C to 95C": *J. Chem. Eng. Data*, vol 5, p 514, 1960.

(10) Davis, J. W., and Collins, A. G., "Solubility of Barium and Strontium Sulfate in Strong Electrolyte Solutions": *Envir. Sci. and Tech.*, vol 5, p 1039, 1971.

(11) Jacques, D. F., and Bourland, B. I., "A Study of Solubility of Strontium Sulfate": *SPE Journal*, April, 1983, p 292-300.

(12) Weintritt, D. J., and Cowan, J. C., "Unique Characteristics of Barium Sulfate Scale Deposition": Journal Petroleum Technology, vol 19, p 1381, 1967.

(13) Ostroff, A. G., "Introduction to Oilfield Water Technology": NACE, Houston, TX, 1979.

(14) Patton, C. C., "Oilfield Water Systems": Second Edition, 1977, Cambell Petroleum Series, Norman, OK, U.S.A.

3 Practical Microbiology

3.01 Terminology and Classifications

Microscopic life forms are present in essentially all oil field water systems. The descriptive name, *microbes*, can be used for all of them. More commonly, they are referred to as bacteria or simply as "bugs."

Scientifically, it is necessary that they be classed by family, genus and specie. A *genus* (plural: genera) is the main subdivision of a *family*. Each genus generally includes many *species*. The genus name is capitalized and precedes the species name, which is not capitalized, i.e.,:

Desulfovibrio desulfuricans
Streptococcus lactis

Each part of these scientific names have meaning to microbiologists. Vibrio, for example, indicates a curved rod shape whereas streptococcus means a chain-like cluster of spherical (coccus form) organisms. A description of a straight, rod-shaped bacteria will include the word bacillus. Spiral shapes are called spirilla.

Microbes can be either plants or animals. The plants are most prevalent and can be subdivided under names such as bacteria, algae, fungi (or fungus), molds, yeasts, etc. Algae contains green chlorophyll which allows them to utilize the energy of sunlight to convert carbon dioxide to complex organic compounds that make-up their cells. The others can grow without sunlight and obtain their energy by other means.

In oil production operations, it has become common practice to refer to all types of microbes as bacteria and to group them according to oxygen requirements as follows:

Aerobic organisms (aerobes)	Require oxygen
Anaerobic organisms (anaerobes)	Inhibited by oxygen
Facultative	Oxygen not required but most grow best when some O_2 is present. Generally counted as aerobes.

Table 2.3-1 is a practical listing of the most troublesome microorganisms.

3.02 Nutritional Requirements and Energy Sources

In common with other forms of plant life, the basic food elements for bacteria are nitrogen, phosphorous, potassium and small amounts of other minerals such as magnesium, calcium and iron. A source of carbon is also required, and this ranges from carbon dioxide to hydrocarbons such as methane and heavier.

In the absence of sunlight, bacteria generally derive the energy required to grow and reproduce from various types of electrochemical oxidation-reduction reactions (see Section 1, Item 1.07). Probably the most important of these in production operations is the reduction of $SO_4^{=}$ ion in water to H_2S. Another is the oxidation of ferrous iron (Fe^{++}) to ferric iron (Fe^{+++}).

3.03 Bacteria Related Problems and Mechanisms

The two major problems that result from growth of bacteria in oil field water systems are corrosion and fouling.

1. The major contribution of bacteria to corrosion in water injection systems is generation of hydrogen sulfide (H_2S) in water that is normally sulfide free (not naturally sour). Bacterially generated H_2S is also a major cause of corrosion on buried pipelines and on the external surfaces of well casing (Section 1, Item 8.04). As noted in Table 2.3-1, Desulfovibrio sulfate reducers are the most notorious generators of H_2S, and their contribution to corrosion of steel is two-fold. This can be most readily explained with the help of the following simple chemical equations.[1]

(a) $SO_4^{=}$ (sulfate ion) + $2H^+$ (hydrogen ion) + $4H_2$ (molecular hydrogen) $\rightarrow H_2S + 4H_2O$

Table 2.3-1
Microorganisms Found in Oil Field Waters

1. *Protozoa*

   Small animals inhabiting contaminated, open waters. Easily identified microscopically because of relatively large size. Protozoa live on smaller organisms and their presence indicates the water has a high level of microbial activity. Controlled with biocides.

2. *Algae and Diatoms*

   Plants that grow only in open, sunlit waters (Figure 2.3-1). Recognized by green (sometimes dark green) slime on water surface or basin walls. Controlled by copper compounds. Oil and H_2S inhibit growth.

3. *Iron Bacteria*

   Occur as slimey chains or filaments (Figure 2.3-2). Composed of bacteria cells surrounded by gelatinous iron oxide excretion. Common in shallow well waters from sands containing iron. Identified by microscopic observation after solution of iron oxide. Control requires chemical treatment of producing sand and detergent-acid cleansing. Frequently found genera are Gallionella, Crenothrix and Sphaerotilus.

4. *Slime Formers*

   This is a general grouping of aerobic and facultative bacteria and fungi that form slime deposits sufficiently thick to be felt and seen. Typical genera are: Pseudomonas, Flavobacterium, Micrococcus, Aerobacter, Aspergillus and Nocardia. Controlled by elimination of dissolved oxygen and treatment with biocides.

5. *Sulfide-Producing Bacteria*

   Most important of these are the sulfate reducers that metabolize sulfate ions ($SO_4^=$) in water to form H_2S. Desulfovibrio desulfuricans is the most prevalent specie and is described as motile obligate anaerobes (meaning they move freely and do not like oxygen). Detected by black colonies in deep tube, nutrient agar medium containing sulfate and iron (Fe^{++}) salt. Another sulfide producer is Clostridium nigrificans, which is thermophilic (prefers elevated temperatures).[1] Clostridium tetani (causes tetanus) has been isolated from oilfield waters.[2] Controlled by limiting aerobic bacteria growth and biocide treatment.

6. *Sulfur Oxidizers*

   Aerobic bacteria that oxidize sulfur and produce sulfuric acid. Beggiatoa and Thiobacillus are examples. The organisms proliferate in open conduits handling sour produced water and in soils near sulfur plants. The organisms can cause acid pollution problems and promote disintegration of concrete structures. Removal of sulfur nutrient is best means of control.

and

$$\text{(b) } \underset{\text{(iron ion)}}{Fe^{++}} + H_2S \rightarrow \underset{\text{(iron sulfide)}}{FeS} + 2H^+$$

According to Equation (a), both hydrogen ions and molecular hydrogen are utilized by sulfate reducers to convert $SO_4^=$ to H_2S. The source of molecular hydrogen is the cathodic reduction reaction (Section 1, Item 1.06), and consumption of hydrogen promotes continued corrosion by preventing cathode polarization (Section 1, Item 1.10). In other words, the sulfate reducers cause corrosion on steel surfaces exposed to *air-free* salt waters by both production of H_2S and by cathode depolarization.

Sulfide generation and cathode depolarization on steel

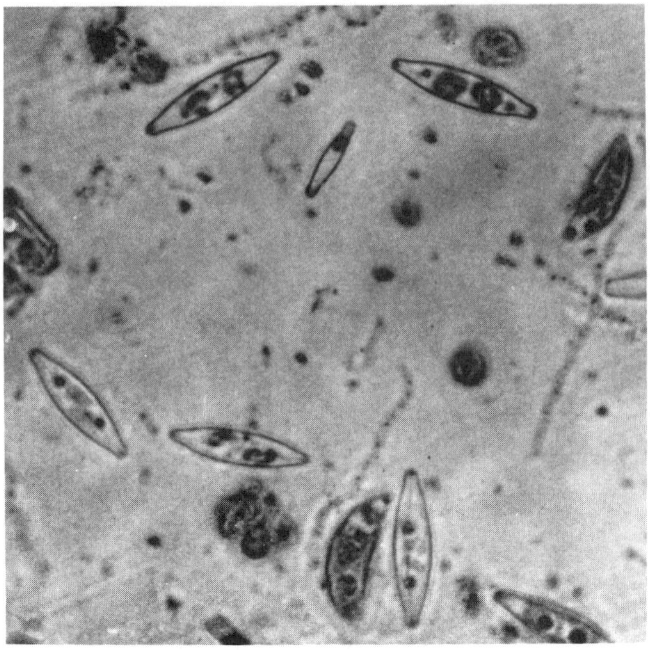

Figure 2.3-1. Diatoms and algae filaments. Approx. 50×.

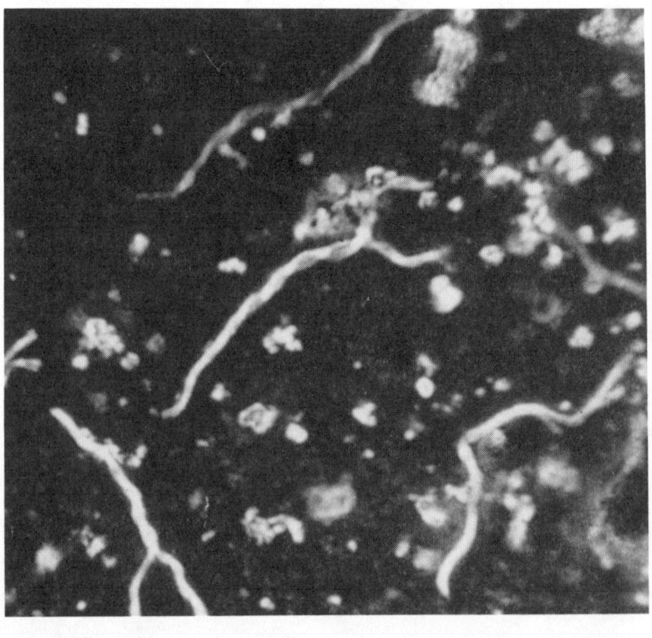

Figure 2.3-2. Iron bacteria filaments (primarily iron oxide residues). Approx. 100×.

surfaces can also be shown by an alternate equation.[3]

$$SO_4^{=} + 8H^{\circ} \text{ (atomic hydrogen)} \rightarrow S^{=} + 4H_2O$$

and

$$Fe^{++} + S^{-} \rightarrow FeS$$

The atomic hydrogen, H°, formed at metal cathodes is utilized directly for sulfate reduction and molecular hydrogen, H_2, is not involved. This appears to be the most likely mechanism.

Concentrations of 2 to 20 mg/L of bacterially generated H_2S have been detected in normally sweet injection water after passage through long, bare steel flowlines. This is sufficient to cause cracking of high-strength (high-hardness) steel equipment (Section 1, Item 7.05).

The slime-forming bacteria (Table 2.3-1) can promote metal pitting in *aerated* salt waters by differential oxygen concentration cell action (Section 1, Item 1.06).

2. Fouling by heavy growths of microorganisms reduces water through-put rates in piping and contributes to injection well plugging. Bacterially generated slimes in combination with corrosion products, clays, scales and other particulate matter, form a mixture that can readily seal injection formation faces. Vigorous growth of sulfate-reducing bacteria produces not only corrosion and pitting, but can generate enough iron sulfide particles to damage some injection wells.

   Shallow water supply wells producing low salinity water with traces of oxygen from sand formations frequently become clogged with iron bacteria (Table 2.3-1).

3.04 Occurence and Growth Habits

Being living organisms, bacteria are difficult to predict with respect to where and to what extent they will grow. Water analyses do not indicate the specific probability of microbial problems in water systems. The opportunity for infection is always present but complex combinations of environmental factors determine whether or not serious growths develop. Following are some conditions that favor growth:

1. Mixed salts in solution, up to about 7 wt %. Some organisms tolerate higher salinities but growth is limited. Presence of magnesium, sulfate and trace metals is conducive to growth. The mixing of produced brines and fresh waters prior to injection often promotes bacterial problems. Traces of organic substances may be involved in this situation.
2. Oxygen in concentrations from a few parts per billion to full saturation. Severe bacterial problems can result from intermittent oxygen entry followed by periods of absence.
3. Temperatures between 30 and 55°C. Some organisms survive outside this range. Significant growth ceases above about 80°C.
4. Low flow velocities. Heavy growth is always on surfaces, viz., colonies are sessile. Nearly stagnant conditions under sludges and deposits in tanks and vessels are favorable.
5. Mixed populations of aerobic, facultative and anaerobic organisms. Many organisms grow best in close association with dissimilar types. This effect is called symbiosis.

The role of dissolved oxygen is important and often not clearly understood. *Serious bacterial infections in oilfield water systems are most frequently associated with trace oxygen contamination*, even intermittent.

Sulfate reducers, known to be strict anaerobes, can thrive under overlying layers of aerobic slime formers in fully aerated waters. Proof of this can be readily seen in recirculating cooling water systems. After scraping away the thick aerobic slime deposits on cooling tower hardware (bolts and braces), black iron sulfide is usually found. The iron sulfide is formed from corrosion of the iron by bacterially generated H_2S. Figure 2.3-3 illustrates the growth conditions on the metal surface. Similar conditions often exist in water injection lines and tanks where there are continuous traces of O_2 or occasional slugs of higher concentration.

3.05 H_2S Generation in Producing Reservoirs

Several cases have been reported where oil and gas production from whole reservoirs has turned sour after initiation of secondary recovery waterfloods. An example is the Wilmington Field in California where seawater was used for flooding. In all reported instances where this has happened the production was initially sweet (sulfide free). There appears to be no doubt that souring resulted directly from the activity of sulfate reducing bacteria within the reservoir rock. A question that remains, however, is whether the bacteria were introduced with the floodwater or were present as dormant cells (spores) within the earth.

Several studies have demonstrated that bacteria of the Desulfovibrio genus can survive in petroleum reservoirs, and at least one patent has issued on their use for enhanced oil recovery.[4] In practice, the presence of H_2S is detrimental due to reduction in value of oil and gas, necessity for gas sweetening, and increased corrosion of equipment.

Case histories of four fields in the U.S. and Canada where the problem has occurred reveals the following similarities:

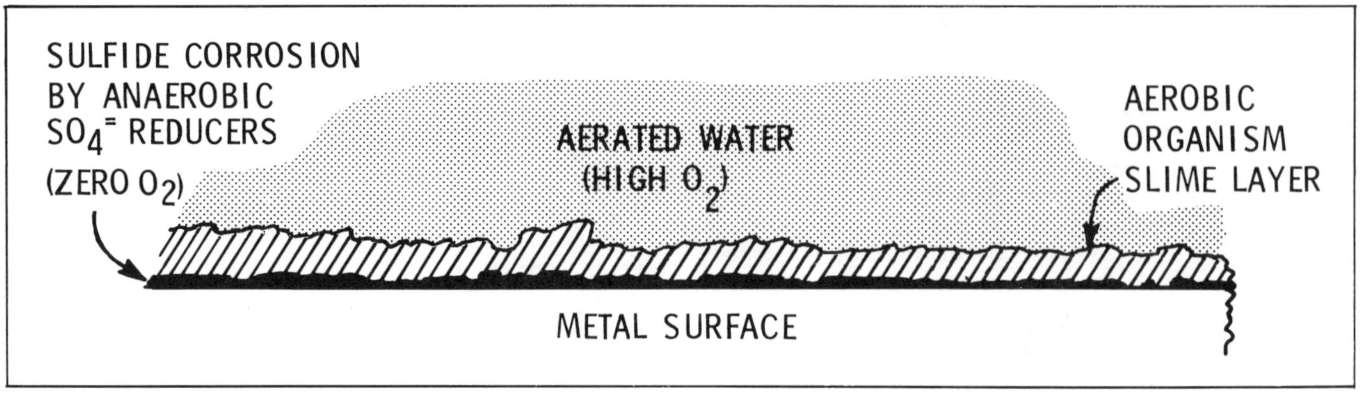

Figure 2.3-3. Anaerobic conditions created at a metal surface by shielding effect of aerobic organism slime layers.

1. Reservoir temperatures were all less than 82°C (180°F). From this observation and the general knowledge that bacterial growth diminishes at higher temperatures, it appears that 80-82°C is the maximum reservoir temperature range above which sulfide generation should not occur.
2. In all four cases, the injected floodwaters contained high concentrations of sulfate ion ($SO_4^=$) relative to the native connate water in the reservoirs.
3. Material balance calculations (sulfur in—sulfur out) indicate the typical efficiency of bacterial conversion of $SO_4^=$ to H_2S is in the range of 60 to 70%. From this, the following correlation was drawn:
   Maximum H_2S concentration (ppm) reached in the produced gas = mg/L (or ppm by weight) of $SO_4^=$ in floodwater × 1.5.
4. Starting from the time that H_2S first appeared in the produced gas, the H_2S concentration increased to the ultimate maximum at a fairly uniform rate over a period of about five years. From this, the following correlation was drawn:

   Maximum rate of increase (ppm/year) of H_2S in produced gas = ultimate maximum concentration (computed from $SO_4^=$ concentration in floodwater) × 0.2.

These observations and guidelines may be useful for engineering and design calculations but should not be construed as infallible. They are presented as information of interest and not as recommendations for systems design.

A question that is often asked is whether or not reservoir souring can be prevented by chemical treatment of the floodwater when conditions are conducive to bacterial growth. This approach does not appear economically feasible due to (1) the large volumes of floodwater to be treated, (2) the high total cost of chemical sterilizing agents and (3) probable loss of chemical in the reservoir. Use of low sulfate content floodwater should be considered but this is often not practical.

3.06 Detection and Assessment of Bacterial Activity

Direct microscopic counts can be made for protozoa, iron bacteria and certain other organisms filtered from water with a membrane filter.[5] However, direct counting of general aerobic populations and sulfate reducers is not feasible for oilfield waters.

Sophisticated laboratory procedures have been developed for growing and observing bacteria. However, many organisms that grow abundantly in water systems cannot be cultured satisfactorily in the laboratory. Further, all organisms do not grow on any one culture medium. Despite the limitations, plate and tube culture tests are the best available means for measuring populations of most microorganisms.

In general, bacteria are cultivated in a substrate consisting of agar (a plant extract with properties similar to gelatin) and the necessary nutrients. The culture medium is inoculated with measured portions of water being tested and then incubated for several days. After plate culturing, visible colonies develop as spots on the agar gel, and each is presumed to have developed from a single organism. By counting the number of colony spots, the number of organisms per milliliter of original water is estimated.

Alternately, the original water is diluted by a series of transfers from tube to tube of the inoculated culture medium, until the last tube shows no growth.[6] Calculations from the number of dilutions, in multiples of 10, indicate how many times the original water was diluted to the point that no bacteria cells were transferred. By simple back calculation, the original population, as cells per milliliter, is established.

Various laboratories use different culture medias and some report counts for specific genera and species. Probably the most widely used media for testing oil field waters are patterned after those described in American Petroleum Institute (API) Recommended Practice RP38.[7] Modified API media are commercially available in special serum bottles for inoculation in the field and these are frequently used by oilfield service companies. Two types are available: (1) clear yellow for sulfate-reducing bacteria and (2) phenol red broth for general aerobic bacteria counts. Disposable, sterile, plastic syringes (available from hospital suppliers) are used to inject test water through a rubber seal on the bottles.

For each type of medium, the original water sample is serially diluted until the dilution factor is at least 10^{-5}. The inoculated bottles are then held at a temperature of 25 to 38°C, or preferably within 5°C of the original water temperature in the system being tested. Bacteria growth in the sulfate-reducing broth bottles is indicated by the solution turning visibly dark. Growth in the phenol red aerobic broth is indicated by a color change from red to yellow *and* turbidity. The highest dilution in the bottle series that shows definite growth (in duplicate) defines the range of the bacteria cell count in the original water.

The broth bottle serial dilution technique is sufficiently reliable and so easy to use that a detailed use procedure is given in Appendix 4A at the back of this book.

3.07 Interpretation of Bacteria Counts

The numerical value of bacterial counts cannot be rigidly interpreted. This is true not only for the counts obtained with the field serial-dilution test method (Method 1, Appendix 4A) but also for counts made on water samples tested in microbiological laboratories. The principal reason is that the numbers of free floating bacteria in water samples do not always correspond to the absolute rate of bacterial growth on the metal surfaces in the water handling system.

Several factors influence the ratio of free floating cells in the water to the surface clinging cells. However, the number of free floating cells, viz., the bacteria count in the water, generally increases when the level of bacterial activity in the system increases. Consequently, bacteria counts are primarily useful for indicating whether activity is increasing, decreasing or remaining fairly constant over a period of time (at same sample point in a system) and for reflecting differences between sample points. A consistent count increase with time at a single point or a large increase between test points indicates possible trouble. For example, if counts are consistently higher downstream of a tank than upstream, you may assume bacteria are growing in the tank. Air leakage into a tank or pump will often be indicated by increasing downstream counts.

While precise interpretation of count numbers is not possible, the count ranges are useful as an aid in deciding on a course of action. Table 2.3-2 provides a general interpretation guide for use when judgements must be made from a limited amount of data. The guide has value for decision making, but it is generally preferable to run repeated tests and look for differences as explained previously, rather than to rigidly interpret the count values.

Table 2.3-2
A Guide for General Interpretation of Bacterial Count Data

Type of Bacteria	*Count No. (Cells/ml)*	*Probable Significance*
Sulfate Reducing	1–10	*Low Count.* Not generally a problem.
	$10–10^2$	*Significant.* Repeat tests for trends.
	$10^2–10^3$	*Significant.* Repeat tests and inspect system for slimes and corrosion.
	$10^3–10^4$	*Serious.* Probability of corrosion and fouling is high. Inspect system. Prepare to initiate remedial measures.
	$10^4–10^6$	*Very Serious.* Initiate remedial measures immediately.
Aerobic	$10–10^3$	*Low Count.* Not generally a problem.
	$10^3–10^4$	*Moderate.* Repeat tests.
	$10^4–10^5$	*Significant.* Slime growth probable. Inspect system. Initiate remedial measures if slimes are found.
	$10^5–10^6$	*Serious to Very Serious.* Inspect system to determine extent and location of slime deposits. Initiate remedial measures.

3.08 Direct Examination of Deposits

When cell count data imply significant or serious bacterial infections are present, the next best step is a visual inspection of affected equipment. Even consistently high counts do not necessarily mean that an expensive treatment program is justified. However, if sensibly heavy slime deposits or corrosion are evident when counts are high, consideration should definitely be given to remedial measures.

Direct examination of surface deposits requires access to the interior of flowlines and water tanks. Heavy growths are easily visable and samples can be scraped off for analysis.

At low velocities, bacterial growths may build to 1.0 cm (0.25 in.) thickness on pipe and vessel walls. The deposits may be brown, black or grey and always feel slick. Dried bacterial slime will burn with an acrid odor leaving little or no residue.

Corrosion test coupons and pipe nipples (see Section 1, Item 10.03) should always be examined for slime deposits when first pulled. Retrievable coupons that are mounted flush to the interior of large volume pipelines can be used for slime detection in situations where it is not feasible to shut in the whole line for direct examination.

Debris collected by scraping pigs should also be analysed for non-hydrocarbon organic matter that may signal heavy bacterial growth. The presence of iron sulfide in scrapings from large water lines handling normally sweet water (i.e., seawater) is a reliable indicator of sulfate reducing bacteria growth.

Table 2.3-3
Some Bacterial Control Agents

Alkyl—Aryl quaternary ammonium chlorides
Chlorine—Hypochlorites
Fatty amines and salts
Alkyl thiocarbamates
Acrolein
Aldehydes
Chlorinated phenols

3.09 Bacterial Control Chemicals

Only a limited number of chemical compounds are used for inhibiting bacterial growth in oilfield water systems. Many potentially effective materials are either too toxic to humans or too expensive for use in large volume, once through operations. Table 2.3-3 lists the most common classes of bacterial control chemicals used in the industry.

Chemicals that slow or inhibit bacterial activity are generally referred to as "biocides." Actually, complete sterilization is rarely if ever achieved at normal use concentrations.

Strong chlorine solutions, such as 0.5 to 2.0 wt % calcium hypochlorite, can be circulated as a slug through flowlines and tanks to decompose slimes and effectively sterilize. Slug volume is adjusted to provide ten minute contact time at the pumping rate. Chlorine (hypochlorite) slugs are used in similar fashion for sterilizing water supply wells.

Continuous application of chlorine is not normally advisable for produced water systems because it is a powerful oxidizing agent and tends to create solids. Electrolytically generated hypochlorite is used for treating raw seawater upstream of deaerators.

3.10 Field Application of Biocides

Commercial oilfield biocides, such as the quaternary ammonium chlorides, fatty amines or acetate salts, thiocarbamates and complex aldehydes, are often injected into water systems continuously at 5 to 20 mg/L (ppm). The chemicals may be injected directly with a chemical metering pump, or a stock solution of the biocide in low salinity clean water is prepared and used to provide increased volume for more accurate metering.

A generally preferred procedure for more economical treatment of large volume water systems consists of injecting 50 to 200 mg/L for four to six hours every three to five days and alternating the type of chemical every two to three weeks. Dosage concentration and frequency is adjusted from cell count data. A model program might alternate between a quaternary, a thiocarbamate and a fatty amine.

Successful control of (primarily) sulfate reducing bacteria has been reported by Aramco in the world's largest seawater flood system (Saudi Arabia's Ghawar Field).[8] Current bacterial control treatment consists of alternate injections of glutaraldehyde at 1500 mg/L for 1 hour and a quaternary ammonium chloride at 500 mg/L for 1 hour. The 1 hour treatments are applied twice each week. This program is believed to be more effective than twice weekly treatments at 150-200 mg/L for 3.5 to 4 hours. The lines are pig scraped as frequently as once each week to remove deposits that harbor bacteria colonies.

REFERENCES

(1) Sharpley, J. M., *Applied Petroleum Microbiology*, Buckman Laboratories, Inc., 1961.

(2) Tant, C. O., "The Isolation of Clostridium Species From Secondary Recovery and Salt Water Disposal Systems": Society of American Bacteriologists, Dallas, TX, April, 1960.

(3) Ostroff, A. G., *Introduction to Oilfield Water Technology*, National Association of Corrosion Engineers (NACE), 1979.

(4) Zobell, C. E., U. S. Patent 2,413,278, 1946.

(5) *Standard Methods For Examination of Water and Wastewater* (APHA, AWWA, WPCF): American Public Health Association, New York, NY, 1980.

(6) Whitesell, L. B., "Field Evaluation of Microbial Problems and Their Control": SPE-AIME Paper No. SPE-64, 1961.

(7) API RP38, "Recommended Practices for Biological Analysis of Waterflood Injection Waters", 1959.

(8) DeWar, E. J., "Control of Microbiologically Induced Corrosion and Accumulation of Solids in a Seawater Flood System," *Materials Performance* (NACE), vol 25, no 7, July, 1986, p 39-47.

4 Water Quality for Injection

4.01 Introduction

Water quality requirements vary with water use. For example, water quality required for human consumption is different from that needed for various industrial purposes. Quality is a measure of how well the water conforms to the requirements of the specific application. Since most of the water handled in the oil industry goes back into the ground, the character of the individual earth formations primarily determines the water quality required. More specifically, it can be said that the primary measure of water quality for injection purposes is the concentration *and* nature of suspended matter (solids) that tend to clog rock pores and flow channels.

An accurate but often surprising observation is that in laboratory tests *a core sample from an injection formation can nearly always be severely plugged by a few pore volumes of the injection water being used in the field*. Unless meticulously prefiltered, even carefully collected and preserved field water samples will seal the face of the rock core and drastically reduce water throughout during the flow tests. Laboratory flood tests using cores are generally unreliable for prediction of well plugging by suspended solids.

In normal operation, injection wells are rarely if ever completely plugged. Conversely, some plugging (restriction of intake capacity) probably always occurs regardless of the treatment given the water.

4.02 The Fate of Solids Entering Rock Interstices

All injection waters contain some particles of suspended insoluble matter. These are typically clays, siliceous silt, fine sand, hydrated iron oxides, iron sulfide, calcium carbonate, bacterial slimes and microorganisms. Produced brines and some well waters contain heavy hydrocarbons and lighter oils that often wet and coat the other solids.

When considering the problem of injection well plugging (decreased intake capacity) one must be concerned with (1) the concentration of the solids, (2) their capability for plugging small pores or flow channels and (3) the capacity of various rock formations to accept and store the solids. This raises the question of what happens to the suspended solid particles when the water enters the rock.

An injection formation may be visualized as a highly efficient filter. This concept is confirmed by the following summary of numerous studies concerning invasion of rock by fine solids:

1. Effective size of rock matrix pore openings in consolidated rock, usually ranges from 5 to 15 micrometers (μm) regardless of rock nature (petrography).
2. To cause plugging, solid particles must be small enough to enter the rock pores. Otherwise, an external filter cake is formed that has a higher permeability than that of the rock matrix.
3. Rock pore flow channels are tortuous and solids that are small enough to enter quickly encounter constrictions that impede deep migration.
4. In core flow tests with water containing small ($<5\mu m$) uniform solids in suspension, particle penetration is usually no more than 4 to 5 cm, and depth of maximum penetration is independent of pressure (below fracture pressure) or core permeability up to about 350 millidarcies (md).

When mixed size solids are present (typical of injection waters), a filter cake develops from particles lodged within but near the rock surface and ultimately on the surface. Small solids finally become trapped within the surface cake.

The permeability of the "inner filter cake" (particles lodged within the pores near the surface) is usually equal to or greater than that of the filter cake on the surface.

Thus the factor that largely controls the rate of continued entry of water into the rock matrix is the *permeability of the surface filter cake or skin* (neglecting saturation effects, capillary and viscous forces). Due to bridging of pore openings by invading particles, only a fraction of the total available pore space is filled by the volume of the particles. It is doubtful that the volume of particles actually in the rock matrix ever exceeds the volume of the surface cake when significant plugging has occurred.

With the mechanics of rock face plugging established, the question may be asked as to why some formations accept large volumes of poor quality injection water while others quickly suffer reduced intake capacity. The most appropriate explanation is that the effective area for water entry into the rock matrix (rock filter surface area) is much larger in some wells than in others. Naturally fractured dolomites, for example, accept "dirty" waters better than uniform sands even when the measured sand permeability is much higher. Injection wells completed in uniform sands have only the area exposed to the borehole for water entry. Any large cracks or fractures connecting to the injection well borehole greatly enlarge the rock filter surface face that can exist in injection wells. Figure 2.4-1 illustrates the concept of variation in area of rock surface face that can exist in injection wells.

4.03 Sources and Properties of Suspended Solids

Depending on original source and prior exposure, injection waters commonly contain a variety of solids as listed previously. (Item 4.02). Corrosion products, clays and scale particles occur frequently in combination with organic matter from bacterial activity. Crude oil and semi-solid hydrocarbons are often present in varying amounts in produced waters. The mixed materials tend to agglomerate or stick together, and it is extremely difficult to obtain reliable particle size measurements. Under a microscope, one can often observe particle clumps ranging in size from larger than 100 μm to less than 1 μm. The particle sizes are prone to change with time and with changes in water alkalinity after a water sample is collected.

Compacted bulk density of the mixed solids range from about 0.9 to 2.0 grams per cubic centimeter (g/cc) with an average of about 1.3. This is equivalent to a filter cake bulk volume of 0.77 cc/g or 770 cc/kg. Since 1 milligram per liter is equal to 1 kilogram per 1000 cubic meters (assuming a water sp. gr. of 1), then:

total suspended solids (TSS) as mg/L × 770
= volume of filter cake as cc/1000 m^3 of water.

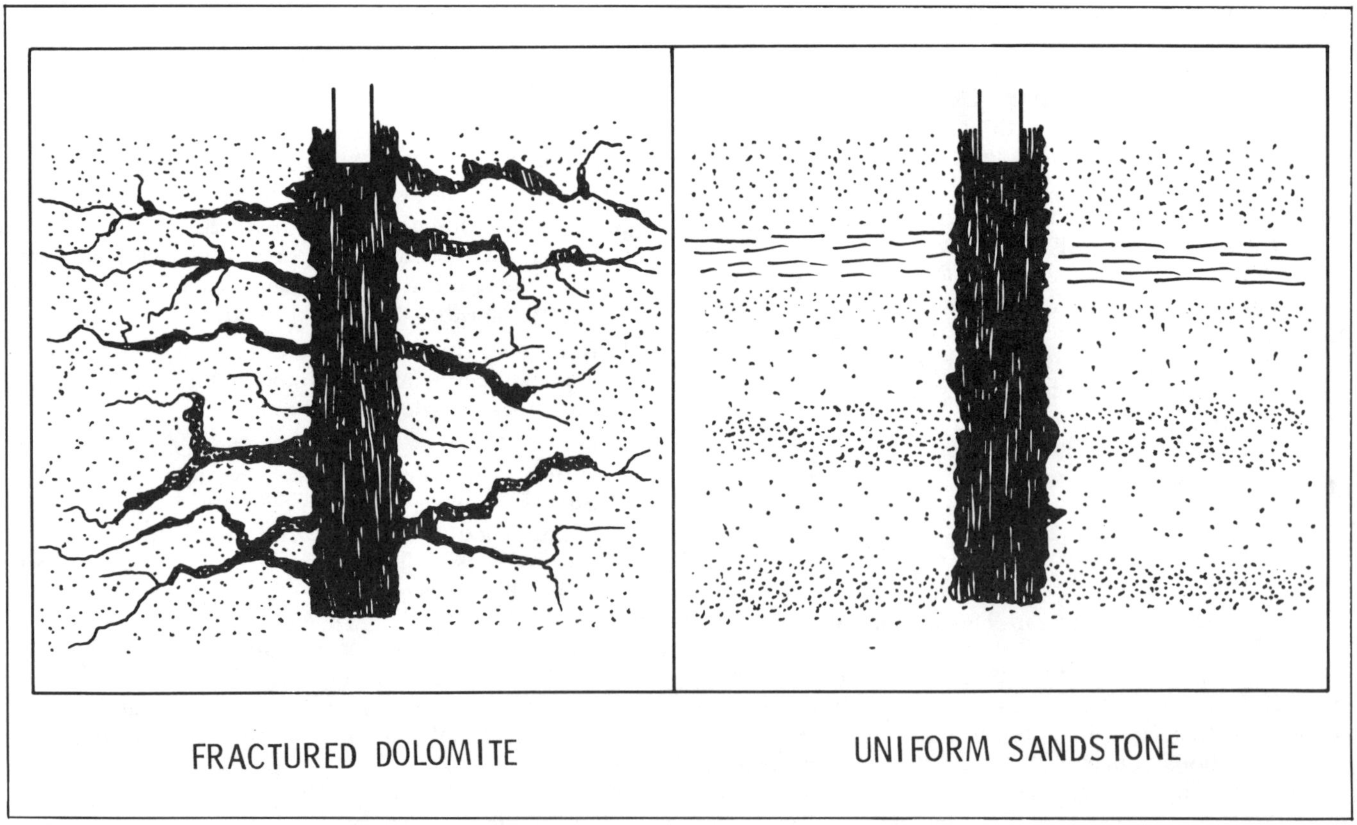

Figure 2.4-1. Illustration of differences in rock face area in injection wells.

At an injection rate of 100 m^3/day, a water containing 36 mg/L TSS will place approximately 1.0 m^3 of compacted solids into the rock interstices in one year.

The measured permeability of filter cakes formed by suspended solids most often found in injection water is usually less than 1.0 md. Fine solids (< 30 μm) that are coated with oil or bacterial slimes often produce compacted filter cakes with permeabilities of 0.1 md and less.

4.04 Water Quality Rating Methods

Technology for predicting with certainty the water quality required for long term injection into individual wells has not been developed. The geometry and capacity of rock flow channels to accept solids cannot be deduced from standard core analyses nor well log data.

A general guide based largely on experience, is that fractured dolomites and limestones can usually accept waters containing 10 to 30 mg/L of suspended matter but water going into uniform sands should contain no more than 3 to 5 mg/L.

Many attempts have been made to devise test methods for quality requirement predictions and for comparatively rating relative injection quality of various waters.[1,2,3] The most widely used method that has evolved consists of a filtration rate test using a porous plastic membrane as the standard test medium. Details of the test procedure are given in NACE Standard TM-01-73.[4] A copy of the test method is reproduced in Appendix 5A. Some companies use modified procedures and apparatus, but the technique of NACE Standard TM-01-73 appears to be generally acceptable. Millipore Filter Corporation, Bedford, Mass., supplies membrane filters and holders that meet requirements of the NACE Standard. The membrane filter discs are 47 mm in diameter and have a mean pore size of 0.45 μm.

The filter test method consists of passing a measured volume of injection water through a membrane filter disc under constant pressure and measuring the flow rate at selected volume intervals. A graphical plot of flow rate vs. water throughout volume produces a curve whose slope is indicative of test filter plugging rate. By using a preweighed membrane disc (or a stack of two matched weight discs), the amount of solids retained from the volume of water filtered can be determined.

Figure 2.4-2 shows a set of curves produced by plotting membrane filter flow rates vs throughput volumes using laboratory prepared waters. In these tests the only solids present were various amounts of precipitated iron sulfide particles. In this test series, the negative slope numbers (m, linear displacement from horizontal per unit length of the line) increased with increasing concentration of suspended particles. A pure water with no solids would yield a horizontal plot line.

The filtration curve slope, m, does *not* always correspond to solids concentration, as demonstrated by the data of Figure 2.4-3. In this case, 0.6 mg/L of small (< 5 μm) particles produced a steeper data plot slope than 4 mg/L of mixed large and small particles.

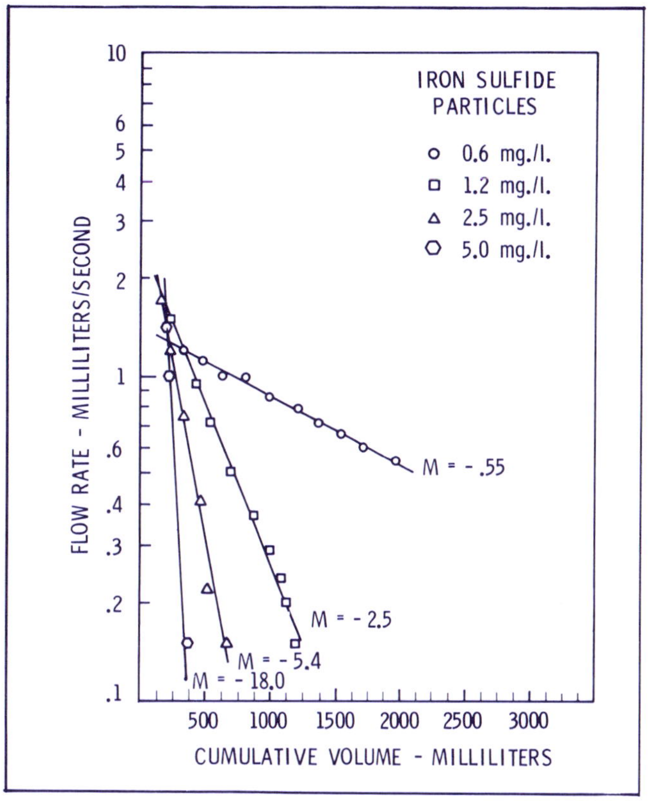

Figure 2.4-2. Effect of particle concentration on membrane filter test plots.

Data plots from the membrane filter tests are actually indicators of the permeability of the filter cakes formed by the solids in the water being tested. The thickness (or length) of the final filter cakes is a function of *solids concentration*, but permeability is also influenced by the *nature* and *size* of the particles. As illustrated by Figure 2.4-3, small particles can pack closer and form more dense (less permeable) filter cakes than mixtures of large and small particles (or large particles alone). For this reason, membrane filter tests run downstream of a field water filter often show steeper data plots than upstream tests even though the downstream solids concentration is lower than upstream.

Gelatinous or soft solids such as iron oxides or slimes produce less permeable filter cakes (steeper data plots) than hard crystalline solids such as scale solids.

In general it can be concluded that the slope of a membrane filter test data plot is a *qualitative* measure of the *ability of solids to plug small pore openings* but provides no quantitative information on the *total plugging capability*. Obviously a high concentration of solids with strong plugging tendency is worse than an equal concentration of a less powerful plugging material. NACE TM-01-73

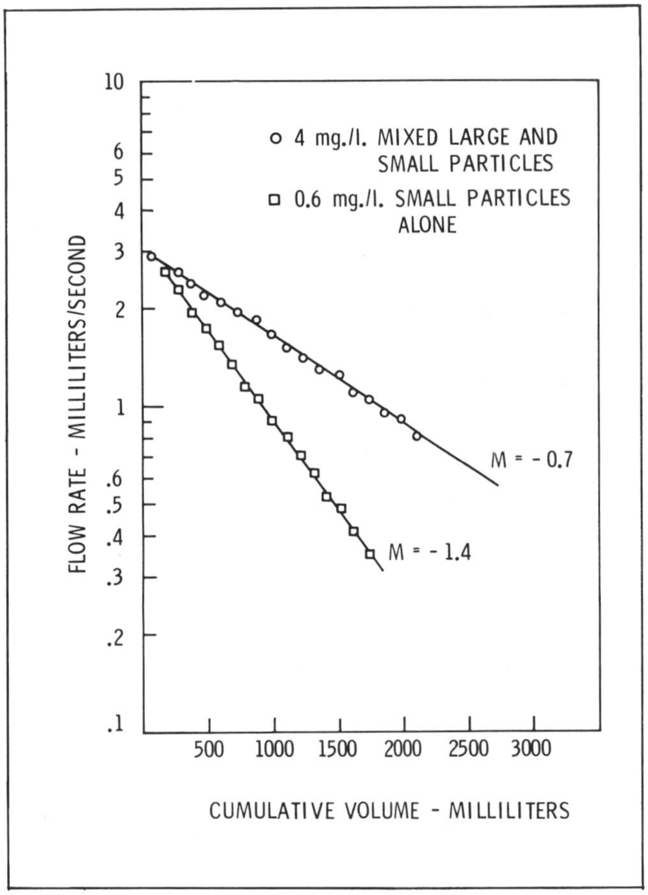

Figure 2.4-3. Effect of particle size distribution.

simply states that the membrane filter test data plots give a general indication of the quality of the injection water.

4.05 An Empirical Water Quality Rating Guide

Graphical plots of membrane filter test data can be helpful for characterizing the plugging nature of suspended solids. However, for relative comparisons of water quality (different waters or point to point tests in an injection system), quantitative determinations of solids concentrations are vital and generally should be the primary test for routine water quality monitoring. The NACE Test Method TM-01-73 provides a suspended solids test method (Procedure 7.5).

Under some circumstances there is need for a more definitive method of water quality rating that provides a numerical value of the total plugging capability. One such circumstance might be a program of correlating measured water quality with injection well performance prior to a waterflood expansion. Ideally the water quality rating value would consider both concentration and relative plugging capability of the solids. Following is a suggested procedure designed for this purpose:

1. Perform a membrane filter test by the general method described in NACE Standard TM-01-73 (Appendix 5A). Make only two determinations of flow rate at specific volumes instead of a whole series as prescribed.

   The first flow rate measurement should be made after some water has been filtered and the change in flow rate/unit volume appears to have stabilized. The amount of water that should be filtered before taking the first flow rate measurement is about 30% of the total volume that will be filtered.

   The second flow rate measurement is made just before the test is terminated. Be sure to *measure the total of all water* run through the test filter.

   Example: Assume that a fairly dirty water is being tested and only about 1.0 L can be filtered before the flow rate decreases to about 0.2 ml/sec. Start the test and flow 250 ml of water through the filter. Stop the flow, empty the cylinder and restart the flow. Using a stop watch, measure the time required to filter an additional 100 ml. The watch may be started at the 50 ml mark on the cylinder and stopped at the 150 ml mark without interrupting flow.

   Calculate the flow rate as ml/sec. This is considered to be the flow rate (QA) at the cumulative volume (CVA) at which the watch was stopped. In the example of the preceding paragraph, the cumulative volume is 350 ml (250 ml initial plus 100 ml during the rate measurement).

   Continue the test, keeping track of the total amount of water filtered. Make another flow rate measurement near the end of the test. For example, start the watch at 850 ml cumulative volume and stop it at 950 ml. Calculate the second flow rate (QB) as ml/sec at the final cumulative volume (CVB). In this example, CVB is 950 ml. Some additional water may be filtered after CVB and this should be measured so that the total volume through the filter is known.

2. Use the following equation to calculate an empirical slope number (ESN):

$$\text{ESN} = \log \frac{\text{QA}}{\text{QB}} \times \frac{2500}{\text{CVB-CVA}}$$

3. Record ESN and total volume of water filtered.

4. Determine weight gain of the membrane disc and, using the measured *total* volume throughput, calculate mg/l of suspended solids in the water (TSS).

5. Add the numerical values of ESN and TSS. The sum can be used as an *index* of relative water quality for injection with the the following implications:

Index Number	General Quality Rating
<4	*Excellent*—Water suitable for most formations
4–12	*Good to Fair*
12–20	*Questionable*—Water may plug sandstones
>20	*Poor*—Water may be suitable for fractured dolomites but generally should be clarified.

The empirical water quality rating guide is described here for information and instruction. It is believed to be useful for the purposes discussed previously, but any application of it is left to the user's discretion and no responsibility can be accepted for the results.

4.06 Clay Swelling and Peptization

The intake capacity of injection wells can be reduced if (1) the formation rock contains a sufficient amount of certain clays (especially montmorillonite) and (2) the injection water is low in divalent cations (calcium and magnesium). The occurrence of this effect to a serious magnitude in waterfloods appears to be rare but the danger should be recognized when injecting into sandstone formations. Low salinity fresh water will generally not damage permeability of clay-containing formations if 10% or more of the cations dissolved in the native formation water *and* in the injection water are calcium and magnesium,[5] viz.:

$$\frac{100\,(\text{mg/L Ca}^{++} + \text{Mg}^{++})}{\text{mg/L Na}^{+} + \text{K}^{+} + \text{Ca}^{++} + \text{Mg}^{++}} = 10\%$$

Mineral analyses of the formation and injection waters should be reviewed prior to starting injection to determine if each water cations contains the minimum desired ratio of divalent cations to total cations. If doubt exists about the clay sensitivity of a particular formation, core tests can be made to determine if permeability loss is likely from flooding with a low salinity water.

REFERENCES

(1) Doscher, T. M., and Webber, L., "The Use of the Membrane Filter in Determining Quality of Water for Subsurface Injection": *Drilling and Production Practice*, API, 1957, p 169-179.

(2) Wright, C. C., "Rating Water Quality and Corrosion Control in Water Floods": *Oil and Gas Jour.*, vol 61, no 20, 1963, p 154.

(3) Barkman, J. H., and Davidson, D. H., "Measuring Water Quality and Predicting Well Impairment": *Jouarnal Petroleum Technology*, vol 24, 1972, p 865-873.

(4) NACE Standard TM-01-73 (1984 Revision), Item No. 53016, "Methods for Determining Water Quality for Subsurface Injection Using Membrane Filters": National Association Corrosion Engineers (NACE), Houston, TX.

(5) Jones, F. O., "Influence of Chemical Composition of Water on Clay Blocking of Permeability": *Journal Petroleum Technology*, April, 1964, p 441-446.

5 Solids and Oil Removal

5.01 Introduction

The major expense for water clarification in oil production operations is in the preparation of water for subsurface injection. Removal of dispersed oil and oily solids prior to surface (land or sea) disposal is also a difficult and costly problem in locations where strict pollution control regulations apply.

Chapter 4 described the general nature of suspended substances in oil field waters, their role in injection well plugging and the nebulous problem of defining required injection water quality. This chapter summarizes the state of the art for removal of suspended solids and oil as practiced in the petroleum producing industry. However, despite intensive design efforts and numerous innovations that have appeared during the last 25 years, there is still no clear definition of the best available technology, especially when economic constraints are considered.

Research on oil separation techniques has been conducted by major oil producers, American Petroleum Institute (API), several universities and U.S. Naval Research. Numerous private concerns that develop and market water treating equipment have also been involved.

A vast amount of knowledge has accumulated in the field and no attempt will be made here to describe all the water conditioning equipment that is available. The examples that will be discussed were selected primarily to illustrate the types of processes that have reached widespread use in production operations. Equipment and chemicals to carry out each of the various unit operations such as filtration, solids contact clarification and flotation are supplied by numerous companies. Identification of specific companies or tradenames is not intended as an endorsement nor as a recommendation for or against any organization or product.

5.02 Solids Removal from Surface Supply Waters

Supply or make-up water for inland flooding operations most often originates from water wells but water is occasionally taken directly from rivers or lakes. Water supply wells are preferably completed in comparatively deep aquifers bearing brackish or salt water that is unfit for human and animal consumption or for agricultural use. Fresh waters are occasionally used where supply is ample. Seawater is normally used offshore and along coastlines.

Some deep well waters contain less than 5 mg/L total suspended solids (TSS) and are free of hydrocarbons and dissolved oxygen. These waters can be injected directly without solids removal treatment but should be handled in totally closed systems to prevent air contamination (see Section 1, Chapter 6).

Shallow well water from sand aquifers is often visably turbid initially with more than 10 mg/L of clays, siliceous silt and iron oxides. Water supply wells drilled along shore lines of rivers, lakes and the sea can also produce dirty water when first started up. The concentration of suspended matter usually declines with time as the wells are produced. The initial surge of high TSS should not be allowed to enter the injection wells. A good practice is to allow the wells to produce for several weeks until the TSS content stabilizes before introducing the water into the injection system. Alternately, temporary filters should be installed.

As a general rule, shallow and surface water sources contain dissolved oxygen but the O_2 content may be below saturation levels. When any amount of dissolved O_2 greater than 0.3 mg/L is naturally present, it is generally permissable to clarify the waters in open treating systems without protection from additional air entry. The logic here is that oxygen will ultimately have to be removed especially if the supply waters are to be mixed with produced water prior to injection.

5.03 Open System Clarification

Open system treatment is particularly advantageous for clarifying surface waters containing more than 50 mg/L of suspended solids (continuously or occasionally) with particle sizes predominately less than 20 micrometers (microns, μm). In general, solids removal is accomplished by coagulation, sedimentation and filtration. Prior settling of large, heavy particles is provided if necessary.

Coagulation is a process for destabilization and initial aggregation of suspended matter. Chemicals such as iron or aluminum sulfate are added at treating rates determined to be optimum by a series of previous on-site tests. Surface charges (negative zeta potential) naturally present on small suspended particles are partially neutralized by the coagulant chemicals thus allowing the particles to move closer together to form aggregates or clumps. Aggregation is further promoted by gelatinous polymeric substances, including hydrated iron oxide or aluminum hydroxide precipitates, which enmesh the small particle clumps into weblike masses called *flocs*. Coagulating and flocculation aids such as activated (hydrated) silica and synthetic polyelectrolyte polymers are often used with the primary iron or aluminum coagulants. In some waters containing natural coagulants, i.e., dissolved iron, the polyelectrolyte polymer additives are used alone without a primary coagulant. Coagulation and heavy floc formation are strongly affected by water alkalinity. For aluminum coagulants the optimum pH is usually 6 to 7 but the pH range for iron can vary from 4 to 10. Control of pH is a vital part of the preliminary testing program and the routine chemical treatment during continued operation.

Another factor that must be controlled in the coagulation-flocculation process is the rate of agitation or mixing. Standard practice is to provide rapid initial mixing for chemical dispersion followed by a period of relatively slow agitation that allows the solids bearing flocs to grow and settle. Specially designed flow through basins constructed of concrete are used in large volume plant water treating. For oilfield use, it is generally preferable to employ compartmented steel vessels called flocculators or *solids contact clarifiers*. Details of their design are available from suppliers such as Petrolite, Infilco, Graver, Permutit and numerous others.

Within the solids contact clarifier units there are provisions for sludge recirculation such that the chemicals and incoming water are allowed to react in the presence of preformed solids, sludge concentration with continuous removal, and removal of treated water. Effluent water is filtered to remove any residual entrained sediment.

5.04 Filters for Surface Waters

Filtration is often used as the sole means of solids removal when TSS concentration is low and majority particle sizes are relatively large (> 10 μm). Some onshore installations for treatment of larger volumes of seawater utilize open gravity filters. These are typically concrete basins fitted with underdrains over which are placed graded layers of gravel topped with sand filter media. The top sand layer is about 30-45 cm (1 to 1.5 ft) thick and effective sand size is typically such that 90% is larger than 0.4 mm. Gravity downflow rates range from 1.5 to 6.5 $L/min/m^2$ (0.04 to 0.16 $gal/min/ft^2$).

Graded bed pressure filters are used for filtering aerated seawater offshore and for surface waters inland, especially in remote locations and where total water volume is relatively low. More detail on design of pressure bed filters is presented later in this chapter.

The efficiency of pressure bed filters for removal of fine particles (> 20 μm) is improved by feeding 0.2 to 2.0 mg/L of an organic polyelectrolyte coagulant aid into the incoming water line upstream of the filter units. This technique works especially well with upflow "deep bed" pressure filters. The point of polyelectrolyte injection should be positioned to allow at least about 1.0 min contact time before reaching the filter.

In a series of efficiency tests conducted on production platforms in the Cook Inlet, Alaska, upflow filters with polyelectrolyte prefeed consistently reduced TSS in the seawater from an average of 80 mg/L to 2-3 mg/L. The suspended solids were primarily glacial silt particles ranging from 1.0 to 10 μm in size.

Some offshore installations clarify seawater with diatomaceous earth (D.E.) filters. The number of inland installations using this method appear to be dwindling especially when handling corrosive municipal and plant waste waters occasionally used as floodwater supply.

D.E. filters employ a body feed of fine grain diatomaceous earth (silica remains of diatom one-cell plants). The filter units contain multiple screen plates or frames covered with woven fabric. Each plate is an integral filter in a group that is manifolded to a common outlet. The fabric covered plates or frames are initially precoated with a slurry of D.E. prior to each run and more is collected from the incoming water to which the body feed was added. Water solids are trapped by the accumulating D.E. filter cake. The amount of body feed is adjusted according to quantity and nature of solids to be removed and the allowable throughput rate (characteristics of the specific filter installation). Typical body feed rates are 5 to 15 ppm fed directly into the raw water ahead of the filter unit.

D.E. filters process about 0.05 $m^3/min/m^2$ (1.25 $gal/min/ft^2$) compared with 0.6 $m^3/min/m^2$ (12 $gal/min/ft^2$) for some high rate upflow pressure bed filters. However, the filtering area in D.E. filters, per unit area of space occupied, is on the order of 20 times greater than that of fixed bed pressure filters.

D.E. filters can effectively remove small particles down to 0.5 μm size when operating properly. However, they have several disadvantages for ordinary oilfield application. The maximum recommended solids load in the water being treated is 30 mg/L, they require frequent operator attention, disposal of used D.E. is a problem and the screens are easily damaged which allows leakage. L'eau Claire Systems, Inc. (New Orleans, LA), has recently introduced a simplified D.E. filter design utilizing tubular screens and either wet or dry filter cake flushing which is reported to eliminate many of the problems traditionally associated with conventional D.E. operation.

Various types of centrifugal separators are used to remove large (> 40 μm) solids. The most popular design is the static cyclone type similar to those used to reduce solids content in circulating drilling fluids. Some particles as small as 5 μm are removed but the removal efficiency for particles smaller than 20-30 μm is very low. The best application for centrifugal separators in water clarification is for solids load reduction upstream of D.E. filters or solids contact clarifiers where space limitations prevent use of settling basins.

Tubular and cartridge filters are available in a variety of designs and retention ratings. Low cost disposable cartridges are used as guard filters downstream of main filters or clarifiers to trap stray carryover particles. They are used beneficially to remove sand grains or other abrasive particles ahead of positive displacement pressure pumps. They are also used as final filters at individual injection wellheads. Low solids holding capacity and the problem of cartridge disposal generally limit use of disposable cartridge filters to low volume systems or to the guard applications mentioned. Cartridge elements are rated for solids size retention ranging from < 1 μm to > 100 μm but the most popular rention range in oilfield use are the 10 to 25 μm elements.

Tubular in-line pressure filters are available with reuseable elements that eliminate need for frequent cartridge replacement and disposal. The elements are backwashed in place either automatically or manually. Figure 2.5-1 shows the external appearance of a three element manual unit (Ronningen-Petter Internal Backwashing Multiplex Filter). In use, unfiltered water enters the bottom inlet and flows up and around the internal filter elements which are stainless steel or fabric screens reinforced by perforated stainless steel backings. Woven fabric sleeves have retention levels down to 1 to 3 μm. Slotted metal and wire mesh screens are available in retention ratings of 25 μm to 1600 μm which provides versatility in media selection for specific needs. After clarification during passage through the tube element walls, the cleaned water moves into the top manifold and on to use. Internal backwash units utilize filtered water from in-service elements to flush and clean one other element at a time as required. The individual filter assemblies typically contain 2 to 16 tubular elements providing up to 3.75 m^2 (40 ft^2) of filter area. Multiple assemblies can be used to provide essentially unlimited throughput capacity.

Figure 2.5-1. External appearance of a three element manual unit (Ronningen-Petter Internal Backwashing Multiplex Filter).

5.05 Clarification of Produced Waters

The techniques and equipment required for optimum cleaning of produced waters containing oil and oily solids are quite different than the methods discussed previously for surface supply waters containing only oil free solids. In fact, large expenditures have been wasted in attempts to use conventional solids removal procedures for oily brine treatment.

Open system treatment of oily produced brines, particularly those containing hydrogen sulfide, can generate large amounts of new solids. Exposure to air promotes precipitation of calcium carbonate, elemental sulfur, iron oxides and iron sulfides. Evaporation of light hydrocarbons and oxidation of heavier fractions increase adherence of oil films to other solids and promotes stabilization of gummy emulsions. Corrosivity is also increased. For these reasons, most produced water handling systems are kept closed to exclude air.

5.06 Closed System Filtration

Closed in-line pressure bed filters are the most widely used type of solids removal equipment for produced waters because of simplicity and low cost. Their efficiency is generally low after several months use because of fouling by hydrocarbons and bacterial growth. A limited number of installations utilize steam and detergent or solvent cleaning on a regular schedule to maintain the filter beds. Such practice is generally not feasible in ordinary oilfield operations.

Figure 2.5-2 illustrates the components of a typical downflow pressure bed filter. Graded sand and anthracite coal are the most popular filter media. Throughput rates are typically 0.1 $m^3/min/m^2$ (2.5 $gal/min/ft^2$) with the fine-grain media illustrated in Figure 2.5-2. Larger grain sizes allow longer filter runs and higher flow rates but efficiency of small (< 20 μm) particle retention is reduced.

Stacked or multimedia filter beds provide coarse to fine filtration (in a downflow flow pattern) with deep bed penetration and longer filter runs at high flow rates. A typical multimedia filter contains a 0.5 m (1.6 ft) layer of 1.0-1.5 mm anthracite over 0.2 m (0.66 ft) of 0.5 mm sand with one or more layers of 2.0 mm and larger garnet or gravel under the sand. Rated throughput capacities up to 0.8 $m^3/min/m^2$ (20 $gal/min/ft^2$) are available.

In all pressure bed filters, media size and available

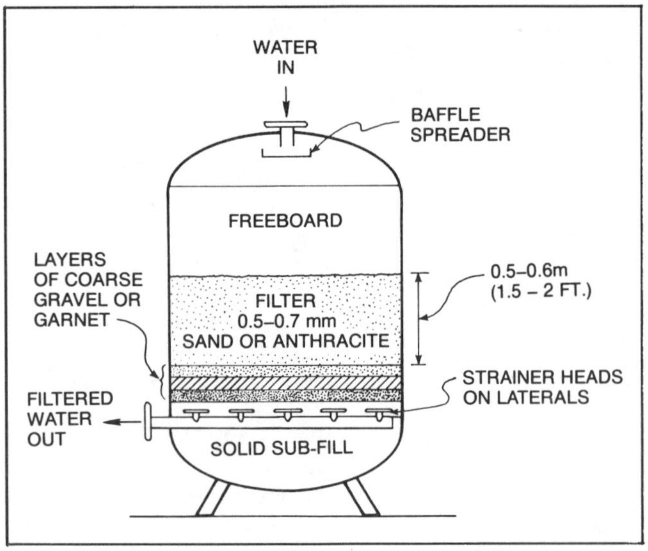

Figure 2.5-2. Typical downflow pressure bed filter.

water pressure (head) primarily determines the flow rate. Flow rate limits the throughput and controls the number of filter units required. As the media becomes clogged with solids, the intergranular velocity at a specific flow rate increases until shear forces compress portions of the bed and hydraulically fractures it. Some of the trapped particles then escape into the effluent.

Normal operating pressure for single and multimedia vertical bed filters is about 500 kPa (72 psi). When pressure drop across the bed reaches about 25 kPa (3.6 psi) the filter run is terminated, flow direction is reversed and the bed is backwashed to avoid solids breakthrough. In high rate filters, allowable pressure drop is 1.5 to 2.0 times greater. Backwash rates are specified according to bed design and must be sufficient to expand the volume of the top layers by about 50-60% upward into the freeboard zone.

Upflow filters utilize multimedia beds and are physically similar to the downflow type. Excessive expansion of the bed during normal upward flow (0.3 $m^3/min/m^2$ or 7.5 $gal/min/ft^2$) is restrained by a steel grid at the top of the bed. Water entering the bottom encounters coarse media first and the fine top media last. With the high upflow rates, fine particles penetrate through the lower bed layers and lodge in the fine grained top layers. Depending on particle size distribution, the whole bed may be substantially filled with solids at the termination of the filter run. Bed cleaning is accomplished by expanding the bed first with air or natural gas and then injecting a mixture of air or gas and water to unload the trapped solids.

As stated under Item 5.04, deep bed upflow filtration with a polyelectrolyte prefeed has been successfully used where small (< 10 μm) solids are being removed. The same type of system with a cationic (positively charged) polyelectrolyte prefeed and relatively large (1-2 mm) sand layers is used as a combination oil coalescer and filter. An example of this type unit is the L'eau Claire oil filter. A detergent is periodically added to the flush water during gas/water purge cleaning to strip adherent oil from the sand media.

5.07 The Problem of Oil Carryover

A closed emulsion treating facility for separation of produced water, oil and gas is depicted in Figure 2.5-3. The treatment train may be more or less complicated than that shown but usually includes some combination of the vessels indicated.

Complete separation of the three fluid phases (gas, oil, and water) is desired but is rarely if ever achieved. Operational control is normally dictated by the maximum allowable water content of the separated oil (usually no more than 0.5 vol %).

The major concern of this discussion is the amount and condition of oil remaining in the water discharge

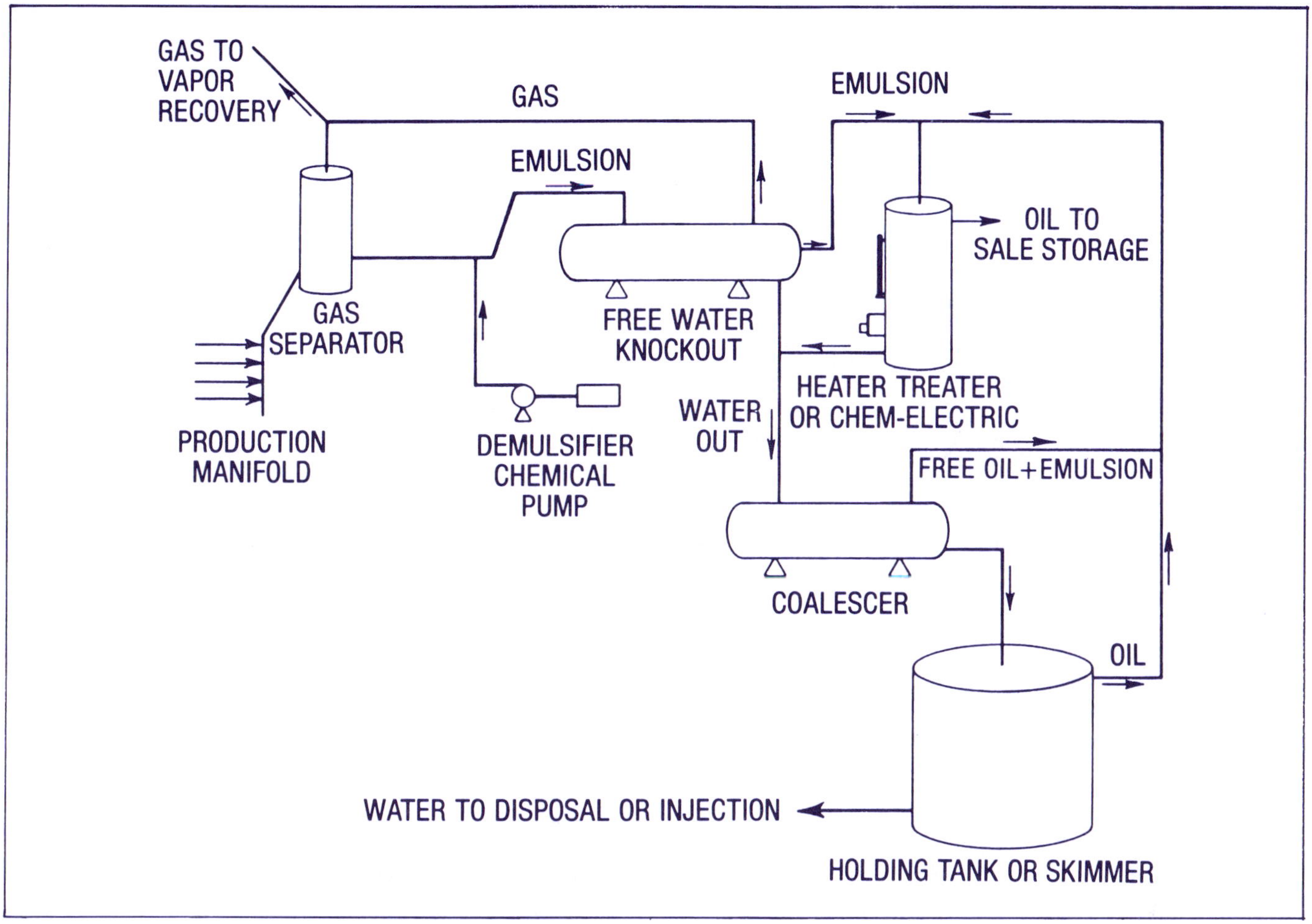

Figure 2.5-3. Closed loop emulsion treating facility.

(carryover oil). Carryover oil in produced water is recognized as a serious problem from both an economic and technical viewpoint. Its worldwide impact is growing for the following reasons:

1. Average produced water-oil ratios are increasing.
2. The number of water injection wells is increasing as more fields are converted to secondary and tertiary recovery. Oil and oily solids are primary contributors to injection well plugging.
3. Public awareness of surface pollution and restrictive legislation are limiting surface disposal options.

Residual oil in the produced water phase usually exists as free entrained oil, emulsified oil and oil that is clinging to solid particles. Free oil may be characterized as the portion that will separate and float to the surface of the water when held in a holding tank or skimmer vessel. Emulsified oil exists as a more or less stable dispersion of small droplets. Each discrete emulsion droplet has an outer layer or skin of emulsion stabilizing agent. The surface layer exhibits a net electrical charge that is sufficiently strong to produce mutual repulsion to other like charged droplets and thus retards coalescence or the merging of several droplets to form one larger mass.

The rate of gravity separation of dispersed oil from water is determined primarily by the density difference between the two immiscible liquid phases, the size of the oil droplets and the viscosity of the water.

When all other factors remain constant, the rise or fall rate of a small (1 to 100 μm) spherical particle in water is approximately proportional to the square of the mean radius. The stabilized terminal velocity (V) is given by the following equation (Stokes Law):

$$V = \frac{2g\,a^2(d_1 - d_2)}{9\eta}$$

where a is the radius of the sphere in cm, d_1 and d_2 the densities of the water and the sphere respectively in g/cm^3, η the coefficient of water viscosity in poises and g is the gravitational constant in cm/sec^2. Dissolved salts

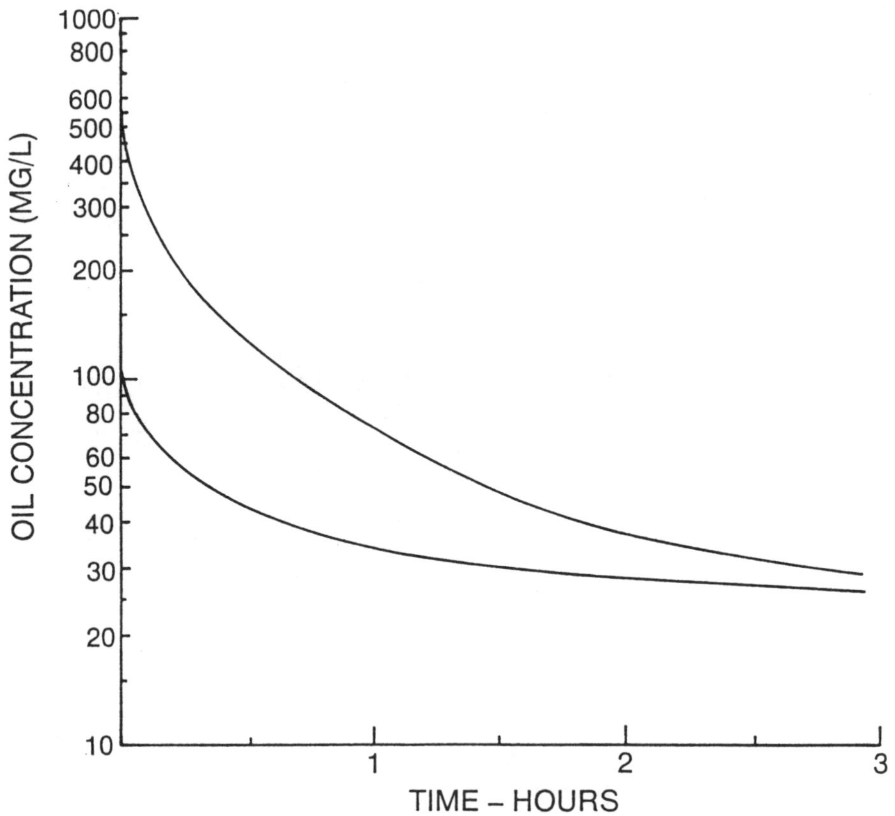

Figure 2.5-4. Gravity separation of crude oils from produced brines—typical rates.

in the water increase density contrast but also increase water viscosity that in turn slows the rise or fall velocity.

Typical rates of gravity separation of crude oil from produced brines in a field holding tank are illustrated in Figure 2.5-4. It can be seen that the separation rate declines rapidly during the first hour and is typically low after two hours. Generally about two hours retention time is sufficient to remove most of the oil that will gravity separate. However, the holding tank must be designed properly with adequate baffling to avoid short circuiting regardless of the size of the tank.

When all other conditions are constant and there is no change in the nature of the oil and water, the residual oil concentration in the water phase after quiescent holding for two or more hours tends to be independent of initial oil content as indicated by the data curves of Figure 2.5-4.

5.08 Factors Influencing Oil Dispersion and Stabilization

Even pure oil can be caused to break up into small droplets and disperse into water. The mechanical energy required is largely dictated by the oil-water interfacial tension (IFT). IFT is the force required to shear an oil/water boundary and is measured as dynes per centimeter.

Crude oils are never pure in the sense that they contain a variety of substances that are not simply hydrocarbons. Some components contain oxygen and nitrogen atoms in their molecular structure that gives them an affinity for water. These so-called polar groups act as natural surfactants and lower both surface and interfacial tension, making the mutual blending of oil and water easier.

Depending on the nature of surfactants present and to some extent the volumetric ratio of the principle liquids, crude oils and water can produce water-in-oil (w/o) or oil-in-water (o/w) emulsions. Most natural crude emulsions are of the w/o type wherein water is the internal dispersed phase. Only a limited number of crude oils naturally produce invert o/w emulsions. These can be difficult to break and the separated produced water often retains significant amounts of dispersed oil.

Some crude oil emulsions consist of oil dispersed in water droplets that in turn are dispersed in oil, the external continuous phase (o/w/o). Such complex emulsions probably occur naturally to some extent but the author has observed them on several occasions in production from wells being treated with water dispersible amine base corrosion inhibitors.

Emulsion upsets are known to occur when a group of wells producing to a common tank battery are all treated with corrosion inhibitor over a short time period. A high concentration wave of inhibitor entering the emulsion treating facility produces "bad" oil that must be recycled to reduce water content prior to custody transfer. In this situation, it has been observed that carryover oil in the water phase is also increased.

Laboratory tests have demonstrated that with certain crude oils even ordinary concentrations, e.g., 15-30 mg/L, of corrosion inhibitor significantly lower IFT and promotes dispersion of oil into water. As pointed out in Section 1, Chapter 3, Item 3.02, the amine constituents of ordinary oilfield inhibitors have surfactant properties and chemical emulsifiers are frequently added that promote oil-in-water (o/w) emulsification.

Another source of IFT lowering surfactants are the demulsifiers, e.g., chemicals used to break w/o emulsions in field operations. Again, laboratory tests have demonstrated that even small excesses of demulsifier chemical over that required to break the primary emulsion can cause transfer of oil into the produced water. The problem is particularly severe when inhibitor stabilized o/w/o emulsions are present and when higher than normal demulsifier addition is required to produce saleable oil. The latter situation often develops when volume throughput exceeds design capacity of the oil treating equipment.

The author has recommended for many years that in addition to testing the water content of lease oil, the chemical service companies supplying corrosion inhibitors and demulsifiers should be required to monitor effects of their products on the carryover oil content of produced water downstream of the emulsion treating facility. Such a program has been carried out for more than ten years in several fields in West Texas. The type and treating program for corrosion inhibitors are modified to avoid excess slug concentrations and demulsifier addition is similarly monitored. Results have been a considerable reduction in average oil content of produced water and reduced expenditure for injection well clean out. No loss in corrosion inhibition has been noted and demulsifier treatment costs have not increased.

In addition to the natural and added surfactants in the produced fluid stream, mechanical factors also aggrevate the carryover oil problem. Overloaded centrifugal pumps running at excess speeds can act as homogenizers and promote stabilization of oil in water dispersions. Liquid impingement as sharp angle piping bends and splashing inside surface tanks has a similar effect.

In producing locations where special oil removal equipment is being contemplated or has already been installed, all the contributing factors discussed here should be studied and remedied as feasible to reduce the carryover oil load in the system.

5.09 Removal of Oil and Oily Solids by Coalescence

In Figure 2.5-3, a coalescer vessel is shown taking water from the free water knockout (FWKO) and the emulsion treater. Some coalescers are nothing more than compartmented tanks with internal baffles that reverse flow direction several times to promote separation of loosely entrained oil. Others use granular coalescing media such as medium coarse sand or carbon particles that are intended to temporarily trap and concentrate oil droplets to larger sizes that separate more quickly. Granular media coalescers are prone to fouling by clumps of oily solids in the same manner as filters.

Entrained oil droplets that are too small to rise rapidly for gravity separation can be caused to coalesce into larger drops by passage through porous oleophilic (oil wettable) plastic media such as polypropylene or polyurethane foam or fabric. One device of this type is illustrated in Figure 2.5-5. In use, oily water flows to the interior of the cartridge and out through the walls where coalescence occurs. Multiple cartridges are used in a single housing to handle large water volumes.

Plastic foam coalescing media can process produced water that contains little or no oil wetted solids but fouling by oily solids is usually a problem. An oleophilic foam separator designed to remove both oil and oil wetted solids is depicted in Figure 2.5-6. Coalesced light oil migrates through the foam pack and is automatically drawn off. Heavy oil, paraffin and oily solids are retained in the body of the foam pack by processes of adsorption, coalescence and entrapment. As clogging of the foam pack progresses, pressure drop increases and triggers an automatic regeneration cycle. During regeneration, water flow direction is reversed to close the check valves and compress the foam pack. Water is flowed back and forth to alternately compress and inflate the foam. The foam pack is then purged with gas and rinse water and returned to service. Compressible foam coalescers have been in use in the petroleum industry since 1981.

5.10 Short Path Separators

Under constant conditions, an entrained particle in water tends to rise or fall at a fixed rate called terminal velocity. The time for separation therefore depends on the length of the path through which the particle must move to reach either the surface of the water or the bottom of the container. In ordinary settling tanks or retention basins, the mean water depth is generally at least 1.0 to 2.0 m (3.3 to 6.6 ft). A 10 μm particle of silt with a density of 2.7 g/cm^3 requires more than 3 hours to fall through 1.0 m of water depth.

Short path separators provide closely stacked parallel

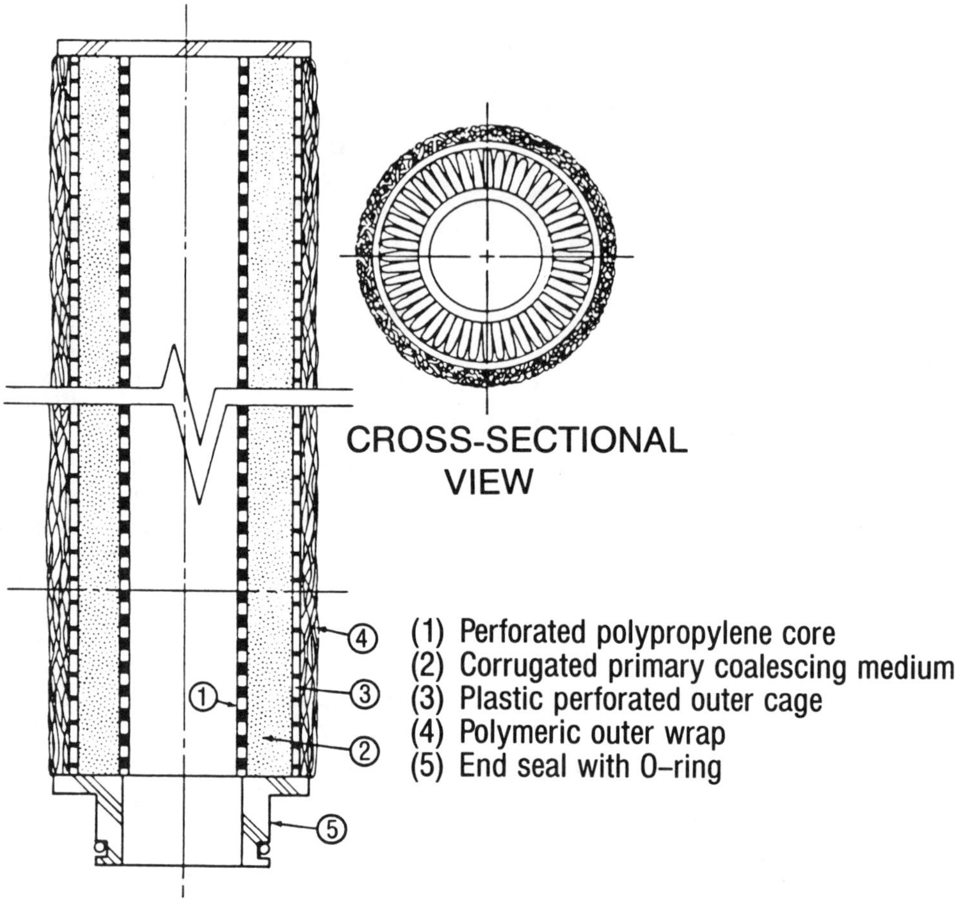

Figure 2.5-5. Representation of liquid/liquid (LL) oil coalescing filter cartridge. (Courtesy Pall Well Technology Corporation).

plates or small diameter tubes through which the water flows without turbulence. Within the confined flow spaces, oil droplets and sludge particles have only a short distance (usually 4 to 10 cm or 1.5 to 4.0 in.) to rise or fall to reach a solid surface where they concentrate. Inclined parallel plate separators are available worldwide from several manufacturers including Monarch Separators Inc., Houston, TX, Pacesetter Separator Co., Corpus Christi, TX, (formerly Pielkenroad Co.) and Facet Enterprises, Tulsa, OK. Various mechanical designs and special features are provided under the trade names of Parallel Plate Interceptor (PPI), Tilted Plate Separator (TPS) and Corrugated Plate Separators (CPS). The essential elements are depicted in Figure 2.5-7 to illustrate the principle of operation. Their primary advantage is that they provide in a small space, with a flow through time of 15-30 min, the same degree of separation achieved in a much larger gravity settling tank in two hours or more. They do not remove emulsified oil unless demulsifier chemicals are added upstream.

5.11 Separation by Flotation

Flotation is a well developed technology that has been adapted to oilfield use for cleaning produced waters. The operating principle depends on increasing bouyancy of entrained oil and solids by attachment to gas bubbles.

Gas bubbles are generated by (1) releasing pressure on water that has been previously charged with dissolved gas or (2) by mechanically dispersing gas into the water in a finely divided state. The latter procedure called induced gas or froth flotation is most widely used in oil production operations.

Figure 2.5-8 shows skeletal features of a four compartment, induced gas flotation machine. Oily water flows through each compartment or cell in sequence. In each cell is mounted a rotating gas induction (cavitator) mechanism with a stationary dispersion assembly as illustrated in Figure 2.5-9. Gas from the upper region of the housing is sucked downward into the induction/dispersion device where it mixes in the form of small bubbles into the incoming water. Treating chemicals added to the oily water ahead of the flotation machine promotes separation (breaking) of oil/water emulsion and attachment of gas bubbles to oil droplets and solid particles. The resulting mixture of gas bubbles, oil and small solids float rapidly to the top of each cell as a froth or foam. Concentration and skimming is provided by a fixed surface baffle that directs the oily foam over an adjustable weir into a skimmings recovery channel taking effluent from all cells. Oily waste is continuously accumulated and drained off, and part of the released gas is recycled.

Some early oilfield installations used air as the flotation gas. Most now employ field gas to minimize corrosion and microbial fouling. Use of air free gas in the sealed flotation machine offers the additional advantage of stripping dissolved oxygen from the treated water. However, explosion proof electric motors and other appropriate safety devices must be provided.

A flotation system design with no mechanical moving parts has recently been put on the market. Gas is introduced by means of an educator nozzle as depicted in Figure 2.5-10. Flotation bubbles are created when the gas/water mixture collides with the striker plate at high velocity. A skimming trough is provided near the top of the vessel to draw off the oily froth. Four cells in series are normally used to provide multiple pass treatment.

Regardless of the bubble creating mechanism employed, all flotation devices are subject to upset and malfunction if concentrated slugs of oil enter the machines. Flotation is therefore normally used as a final or end of the pipe treatment downstream of primary separation equipment such as that depicted in Figure 2.5-3.

5.12 Oil Removal Process Control

Regardless of the type of oil removal process used, reliable data are necessary to determine if the treated water is sufficiently free of residual oil and oily solids to meet requirements for either surface disposal or subsurface injection.

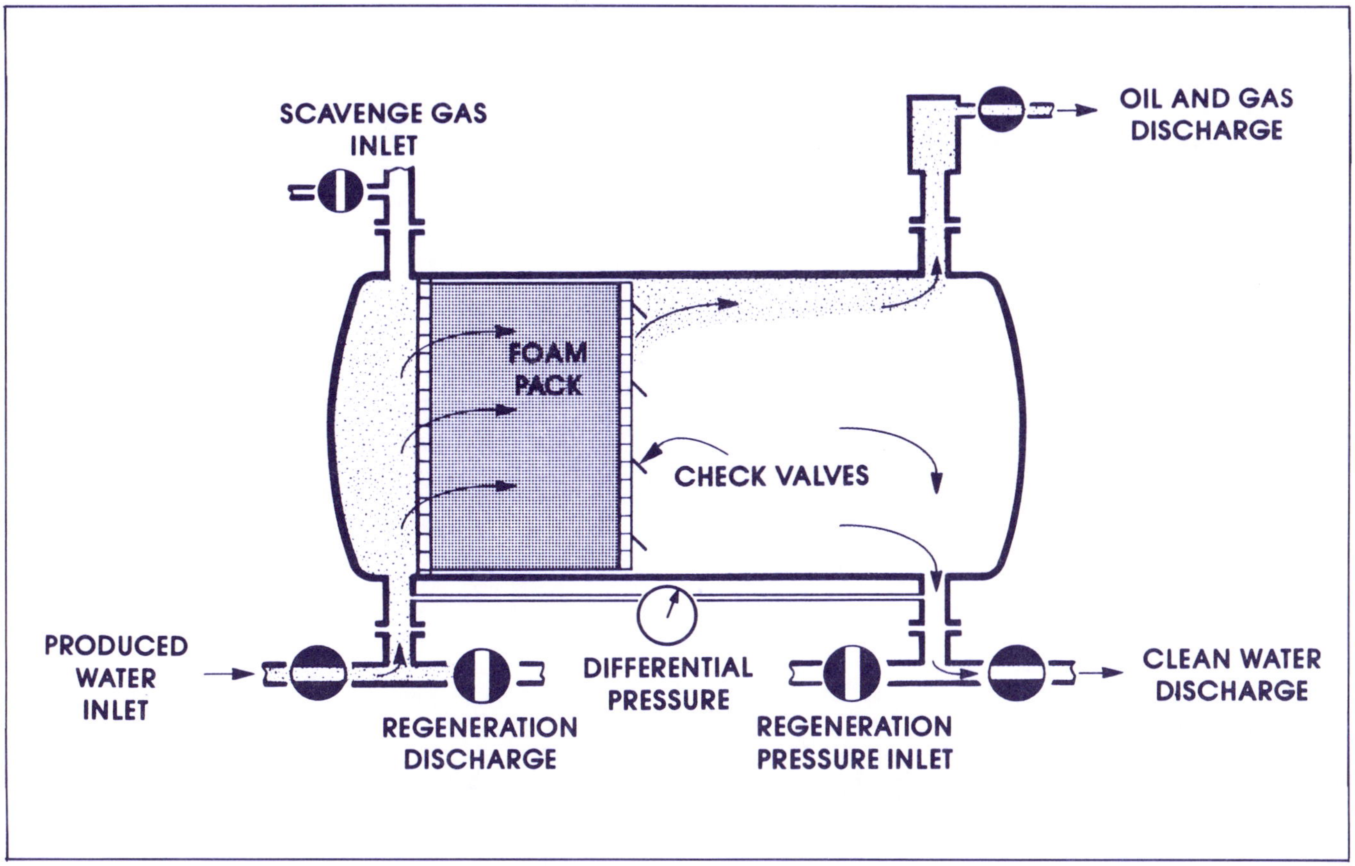

Figure 2.5-6. Oil/water separator using oleophilic foam. (Courtesy Marco Seattle Company).

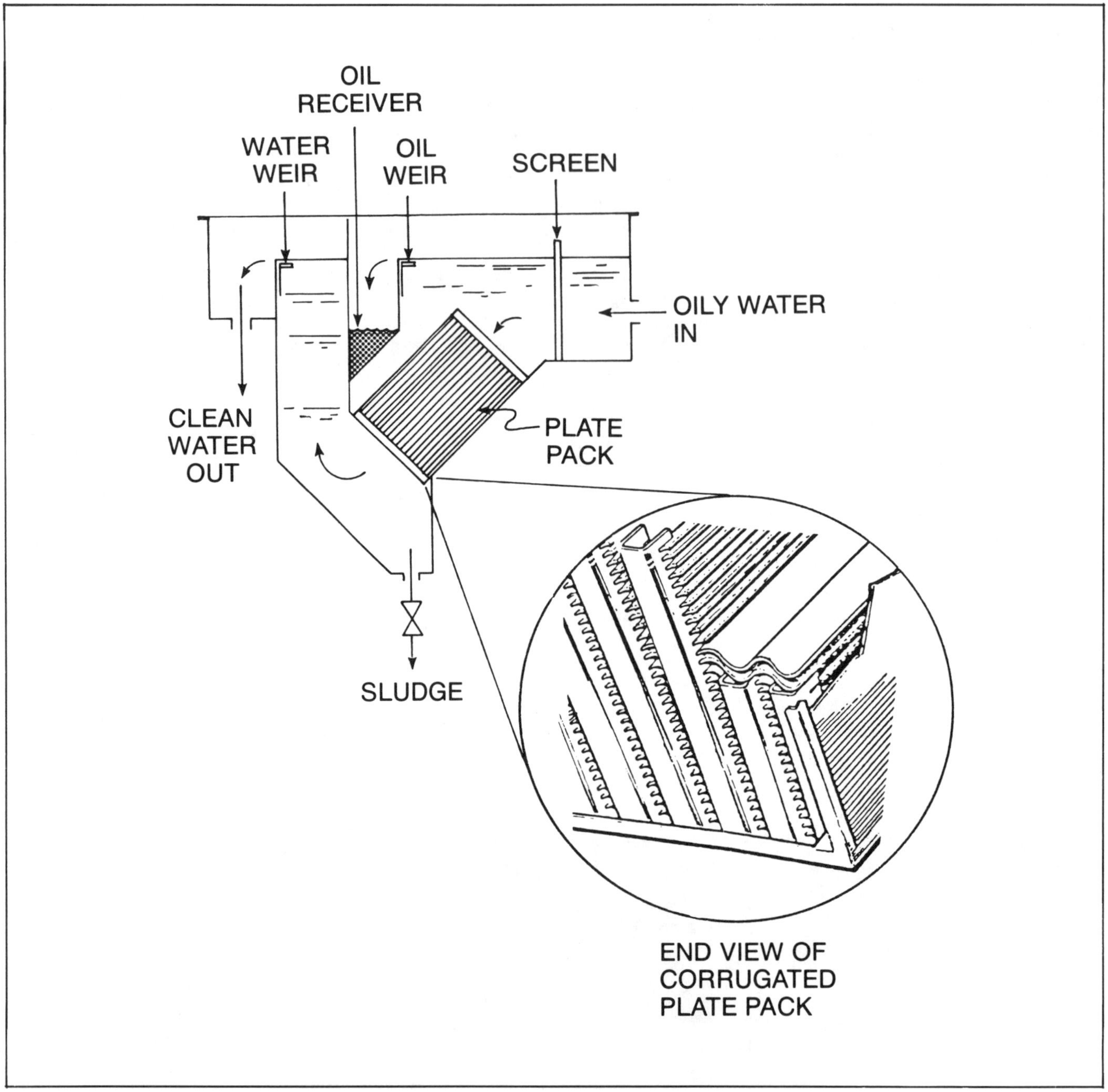

Figure 2.5-7. Elements of a Short Path Separator.

Allowable oil content for surface disposal may be controlled by government regulatory agencies such as the EPA in the United States. Most states in the United States also impose local limits. The applicable laws and regulations should always be obtained prior to planning a surface disposal system.

The discussion in Section 2, Chapter 4 (Water Quality for Injection) provides guidelines for estimating allowable amounts of suspended matter including oily solids in water to be injected subsurface. However, no general concensus has been reached on the concentration of free entrained oil that may be considered detrimental to water injection wells. Some operators set a goal of no more than 50 mg/L (or ppm) as the average daily maximum measured as free floating oil.

When setting any numerical limit, a practical problem

arises as to how to measure oil concentration in water with reasonable accuracy. First, if anything less than total removal is achieved (usual situation), the oil concentration will probably vary significantly during each day. Secondly, portions of water removed from the system for analysis can provide erroneous data because of oil separation in the sample containers or even in automated continuous on-line analyzers.

Considerable improvement has been made in continuous automatic oil analyzers but such instrumentation has several drawbacks for use in typical oilfield operations. These include high initial and maintenance costs, the need for multiple testing units in large fields, and the chronic problem of fouling by oily solids.

General practice is to periodically collect multiple grab samples during an operating day and eventually accumulate sufficient data to establish a statistical daily average oil concentration. Analysis for oil content may be run on site but samples are usually submitted to a local analytical laboratory. Samples should be taken on a regular schedule both up and downstream of any type oil separation equipment to monitor removal efficiency. A recommended sample collection procedure is outlined in Appendix 6A as an aid in improving reliability of grab sample analysis.

5.13 Oily Waste Disposal and Injection Well Cleaning

All of the various methods for separation of oil and oily solids discussed in preceding sections of this chapter produce some volume of relatively concentrated waste. Free oil and associated solids may be diverted to the lease stock tanks and eventually disappear into the pipeline. High water content backwash from filters or coalescers and separated sludges from flotation units must be eliminated in some other way.

Some locations are fortunate in having one or more high capacity disposal wells that will accept heavily contaminated water. The disposal wells can be repeatedly fractured to maintain injectivity if the location, thickness and lithology of the disposal formation permits. However, the water base waste (backwash and flotation overflow) is often simply recycled back through the system and eventually ends up in the primary waterflood injec-

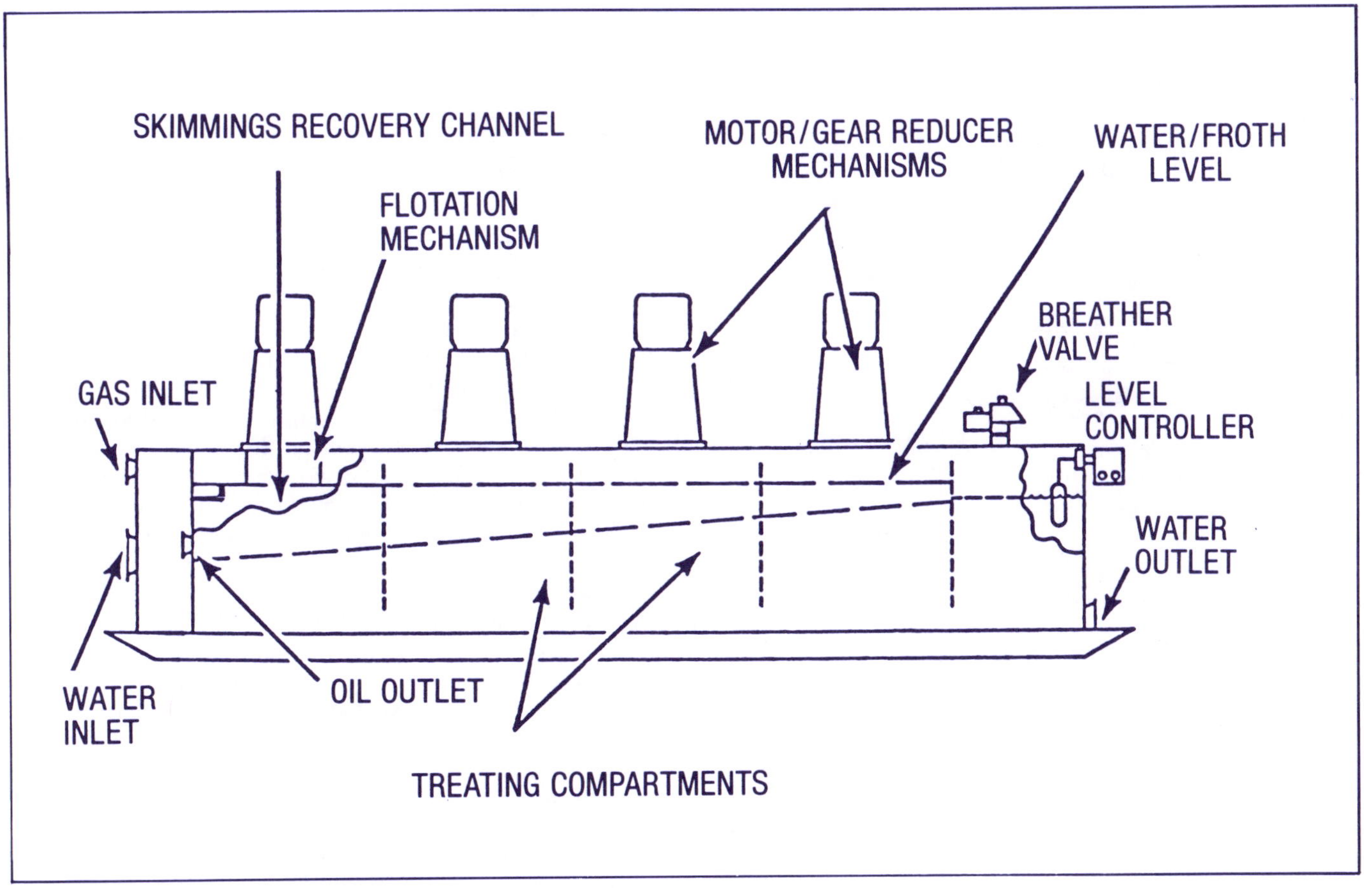

Figure 2.5-8. Induced gas flotation machine for water clarification. (Courtesy Petreco Division, Petrolite Corporation).

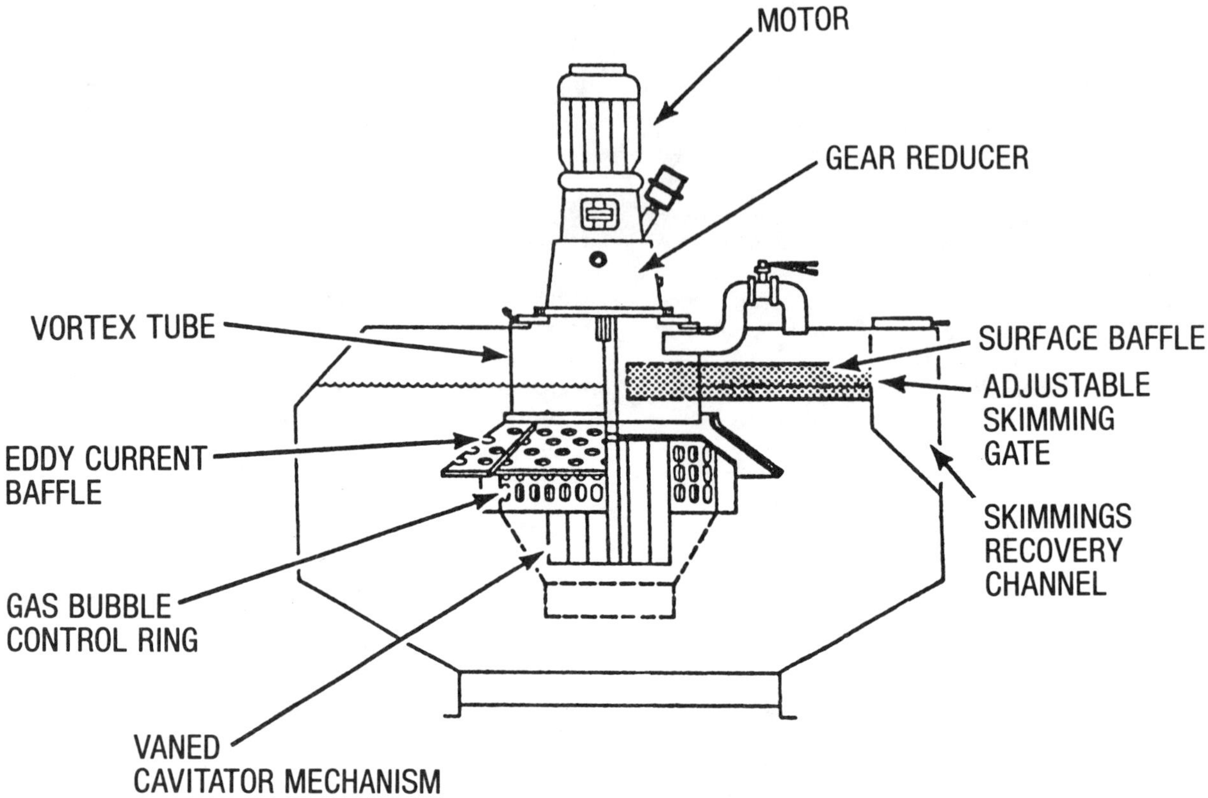

Figure 2.5-9. Induced gas flotation cell mechanism. (Courtesy Petreco Division, Petrolite Corporation).

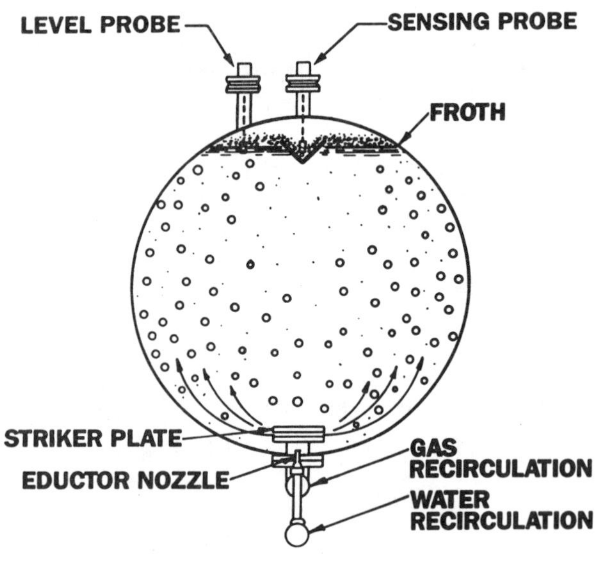

Figure 2.5-10. Static induced gas flotation cell. (Courtesy L'eau Claire Systems, Inc.).

tion wells during periods of upset. About the only alternative is controlled incineration, which is too expensive for most oilfields.

Gradual bleed-off of oily solids into waterflood injection wells is not necessarily all bad, especially if the producing formation is naturally fractured. Slow fillup of conduit fractures may indeed improve the flood pattern and conformance. The situation that should be avoided most is intermittant entry of concentrated slugs of solids that can bridge and plug near the injection wellbores.

When formation face plugging does occur to the extent that injection pressure approaches fracture pressure or intake capacity becomes too low, the injection wells must be cleaned and stimulated. Simple backwashing is often ineffective and may actually prove detrimental. An additional problem is that oil and oily solids in the backwash water must be separated for disposal by some means. Ordinary acid treatments of injection wells do not dissolve heavy hydrocarbons in the accumulated well sludge deposits and frequently provide only temporary improvement.

The best answer appears to be periodic cleaning by injecting limited volumes of acid containing high concentrations (10-15 vol %) of micellar surfactant solvents such as NL Treating Chemicals CheckerSol additive. These materials are capable of solubilizing heavy oils and other organics and dissolving most of the usual residue such as corrosion products and carbonate scale particles. After a soak period (12-24 hours), the solubilized, dissolved and dispersed plugging material is swept away from the injection wells by resumed water injection.

The micellar solubilizing chemicals are now being used extensively at lower concentration (5 vol %) for acid stimulation of producing wells. In badly fouled water injection wells, higher concentrations of the additives (10-15 vol %) promote more complete solubilization of oily sludges and extend the time period between treatments. In fact there is growing evidence that a regular program of injection well cleaning with the acid/surfactant solvents may prove more cost effective than installation and operation of surface water treating facilities such as gas flotation. This is especially true where there are no adequate means for disposing of the concentrated waste discharge from the surface treating equipment.

Conversion Factors

U.S. to metric factors:

To convert from	To	Multiply by
acre	hectare	0.40469
acre	metre^2	4,046.9
acre-feet	metre^3	1,233.5
barrel (42 US gal)	metre^3	0.15899
barrel/day (1 kg/dm^3)	tonne/annum	58.0304
barrel/US ton	metre^3/tonne	0.17525
British Thermal Unit	joule	1,055.1
BTU/bhp-hour	watt/kilowatt	0.39301
BTU/foot^3	kilojoule/metre^3	37.2590
BTU/US gallon	kilojoule/metre^3	278.716
BTU/minute	watt	17.5843
BTU/hr-foot^2 °F	watt/metre^2 °C	5.67826
BTU/pound mass	joules/gram	2.32600
centipoise	pascal-second	0.00100
centistoke	millimetre^2/second	1.00000
darcy	micrometre^2	0.98692
foot	metre	0.30480
foot^3	liter	28.1361
foot^3	metre^3	0.02832
foot^3/day	metre^3/day	0.02832
foot^3/pound	metre^3/kilogram	0.06243
foot-pound-force	joule	1.35582
foot-pound/minute	watt	0.02260
gallon US liquid	liter	3.78531
gallon US liquid	metre^3	0.00379
grain/gallon US	gram/metre^3	17.1181
horsepower	kilowatt	0.74570
horsepower-hour	megajoule	2.68452
inch	millimetre	25.4000
inch^2	millimetre^2	645.160
inch^3	millimetre^3	16.3871
kilowatt-hour	megajoule	3.60000
mile, nautical	kilometre	1.85200
mile, US statute	kilometre	1.60934
mile^2 US statute	kilometre^2	2.58999
mile (US stat.)/hour	kilometre/hour	1.60934
ounce force av	newton	0.27801
ounce mass av	gram	28.3495
ounce US fluid	centimetre^3	29.5735
ounce US fluid	liter	0.02957
pint US dry	centimetre^3	550.611
pint US liquid	centimeter^3	473.177
pound force	newton	4.44822
pound mass av	kilogram	0.45359
pound force/inch^2	kilopascal	6.89476
pound mass/foot^3	kilogram/metre^3	16.0185
pound mass/US gallon	gram/centimetre^3	0.11983
ton force US	kilonewton	8.89644
ton mass US short	kilogram	907.185
ton mass US short	tonne	0.90718

Equivalent metric to U.S. factors:

To convert from	To	Multiply by
hectare	acre	2.47105
metre^2	acre	0.00025
metre^3	acre-feet	0.00081
metre^3	barrel (42 US gal)	6.28981
tonne/annum	barrel/day (1 kg/dm^3)	0.01723
metre^3/tonne	barrel/US ton	5.70602
joule	British thermal unit	0.00095
watt/kilowatt	BTU/bhp-hour	2.54443
kilojoule/metre^3	BTU/foot^3	0.02684
kilojoule/metre^3	BTU/US gallon	0.00359
watt	BTU/minute	0.05687
watt/metre^2 °C	BTU/hr-foot^2 °F	0.17611
joules/gram	BTU/pound mass	0.42992
pascal-second	centipoise	1,000.0
millimetre^2/second	centistoke	1.00000
micrometre^2	darcy	1.01325
metre	foot	3.28084
liter	foot^3	0.03531
metre^3	foot^3	35.3147
metre^3/day	foot^3/day	35.3147
metre^3/kilogram	foot^3/pound	16.0185
joule	foot-pound force	0.73756
watt	foot-pound/minute	44.2539
liter	gallon US liquid	0.26417
metre^3	gallon US liquid	264.173
gram/metre^3	grain/gallon US	0.05842
kilowatt	horsepower	1.34102
megajoule	horsepower-hour	0.37251
millimetre	inch	0.03937
millimetre^2	inch^2	0.00155
millimetre^3	inch^3	0.06102
megajoule	kilowatt-hour	0.27778
kilometre	mile, nautical	0.53996
kilometre	mile, US statute	0.62137
kilometre^2	mile, US statute	0.38610
kilometre/hr	mile (US stat.)/hour	0.62137
newton	ounce force av	3.59694
gram	ounce mass av	0.03527
centimetre^3	ounce US fluid	0.03381
liter	ounce US fluid	33.8141
centimetre^3	pint US dry	0.00182
centimetre^3	pint US liquid	0.00211
newton	pound force	0.22481
kilogram	pound mass av	2.20462
kilopascal	pound force/inch^2	0.14504
kilogram/metre^3	pound mass/foot^3	0.06243
gram/centimetre^3	pound mass/US gallon	8.34541
kilonewton	ton force US	0.11240
kilogram	ton mass US short	0.00110
tonne	ton mass US short	1.10231

(Courtesy Oil, Gas & Petrochem Equipment, July 1986.)

Specific Gravity (sp.gr.) = Approximate g/cm^3

$$\text{Specific Gravity} = \frac{141.5}{131.5 + {}^\circ\text{API}} \qquad {}^\circ\text{API} = \frac{141.5}{\text{sp.gr.}} - 131.5$$

Temperature

APPENDIX 1A

Simplified Outline of Corrosion Theory

Introduction

This brief section is presented for benefit of persons desiring a somewhat more theoretical treatment of corrosion processes than that presented in the main text. Being a summary, no attempt is made to teach fundamentals of the mathematics or physical chemistry involved.

1A.1 Basic Definitions

Table 1A-1 defines several of the electrical, physical and chemical quantities that will be considered in the following discussion. The definitions are primarily for review and orientation. The reader is referred to standard physical chemistry textbooks for more complete explanations of the quantities. References for further study of the principles described in this section are provided.

Table 1A-1
Definitions

Quantity	*Unit*	*Symbol*	*Description*
Electromotive Force (EMF or Potential)	Volt	E	Force or pressure that drives an electric current through a closed circuit. $E = IR$
Current	Ampere	I	Amount of current sent by one volt through a resistance of one ohm. $I = \frac{E}{R}$ (Ohm's Law)
Resistance	Ohm	R	Property of a circuit that retards or reduces current flow. $R = \frac{E}{I}$
Electrical Quantity	Coulomb	C (or Q)	Amperes × Time (seconds)
	Faraday	F	96,500 Coulombs. Quantity of electricity that will deposit or dissolve one equivalent weight of any substance such as a metal.
Equivalent Weight	(Ratio)	e	Atomic or molecular weight divided by valence.
Electrochemical Equivalent	(Ratio)	k	Grams per coulomb that is electrochemically oxidized or reduced at 100% efficiency. $k = \frac{e}{F}$ (for metals)
Redox Potential	Volt	E°	Standard half cell potential under equilibrium (zero current) conditions. See Table 1.1-1.
Corrosion Potential	Volt	E_{corr}	Potential of corroding metal vs. potential of reference electrode.
Corrosion Current (Proportional to Corrosion Rate)	Ampere	I_{corr}	Amount of current flow equal to: $\frac{E_{corr}}{R}$ where R is the sum of all resistances in the cell.
Energy or Work	Joule	J (or W)	Volts × Coulombs. Change in free energy, ΔG, is a measure of electrical energy available from a system.
Gas Constant	(Ratio)	R	8.314 J/deg mole
Absolute Temperature	Degree	K	deg C + 273.16
Effective Concentration	Activity	q (or a)	Molality × Activity Coefficient

1A.2 Fundamentals of Electricity and Electrochemistry

Electric current flow is manifested as a form of energy that can easily be transformed into other, more useful forms such as heat or mechanical movement. The generally accepted physical concept of electricity is that of electrons moving predominantly in one direction, presumably from atom to atom, through a conducting substance. An electrical conductor is by definition a material through which electrons move with relative ease. The direction of electron flow must always be from a point of high potential to a point of low potential, and the potential difference, ΔE, between the two points is the driving force or voltage.

Any property or condition that opposes or diminishes electron flow can be considered and treated mathematically as resistance. Resistance in a conductor can be either ohmic, inductive or capacitive in nature but the Ohm's Law relationship applies precisely with proper interpretation.

Figure 1A-1 illustrates the relationship of E, I and R in a metallic circuit. For any value of E, a variable amount of current I will flow, depending on the slope, m, of the straight line representing R. Metallic conduction produces no change in the chemical properties of the conductor.

When part of the electrical circuit is an electrolyte such as salt water, electric energy moves through the liquid by means of both negative and positive carriers known as ions (electrically charged atoms or groups of atoms). Electron transfer (redox) reactions occur at the liquid/solid interfaces (see Section 1, Item 1.07).

A combination of two electrical conductors (electrodes) immersed in an electrolyte is called a cell. The total of the positive and negative current (resulting from + and − ion movement) in the electrolyte of a cell is always exactly equivalent to the total current carried in the metallic conductors by electrons alone. In contrast to metallic conduction, electrolytic conduction involves chemical reactions that occur at the electrode surfaces and matter changes form. *Electrochemistry is the science which deals with the conversion of electrical energy into chemical energy and vice versa.*

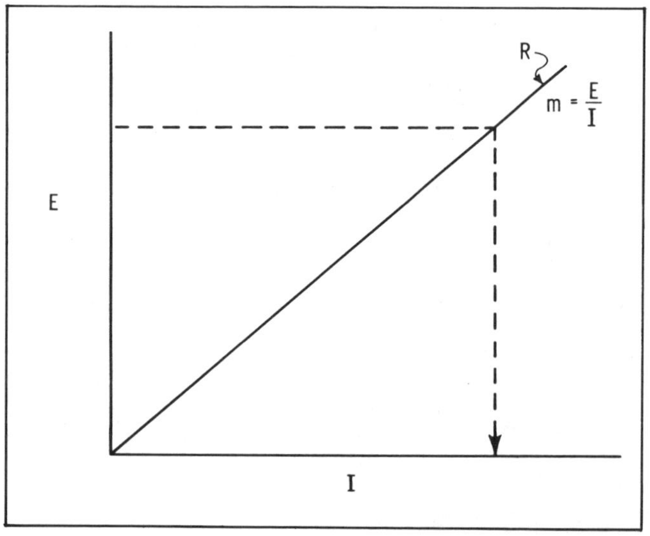

Figure 1A-1. Graphic representation of Ohm's Law.

Electrochemical reactions consist of two or more partial oxidation and reduction (half cell) reactions. The number of electrons released at one electrode is equal to the number consumed at the other electrode and there can be no net accumulation of electrical charge.

In an electrochemical cell such as a metal corroding in salt water, the chemical work done by the system per mol of reactant (metal) consumed is nFE, in which n is the valence change, F is the Faraday and E is the cell potential or emf. At the same time, the free energy of the system decreases by the amount G (Gibbs free energy change), and:

$$\Delta G = -nFE$$

The free energy change must be negative (a net energy loss) for the reaction to be spontaneous. Thus, the sign of the free energy change can indicate if a corrosion reaction is possible. However, such calculations are rarely made for practical corrosion studies since the magnitude of ΔG is *not* an indicator of corrosion rate.

1A.3 Sign Conventions of Electrochomical Reactions

When Luigi Galvani published his studies of electrochemical action in 1791, the structure of matter and the nature of electricity were not known. An arbitrary convention was established which envisioned flow of *positive* current from positive electrode to negative electrode. The cathode was designated as the electrode at which current *enters* from the electrolyte or earth and the anode as the electrode at which positive (conventional) current *leaves* to return to the electrolyte or earth. This direction of current flow is opposite to the real path of movement of electrons, as depicted in Figure 1.1-3, Section 1. *The concept of positive or conventional current flow is still used in corrosion engineering and is always implied when the sign of the current carrier is not designated* (see Section 1, Item 1.09).

Due to the confusion that persists regarding current flow direction, corrosion scientists and electrochemists now prefer to designate the anode of a cell as the electrode at which oxidation (electron loss) reactions predominate. The cathode is the electrode at which reduction or the consumption of electrons occurs (see Section 1, Item 1.07).

In spontaneous local corrosion cells and in galvanic (dissimilar metal) cells, the electrical energy is supplied by release of electrons (oxidation) of the corroding anode metal. As depicted in Figure 1.1-3, the cathode is the positive electrode because its potential is less negative than that of the anode when measured against a reference or third electrode. Electrons flow *through the metal path from anode to cathode* to sustain the electrochemical oxidation (anode) and reduction (cathode) reactions.

In electrolytic or impressed current cells, electrical energy is supplied by an outside source such as the generator illustrated in Figure 1A-2. In this situation, reduction (electron consuming) reactions occur at the electrode connected to the negative pole (electron source) of the outside generator. This electrode is therefore called the cathode. Corrosion (electrochemical oxidation) occurs at the anode connected to the positive pole of the generator. The + and − or positive and negative designations on the poles or terminals of an electrical generator or rectifier are used strictly as conventional electrical symbols. These electrical designations (+ and −) have no meaning with respect to the electrochemical reactions that occur at the electrode surfaces in contact with the electrolyte.

Electrons always *flow from the anode* through the generator *to the cathode* (same as in natural cells) and *conventional* current flows from anode to cathode through the connecting electrolyte or soil.

When the potential of an impressed current cell is sufficiently high, electrons can be supplied by anodic oxidation reactions other than corrosion of the anode metal (see Section 1, Item 1.07). Noble metals such as platinum and conducting non-metals such as carbon can serve

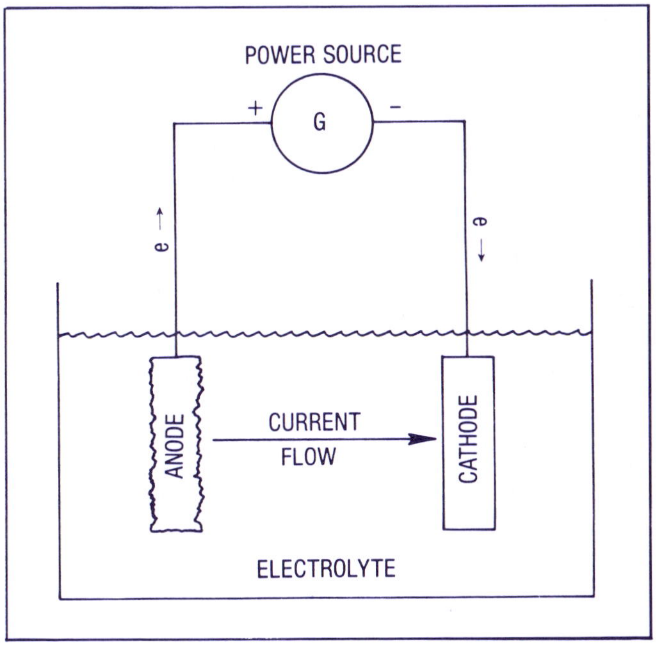

Figure 1A-2. Impressed current or electrolytic corrosion cell.

as anodes without corroding. In fact, impressed current cathodic protection systems often utilize non-corroding anodes such as graphite and platinum coated titanium. Free O_2 and Cl_2 gases are formed at the inert anodes because of electrochemical oxidation of H_2O and Cl^- ions.

1A.4 Measurement of Cell Potential

The oxidation-reduction (redox) reactions in an electrochemical cell are controlled by an electrical potential at each electrode and the potential of the whole cell, E_{cell}, is:

$$E_{cell} = E_{left} + E_{right}$$

where

E_{left} = half cell potential of the left hand electrode
E_{right} = half cell potential of the right hand electrode

The absolute value of E_{cell} is experimentally difficult to determine since the value changes as current is drawn from the cell.

In practice, the emf of a cell is measured with a zero current potentiometer using an internal standard or reference cell whose potential is known. When the potential of the whole cell, E_{cell} and the potential of either electrode is known, the potential of the other electrode (half cell) can be determined.

A common laboratory reference electrode (half-cell of known potential) is the standard hydrogen electrode (SHE). It consists of a platinized platinum electrode dipping into a solution of hydrogen ions whose activity equals 1.000 while hydrogen gas bubbles over the electrode at a pressure of 1 atmosphere. The potential of the standard hydrogen electrode is *arbitrarily* assigned the value of 0.000 volt, and all other reference electrodes and half-cells are compared to it.

1A.5 Standard Half Cell Potentials

Standard oxidation (redox) potentials, E_o, (such as listed in Section 1, Table 1.1-1) refer to the emf of cells in which one of the half cells is the standard hydrogen electrode (SHE). When a metal electrode surrounded by a solution of its own ion at unit activity, is found to be the negative pole of a galvanic cell of which the SHE is the other electrode, the potential of the metal-metal ion half cell is assigned a negative value. It is said to be negative to the hydrogen electrode. Metals above hydrogen in Table 1.1-1 will electrochemically oxidize (lose electrons) and go into solution replacing hydrogen. For example:

(a) $Zn^\circ + 2H^+ \rightarrow Zn^{++} + H_2$

This equation represents the difference in the following two half-cell reactions:

(b) $Zn^\circ \rightarrow Zn^{++} + 2e$ (anodic oxidation reaction)

and

(c) $2H^+ + 2e \rightarrow H_2$ (cathodic reduction reaction)

Experimentally it has been shown that the standard emf of reaction (a) is 0.763 v. The measured emf is assigned a negative (−) polarity.

Metals below hydrogen in the list do not easily force the displacement of hydrogen from solution (equation c) and their standard oxidation potentials are given a positive sign (+) for reference purposes.

1A.6 Non-Standard Potentials

The potential of a system in which reactants are not at unit activity, may be calculated by the Nernst Equation:

$$E = E_0 + 2.3 \frac{RT}{nF} \log \frac{a\,(\text{oxid})}{a\,(\text{red})}$$

where E is the actual half cell potential, E_o the standard redox potential, R is the gas constant, T is the absolute temperature, n is the number of electrons transferred, F is the Faraday constant, a (oxid) and a (red) are activities of oxidized and reduced species. It can be shown by the equation that as the amount of oxidized species, a (oxid), increase; E anode becomes more positive or less negative (see following discussion of polarization effects).

E_o for iron at 25°C, at unit concentration (1 gram-ion in 1000 g H_20), is approximately −0.44 volts vs. SHE (see Table 1.1-1). When Fe^{++} ion concentration is near zero, E can be −0.7 volts vs. SHE. When Fe^{++} ion concentration at the anode surface is very high, E can become more positive and drop to −0.2 volts vs. SHE.[1]

1A.7 Polarization

Electrode polarization is defined as the deviation from the open circuit (zero current) potential resulting from the passage of current. Passage of current alters the effective concentration of a (oxid) and a (red) in the Nernst Equation. As current flows, the anodic and cathodic potentials are displaced from their equilibrium or rest values and approach each other.

Electrochemical polarization is caused primarily by two factors affecting the reaction kinetics:

1. A slow step in the reaction sequence. This is called activation polarization.

and

2. A deficiency of reactant available to the electrodes as a result of slow diffusion.

   This is called concentration polarization and primarily affects only the cathode reduction reactions.

Equation (c) under paragraph 1A.5 above indicates reduction of hydrogen ions to molecular hydrogen. The reaction occurs in two parts:

(d) $2H^+ + 2e \rightarrow 2H^\circ$

(e) $2H^\circ \rightarrow H_2$ (slow step)

As part (d) occurs at the cathode, $E_{cathode}$ is diminished due to the slowness of part (e).

This is an example of *activation* polarization. If, however, a cathode depolarizer such as oxygen is present, part (e) does not occur and instead:

(f) $2H^+ + {}^1\!/_2O_2 + 2e \rightarrow H_2O$

which is a fast reaction. Thus, in the presence of oxygen, corrosion is rapid since $E_{cathode}$ remains close to the value of $E^o{}_{cathode}$. Cathode polarization is small and anode/cathode ΔE remains large.

The effective availability or rate of diffusion of H^+ ions to the cathode (equation d) is also limiting and this is an example of *concentration* polarization. One reason why acidic solutions are highly corrosive to iron is because concentration polarization is minimal at the cathode.

1A.8 Polarization Curves and Mixed Potential Theory

Graphical plots of electrode (half cell) potentials vs. corrosion current illustrate the corrosion process. Figure 1A-3 is a schematical polarization diagram depicting corrosion of pure zinc in hydrochloric acid. Note that I_o, for the H^+/H_2 cathode reaction, corresponds to E_o = 0.00 for the SHE half cell while I_o for the Zn^{2+}/Zn anode reaction, corresponds to the E_o = 0.76 value listed in the electromotive series (Table 1.1-1). This is the initial or open circuit potential for the zinc/zinc ion half cell.

In the overall reaction, zinc metal dissolves (corrodes) and H_2 gas is displaced from solution. The potentials of the anodic and cathodic reaction sites (local electrodes) are displaced as corrosion continues due to polarization represented by the straight lines labeled $Zn \rightarrow Zn^{2+} + 2e$ and $2H^+ + 2e \rightarrow H_2$. At the point of convergence of these hypothetical polarization curves, oxidation and reduction is occurring equally, viz.: zinc dissolution is equal to hydrogen evolution. The cell potential at this point is called the *mixed* or corrosion potential, E_{corr}. The corresponding corrosion current density, I_{corr}, indicates the rates of both zinc corrosion and hydrogen evolution.

The relationship between I_{corr} and corrosion rate can be calculated by Faraday's law:

$$M_w = kC$$

M_w = weight of metal loss, grams
k = electrochemical equivalent of the metal
C = coulombs, or I_{corr} (as amperes) × seconds

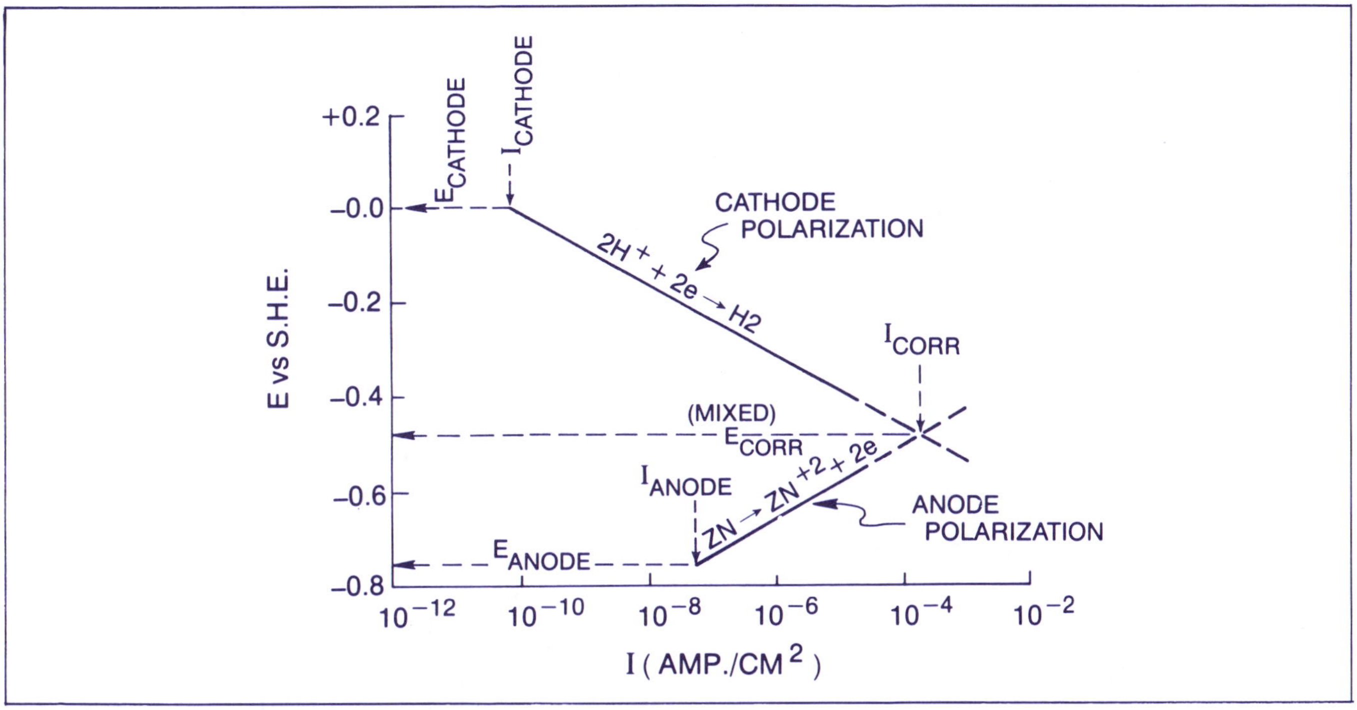

Figure 1A-3. Polarization diagram of zinc corroding in acid.

For zinc, k = 3.4 × 10^{-4} grams/coulomb. The k value for iron is 2.9 × 10^{-4} grams/coulomb. This translates to the value of 9.1 kilograms (20.1 lb)/amp-year discussed under Section 1, Item 1.21, for electrolytic corrosion of steel. A current density, I_{corr}, of 1 μamp/cm^2 roughly corresponds to a corrosion rate of 1 mil penetration/year or 5.5 mdd for most metals.[2]

Referring again to Figure 1A-3, it can be seen that the position of the point I_{corr} is determined by the slopes of the cathode and anode polarization curves. For example, if the slope of the cathode polarization curve is moved sharply downward, it would intersect the anode polarization curve at a lower value of I_{corr} and the rate of corrosion would decrease. This effect can occur if a chemical inhibitor is added that slows the cathodic reaction (H_2 generation). The cell is then said to be under cathodic control and the chemical additive would be classed as a cathodic inhibitor. The slope of either curve is computed as

$$\frac{\Delta E}{I_{applied}}$$

and thus has the mathematical dimension of R as in Ohm's Law (Fig. 1A-1). For this reason, any condition which retards polarization (a change in half cell potential E) is called polarization resistance. For example, it was shown in Item 1A.7 that presence of oxygen *retards* cathode polarization and, in this case, polarization resistance is high since E_{corr} remains high.

1A.9 Experimental Determination of I_{corr}

It has been established that in an active corrosion cell the rate of metal loss is proportional to I_{corr}, or it can be said that the corrosion rate is determined by corrosion current. Therefore, several procedures[3] have been developed to measure I_{corr} in order to prove the mixed potential theory and to rapidly measure corrosion rate under controlled conditions.

In one method, a metal specimen, called the working electrode, is placed in an electrolyte together with a reference electrode. Current is impressed onto the working specimen by creating a voltage difference between it and a third or auxiliary metal electrode in the same solution. The combination of the working electrode and the auxiliary electrode immersed in an electrolyte and connected to an outside power source is, therefore, an electrolytic cell. When the auxiliary electrode is made the anode, the working electrode becomes the cathode and is cathodically charged. Since this is an *electrolytic* cell, the cathodic charge or voltage is assigned a negative (−) polarity sign (see 1A.3). As current is incrementally increased, the voltage or potential of the working electrode is measured against the reference whose potential remains relatively constant. Each change in potential is plotted for each controlled change in current to establish the relationship $\frac{E}{I}$. This is called a potentiodynamic test in the cathodic direction. Application of current is continued until each additional increment of current pro-

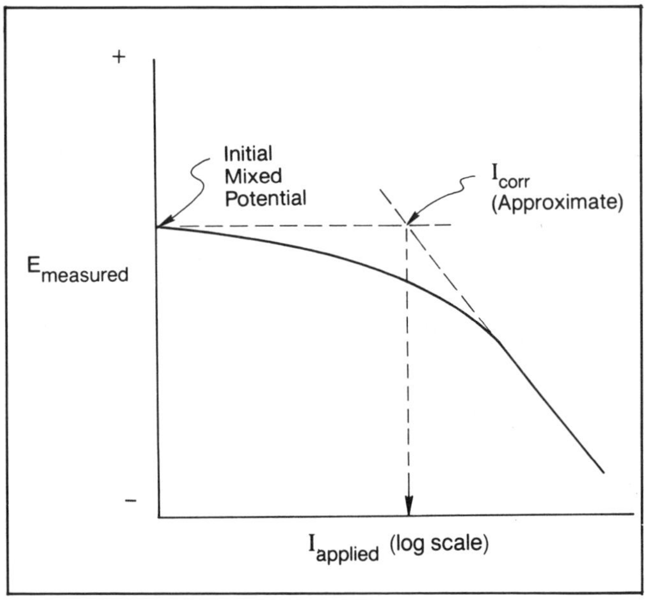

Figure 1A-4. E Log I Curve.

duces a large but consistent change in potential and the plot becomes linear. If E is plotted against the logarithm of applied current I, the result is called an *E log I curve* (see Fig. 1A-4). The bend or knee in the curve where the linear (Tafel) slope commences is taken as the point where $I_{applied}$ is equivalent to I_{corr}.

The E log I procedure has been applied in field tests where a well casing is the working electrode. The casing potential is measured at the well head against a copper-copper sulfate ($Cu/CuSO_4$) reference as current is incrementally applied from a power source such as a welding generator through an auxiliary anode bed. The test data are used to estimate cathodic protection current requirement.[4]

Such data plots represent the cathodic polarization curve for the specific working electrode (casing) and its specific environment. Note that in practice the data curve for $\frac{\Delta E}{I_{applied}}$ is not a well defined straight line and validity of the graphical determination of the value of I_{corr} (and thus the cathodic protection current requirement) is still being debated.[5]

1A.10 Linear Polarization

An alternative (and more accurate) procedure to the E log I method for finding I_{corr} is called linear polarization. This technique (also called polarization resistance) is important to petroleum operations as a very rapid way to measure corrosion rates in the field. A description of test instruments using this method (corrosion meters) is given in Section 1, Item 10.03.

The sensing elements of the corrosion meters may be simple cylinders of metal as shown in Figure 1.10-6. One cylinder acts as the working electrode and one is the auxiliary. A power supply in the instrument forces a voltage difference between the working and auxiliary electrodes. The auxiliary may also serve as a reference (Magna Corrater™) or a third metal reference electrode may be provided (Petrolite PAIR™). The applied current density required to shift the potential of the working electrode by 10-20 millivolts is measured by the instrument and converted to corrosion rate (usually as MPY).

The working electrode offers a certain reluctance or resistance (R) to the forced shift in potential and this effect (within 10-20 mv range) has been found to be proportional or *linear* to I_{corr}. The electrochemical relationship is expressed mathematically as:

$$\frac{\Delta E}{\Delta I} = k\, I_{corr}$$

where ΔE = a small potential change of a corroding specimen (usually 10–20 millivolts)
ΔI = the current required to produce ΔE (milliamperes)
k = proportionality constant, which depends on the valence and density of the metal and on the electrode area and spacing
I_{corr} = corrosion current

The term, $k\, I_{corr}$, has the units of resistance as dictated by Ohm's Law where:

$$R = \frac{E}{I}$$

Most field instruments for measuring instantaneous corrosion rates by the linear polarization or polarization resistance method are preprogramed to convert measured current at a fixed ΔE into mpy (mils penetration per year).

Some research has been done to adapt the linear polarization technique to well casing current requirement for cathodic protection.

1A.11 Passivity and Anodic Protection

In laboratory corrosion studies, polarization plots are derived under carefully controlled conditions for observing corrosion behavior of various metals in various environments. In typical experiments, polarization of the metal specimen is forced by application of either anodic or cathodic current from a second electrode. The data plots or polarization curves are usually not simple straight lines as was depicted in Figure 1A-3.

In the example plot shown by Figure 1A-5, a moderate level of corrosion is indicated by the initial value of I_o. This may be the corrosion rate of clean carbon steel in aerated salt water. As the potential E of the spec-

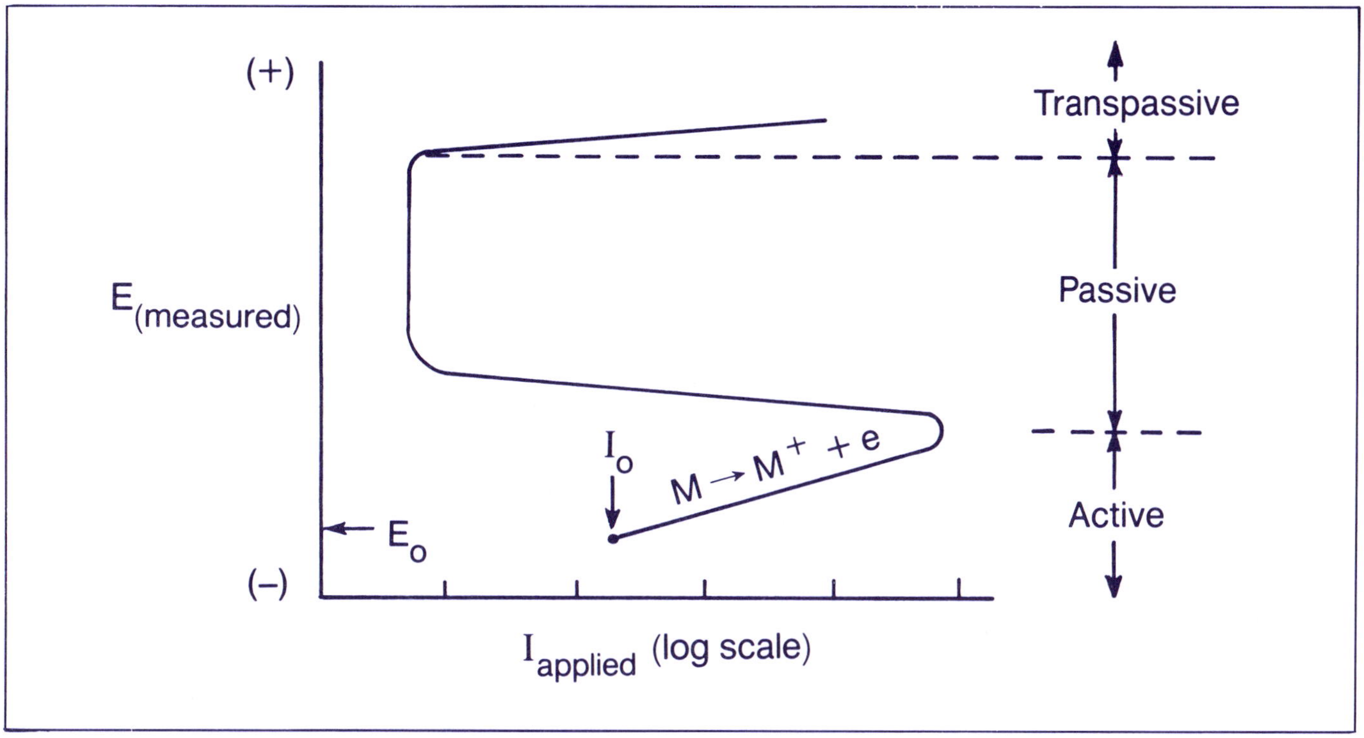

Figure 1A-5. Example of an Anodic Polarization Plot for an Active-Passive Metal.

imen is forced to a less negative (more anodic) range by applied current the corrosion rate increases along the line labeled $M \rightarrow M^+ + e$.

At and above a certain potential, the value of $I_{applied}$ (corresponding to corrosion rate) drops to a low level and remains as long as E stays within a certain range labeled Passive. This means that less anodic current is now required to shift and hold the corrosion potential E_{corr} than was initially required through the active region. An explanation for this effect is that the applied anodic current causes a very dense and complete layer of protective iron oxide to form on the metal surface. The metal is now anodically passivated and corrosion will remain low as long as the potential is held in the required range. This effect is called anodic protection. If the anodic current is interrupted and E drops to the active range, corrosion restarts. If E is forced too high, very aggressive corrosion takes place in the transpassive region.

Anodic protection is being applied to steel tanks holding concentrated sulfuric acid for example. Current and potential are constantly monitored by electronic equipment utilizing an inert electrode (cathode) mounted in the tank. This is an appropriate application of anodic protection because concentrated sulfuric is an oxidizing acid.

Anodic protection is under consideration for corrosion control on offshore drilling/production platforms in lieu of cathodic protection. However, maintenance of the passivating oxide surface films is difficult in sea water and there is real fear of catastrophic corrosion if the protection system fails even for a short period.

Certain stainless steels spontaneously passivate *without applied current* under oxidizing conditions (see Section 1, Item 1.15). A polarization plot for 18-8 stainless in aerated sulfuric acid may look similar to Figure 1A-5 except current density (indicative of corrosion rate) is natural and not applied.

1A.12 EMF-pH (Pourbaix) Diagrams

Thermodynamic calculations have been used to construct diagrams indicating equilibrium forms of metal compounds that are most stable (lowest free energy state) at specified conditions.[6] Figure 1A-6 shows a simplified EMF-pH diagram for the Fe-H_2O system where potential is plotted against pH value (portions omitted). The EMF is related to oxidation state and pH indicates acidity, both of which strongly influence solubility and phase stability.

In the sample diagram, the lower region marked Fe indicates essentially stable iron (no corrosion) as might be achieved by cathodic protection. The left upper portions shows ionic species being the thermodynamically stable forms of iron and thus implies corrosion. The upper right region labeled $Fe(OH)_x$ suggests that over the corresponding ranges of metal potential and solution pH, the iron should be anodically passivated.

In common with other thermodynamic approaches, the

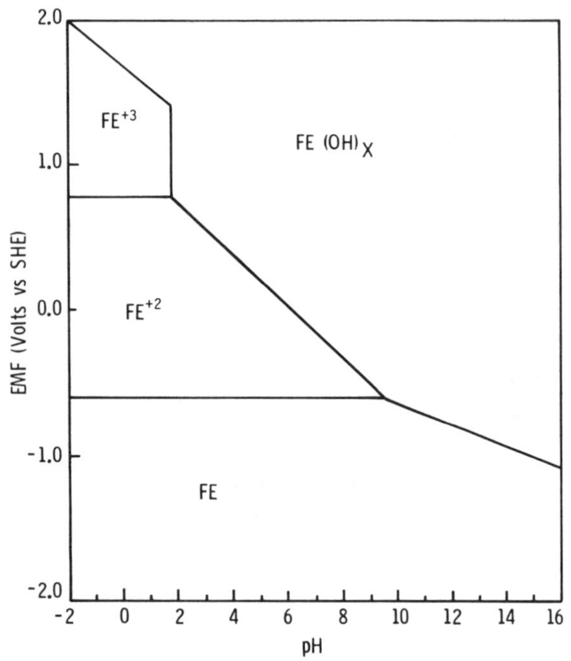

Figure 1A-6. Gross features of an EMF-pH diagram for Fe-H_2O system.

EMF-pH diagrams represent equilibrium conditions and should not be used to infer corrosion rates. However, the diagrams can be an aid in controlled corrosion research experiments. For example, Silverman employed Pourbaix (pronounced *poor-bay*)diagrams in the diagnosis of real in- plant corrosion problems involving titanium exposed to a concentrated acidic salt solution and cadmium in a system containing a corrosive complexing agent.[7]

This limited discussion of Pourbaix diagrams has been included simply to provide the reader a level of recognition when the diagrams are found in technical papers dealing with corrosion.

REFERENCES

(1) LaQue, F. L., and Copson, H. R., *Corrosion Resistance of Metals and Alloys*, Reinhold Publishing Corporation, New York, 1963.

(2) Fontana, Mars G., and Greene, N. D., *Corrosion Engineering*, McGraw-Hill Book Company, New York, 1967. (Revised 1979).

(3) American Society for Testing and Materials (ASTM) G3-74, Recommended Practice for Conventions Applicable to Electrochemical Measurements in Corrosion Testing. See also Recommended Practice:

G5-78 Standard Reference Method for Making Potentiostatic and Potentiodynamic Anodic Polarization Measurements.

G59-78 Conducting Potentiometric Polarization Resistance Measurements.

G69-78 Measurement of Corrosion Potentials of Aluminum Alloys.

(4) Ewing, S. P., and Bayhi, J. F., *Corrosion*, vol 4, no. 6, 1948.

(5) Husock, Bernard, *Materials Performance*, January, 1984, p 39.

(6) Pourbaix, M., *Atlas of Electrochemical Equilibria in Aqueous Solutions*, Cebelcor, NACE, 1974.

(7) Silverman, David C., "Application of EMF-pH Diagrams to Corrosion Prediction," *Corrosion* (NACE), vol 38, no 10, October, 1982.

OTHER SUGGESTED READING

Corrosion Basics, An Introduction, NACE, Houston, TX, 1984.

Corrosion and Its Control, Atkinson and VanDreffelaar, NACE, 1982.

Corrosion and Corrosion Control, Uhlig, John Wiley and Sons, Inc., New York, 1971.

APPENDIX 2A

Additional Information on Corrosion Inhibiting Chemicals

2A.1 Oxygen Corrosion Inhibitors

Because oxygen corrosion or rusting is so common and obvious to everyone, it is natural that some of the first compounds used for corrosion control were materials added to boilers and cooling water systems exposed to air. Early day operators of steam generators and boilers treated their water with wood chips or bark to reduce both corrosion and scaling. The tannic acid and other wood extracts consumed some of the oxygen and also acted as rust inhibitors.

Later it was discovered that complex salts such as sodium chromate (Na_2CrO_4), sodium phosphates (Na_3PO_4) and sodium nitrite ($NaNO_2$) arrest oxygen corrosion especially when alkalinity is increased with caustic (NaOH). Complex phosphates such as sodium hexametaphosphate ($Na_6P_6O_{18}$) were developed about 50 years ago and are still used in cooling and boiler water treatment.

More recently zinc salts of organic phosphonic acids have been employed in the treatment of aerated cooling waters for replacement of chromates. Use of chromates is declining because of toxicity and pollution control regulations. Sodium molybdate (Na_2MoO_4) is also being used as a chromate replacement.

The exact mechanism of inhibition by these materials is not yet understood. Chromates, nitrites and molybdates apparently act by promoting formation of stable iron oxide films on steel. The films are electrically conductive which indicates they do not inhibit by increasing electrical resistance. Further, the oxide films are cathodic to the underlying steel surface. For example, when chromate inhibitor concentration is too low, the chromium containing film actually promotes corrosion at areas (anodes) where the film is disrupted or broken. When chromate concentration is high, the entire coated metal surface behaves as a noble cathode, i.e., it is passivated (see Appendix 1A, Item 1A.11).

Low-hardness waters containing some dissolved iron can be inhibited with silicate salts to prevent oxygen corrosion. In this case, the silicate salts themselves do not form the protective iron oxide film but evidently strengthen it.

The example of zinc salts of orgánic phosponic acids is a case where both the anion (phosphonic acid) and the cation (Zn^{++}) help to form a protective inhibitor film. Zinc is used as an adjunctive inhibitor, not only with the organic phosphonic acids, but with inorganic phosphoric acid as well. Ternary mixtures of zinc, phosphate or phosphonate, and chromate salts form highly effective oxygen corrosion inhibitors in recirculating cooling systems.

An example of organic anionic (negatively charged) oxygen corrosion inhibitors, is the salts of organic acids. Typically the organic acids have in their molecular structure, acid groups such as:

$$R - COOH \text{ (Carboxylic)} \quad \text{or} \quad R - SO_3H \text{ (Sulfonic)}$$

R represents a hydrocarbon chain or ring.

For rust prevention in cutting oils and lubricants the sulfonic acids have proven to be one of the best of this class of chemicals. Their structure can be represented as follows:

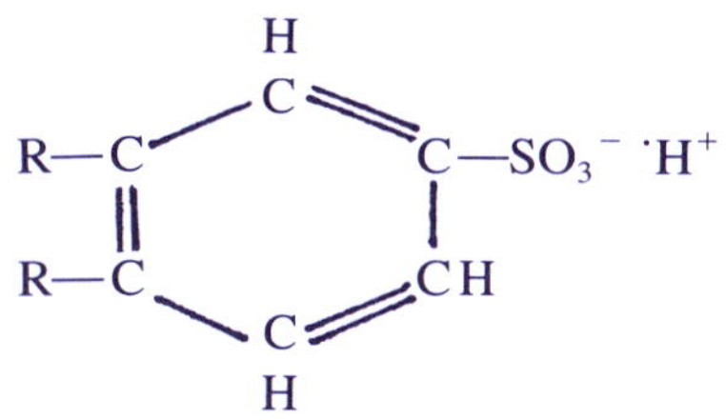

Here, R is a fatty or alkyl chain attached to the ring. Without the $-SO_3^-$ group, the alkyl substituted ring compound is simply an oil or heavy hydrocarbon. When crude oil is treated with strong sulfuric acid (H_2SO_4), some sulfonic acids are formed which can be separated. Their salts are the "petroleum sulfonates." The $-SO_3Na$ group, formed by neutralizing the sulfonic acid with caustic (NaOH), makes the oily molecule more water soluble or hydrophilic. Thus, the final material is made up of a large, oily, hydrophobic (or lipophilic) residue and the water loving hydrophilic solubilizing group. The hydrophobic-hydrophilic balance determines the solubility of the material. In a sense, the material is too oily to stay dissolved in water and too hydrophilic to remain in true solution in oil. Therefore, it accumulates at oil/water interfaces and at liquid/solid interfaces such as the surface of a metal immersed in water.

Another good anionic organic inhibitor of oxygen corrosion is mercaptobenzothiazole or MBT. It is usually employed as a sodium salt:

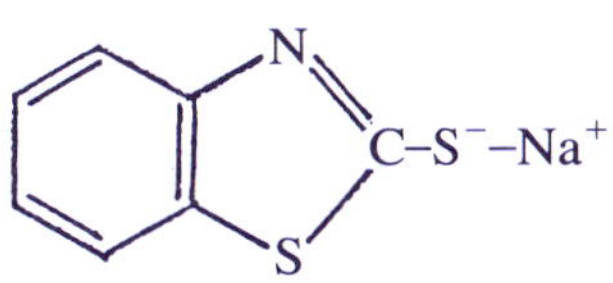

The carbon atoms in the rings are omitted here for sake of simplicity.

In practice, the inhibitor chemicals are rarely used alone.

Literally thousands of combinations in various proportions have been employed in commercial inhibitor preparations.

In open cooling water systems, the total concentration of mixed inhibitors is usually held within the range of 50 to 100 mg/L (ppm) by frequent additions. In plant operations, the cooling water treatment program is normally controlled by a contract water service company.

In closed systems, such as automotive radiators, a one time treatment of as much as 0.5 weight percent is applied.

2A.2 Cationic and Mixed Charge Organic Inhibitors

A majority of the inhibitor compounds used for treating oil and gas wells contain one or more of the amines discussed in Section 1, Item 3.02. The amine functional groups are attached to large alkyl hydrocarbon chains or form a part of imidazoline ring structures to which long alkyl chains are attached:

a. $CH_3(CH_2)_nNH_2$ — Primary monoamine
n = 11 to 17 typically (alkyl chain)

b. $RNH(CH_2)_3NH_2$ — Diamine with primary and secondary amine groups
$R = CH_3(CH_2)_n^-$
n = 11 to 17

c. 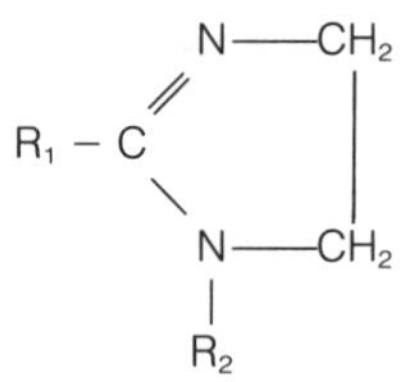 — Imidazoline with tertiary amine group

$R_1 = CH_3(CH_2)_n^-$
R_2 = Same or may be H^+, $-(CH_2)_nNH_2$, etc.

The amine groups are alkaline (basic) and readily take on hydrogens ions (protons) to assume a net positive charge:

$$\underset{\text{(amine)}}{RNH_2} + \underset{\text{(hydrogen ion)}}{H^+} \rightarrow \underset{\text{(protonated amine)}}{RNH_3^+}$$

Hydrogen ions are, of course, readily available from water and acids. The protonated amine is a cationic (positively charged) surfactant which strongly adsorbs to metal surfaces.

A particularly effective inhibitor composition is formed when a long chain diamine is protonated with an organic acid to form an amine-acid salt or soap:

$$[RNH_2\,(CH_2)_3NH_3]^{++} \cdot 2[RCOO]^-$$

Here we have both cationic (positive) and anionic (negative) functionality. Due in part to the anionic functionality, these mixed charge inhibitors can perform better in the presence of oxygen than the unreacted amines alone do.

The amine-acid salts and the imidazolines are the two products most widely used for preparing general purpose oil soluble and water dispersible oilfield inhibitors (see Section 1, Item 3.06). Treating procedures are listed in Section 1, Item 8.11.

2A.3 Reference List of Inhibitor Chemicals

Table 2A-1 lists many of the most widely used inhibitor chemicals according to their group classifications:

Table 2A-1
Inhibitor Chemicals

Group Class	*Compound*		*Applications*
Inorganic Anionic	Phosphates		Boiler feedwater, with caustic (NaOH) and dispersants
	Sodium ortho	(Na_3PO_4)	
	Sodium pyro	$(Na_4P_2O_7)$	
	Sodium tripoly	$(Na_5P_3O_{10})$	Cooling water, with phosphonates, chromates and dispersants
	Sodium hexameta	$(Na_6P_6O_{18})$	
	Chromates		Cooling water, with phosphates, phosphonates and dispersants (Use limited by toxicity)
	Sodium chromate	(Na_2CrO_4)	
	Sodium dichromate	$(Na_2Cr_2O_7)$	
	Sodium molybdate	(Na_2MoO_4)	Use same as chromates
	Silicates		Low salinity (fresh) water distribution systems
	Sodium meta	(Na_2SiO_3)	
	Sodium ortho	(Na_4SiO_4)	
	Borates		Automotive cooling systems, with phosphates, nitrite, and MBT
	Sodium meta	$(Na_2B_2O_4)$	
	Sodium tetra	$(Na_2B_4O_7)$	
	Nitrite		Same as Borates
	Sodium	$(NaNO_2)$	
Organic Anionic	Sulfonates Alkyl Aryl $(R \cdot SO_3^-$, Na^+ or $Ca^{++})$		Cutting/cooling oils Lubricants Hydraulic Fluids
	Mercaptobenzothiazole (MBT) Sodium MBT		Automotive (see Borates) Glycol dehydration systems (see Section 1, Chapter 8, Item 8.15)
	Triazoles Benzo Tolyl		Antifreeze inhibitors for protection of copper
	Thioureas Alkyl	$(RNH_2)_2CS$	Drilling fluid additives Detergents (concentrated)
	Acetylenic Alcohols Propargyl Hexynol	$(RC \equiv CROH)$	Acid inhibition, with complex amines
	Phosphates Amino, polyalkyl -sodium salts	$N(RPO_3^{=} \cdot Na_2^{+})_3$	Cooling water, with chromates, phosphates and zinc salts (primarily for scale inhibition and silt dispersion)
Organic Cationic	Amine/Acid Salts		Base materials for oilfield inhibitors
	Typically, high molecular weight diamines	$RNH(CH_2)_3NH_2$	
	combined with high molecular weight organic acids	$R \cdot COOH$	
	Imidazolines	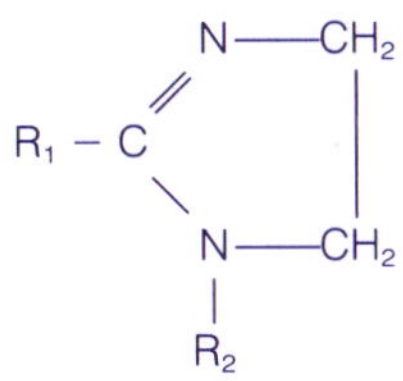	
	Amides	$R \cdot CONH_2$	
	Quaternary ammonium salts	$[R_4N]^+[Cl]^-$	

APPENDIX 3A

Scaling Tendency Calculation Examples

3A.1 Calcium Carbonate

(a) Refer to the water analysis of Sample 1, Table 2.1-6, Chapter 1, and the calculation procedure under Item 2.05, Chapter 2, Section 2.

CALCULATE

$$SI = pH - pH_s = +1.0 \text{ at } 50°C\ (122°F)$$

The value of pH_s for this water is computed as follows:

STEP 1—compute ionic strength (use Table 2.2-2)

Ion	mg/L	Ionic Strength
Na^+	1372	0.03 (includes K^+)
Ca^{++}	101	0.005
Mg^{++}	28	0.0023
Cl^-	2265	0.032
$SO_4^=$	24	0.0005
HCO_3^-	165	0.0013
	Total Ionic Strength =	0.0711

STEP 2

K from Figure 2.2-1 = 1.7
pCa from Figure 2.2-2 = 2.6
pAlk from Figure 2.2-2 = 2.6
pH_s = 6.9

STEP 3

$$SI = \underset{(pH)}{7.9} - \underset{(pHs)}{6.9} = +1.0$$

With an SI value of +1.0, the water (as represented by the analysis) has a definite tendency to precipitate $CaCO_3$ scale.

(b) Assume that the water is saturated in the water supply system with a gas containing 4 mol percent CO_2 at 5 bar (72.5 psi) total pressure.

What is the actual pH in the water supply system at 50°C?

What is the SI for $CaCO_3$ in the system?

See calculation procedure under Item 2.05, Chapter 2, Section 2.

STEP 1

$$R = \frac{\underset{(mg/L\ HCO_3^-)}{165} \times 0.82}{\underset{(mol\ fraction\ CO_2)}{0.04} \times \underset{(S_f\ from\ Fig.\ 2.2\text{-}3)}{1430}} = 2.37$$

STEP 2

Applying R to Figure 2.2-4, system pH is about 6.6 compared to the value of 7.9 shown by the analysis. Substituting this pH value for the analysis pH and using pH_s from (a) above

$$SI = \underset{(pH)}{6.6} - \underset{(pH_s)}{6.9} = -0.3$$

The water should not deposit $CaCO_3$ scale in the water supply system at 50°C.

(c) The following is an alternate digital method for calculating pH_s without use of charts:

Table 3A-1
Determination of Saturation pH by Hand Calculator[1]

Calculation Procedure

1. Enter ppm Ca as $CaCO_3$
2. Multiply by total alkalinity as ppm $CaCO_3$
3. Take log of the product, store in memory, and clear
4. Enter water temperature, degrees Fahrenheit
5. Multiply by temperature coeff.: $T < 77° = 0.0142$, $T > 77° = 0.00915$
6. Add memory to product, store sum in memory, and clear
7. Enter total dissolved solids in mg/L, and take log
8. Multiply by 0.10, add constant per temperature: $T < 77°$, 12.65; $T > 77°$, 12.27
9. Subtract memory, round off to nearest hundredth as pH_s

Example

Ca = 132, Total Alk. = 107, T = 55° F, TDS = 296 mg/L
log 132 × 107 = 4.15
55 × 0.0142 = 0.78 (+4.15 = 4.93)
log 296 × 0.1 = 0.25 + 12.65 = 12.90 − 4.93 = 7.97; = pH_s

[1]Excerpt from the titled paper by J. R. Hefler, NACE *Materials Performance,* vol 22, no 4, April, 1983, p 20–22. (Courtesy NACE).

3A.2 Calcium Sulfate

Refer to the water analyses of Samples 2, 3, and 4, Table 2.1-6 and the calculation procedure under Item 2.07, Chapter 2, Section 2.

(a) Calculate SI values at 60°C for Sample 2 (seawater) and Sample 3 (connate water). Use meq/L.

	Sample 2	*Sample 3*
$[CaSO_4]$	$= 23.2$	8.5
ECl	$= 54.1 - 23.2 = 31$	$933.1 - 8.5 = 925$
NM	$= 484.5 + 111.0 = 596$	$1374.0 + 232.6 = 1607$
R	$= \frac{596}{31} = 19$	$\frac{1607}{925} = 1.7$
K_s	$= 23 - 19 = 4$	$8.5 - 1.7 = 6.8$
	Undersaturated at 60°C	Approximately saturated at 60°C
M	$=$	$\frac{1.7}{8.5} = 0.21$
S_{60}	$= 23.2\left(\frac{936}{596} + \frac{596}{2808}\right) = 41.3$	$1.2(6.8) = 8.2$
SI at 60°C	$= 23.2 - 41.3 = -18.1$	$8.5 - 8.2 = +0.3$

(b) Calculate SI value for a mixture containing 20 vol % seawater (Sample 2) and 80 vol % connate water (Sample 3). Calculate SI value for Sample 4 (produced water).

Composition of 20/80 mixture:

Na^+	$0.2(484.5) + 0.8(1374.0)$	$= 1196.1$ meq/L
Ca^{++}	$0.2(23.2) + 0.8(933.1)$	$= 751.1$ meq/L
Mg^{++}	$0.2(111.0) + 0.8(232.6)$	$= 208.3$ meq/L
Cl^-	$0.2(562.1) + 0.8(2521.1)$	$= 2129.3$ meq/L
$SO_4^=$	$0.2(54.1) + 0.8(8.5)$	$= 18.0$ meq/L

Note that the meq/L Cl^- of theoretical mixture compares with that of Sample 4 but meq/L $SO_4^=$ is twice that of Sample 4. The low $SO_4^=$ content of Sample 4 was noted in the discussion of Figure 2.1-5, Item 1.10, Section 2.

	20/80 Mixture	*Sample 4*
$[CaSO_4]$	$= 18$	9
ECl	$= 751 - 18 = 733$	$700.5 - 9 = 692$
NM	$= 1196 + 208 = 1404$	$1241 + 202.5 = 1444$
R	$= \frac{1404}{733} = 1.9$	$\frac{1444}{692} = 2.1$
K_s	$= 18 - 1.9 = 16$	$9 - 2.1 = 7$
	Over saturated at 60°C	Approximately saturated at 60°C
M	$= \frac{1.9}{18} = 0.11$	$\frac{2.1}{9} = 0.23$
S_{60}	$= 0.68(16) = 11$	$1.2(7) = 8$
SI at 60°C	$= 18 - 11 = +7$	$9 - 8 = +1$

(c) Formation temperature is 90°C (184°F). Compute solubility of $CaSO_4$ and SI in these waters at 90°C using Table 2.2-3.

20/80 Mixture	*Sample 4*
$0.82(11) = 9$	$0.82(8) = 7$
SI at 90°C $= 18 - 9 = +9$	SI at 90°C $= 9 - 7 = +2$

Precipitation of 9 meq/L $CaSO_4$ from the 20/80 mixture at formation temperature leaves $18 - 9 =$ 9 meq/L $CaSO_4$ which compares with the 9 meq/L $CaSO_4$ found by analysis in Sample 4, the actual produced water. With these calculations, it is apparent that the 20/80 mixture is supersaturated with $CaSO_4$ and that $CaSO_4$ scale had previously precipitated from the produced water (Sample 4) prior to analysis. (See also comments in the next to last paragraph of Item 1.10, Section 2.)

APPENDIX 4A

Field Determination of Approximate Bacteria Populations in Water (Modified from API RP38)

Procedure:

1. Obtain a supply of filled nutrient broth bottles[1] (sulfate-reducing and general aerobic types) and a supply of sterile disposable syringes (3 ml capacity).[2]

   Each complete test requires at least 12 bottles of each type and four syringes. If desired, the sulfate-reducing or general aerobic bottles may be used separately, depending on the type of organism that is of greatest interest.

2. Arrange the bottles in duplicate rows of 6 each. From left to right, label the bottles 1, 10^{-1}, 10^{-2}, 10^{-3}, 10^{-4}, 10^{-5} to indicate the dilution factor. Be sure to duplicate so there are two bottles for each dilution. If both sulfate reducing and general aerobic counts are being made, you will have 4 rows of 6 bottles each.

3. Pull the small circular metal tab off the top of each broth bottle. Leave the stopper seal ring in place. Do *not* open the bottles or break the rubber seal.

4. The water sample to be tested should be carefully collected to avoid contamination. Allow the water tap to flow at least 3 minutes to be sure a representative sample is taken.

   Collect the sample in a clean *sterile* bottle and label as to source, date, time and appearance. Prescription bottles are suitable.

   If sterile bottles are not available and you want to run the test on location, you may collect the sample directly into a sterile syringe from the flowing water stream.

5. To use the syringe, open the paper package and remove the syringe (with the plastic guard covering the needle). Tighten the needle with a clockwise twist. Pull the plastic guard straight off the needle. Do not twist. Do not touch the needle. *Be cautious*. The needle is very sharp and dangerous.

6. Draw water into the syringe from the water sample container or directly from the flowing stream. Collect at least 2 milliliters (cc) of water in the syringe.

7. Point the needle upward at eye level so the plunger in the syringe is visible through the clear plastic body. Tap the body of the syringe, if necessary, to cause all gas bubbles to rise. Slowly move the plunger upward displacing all gas and some water out of the needle. Stop the top of the plunger at the 1 ml (or cc) mark.

8. Carefully insert the needle through the exposed rubber seal of the first bottle labeled "1". Push the needle all the way in so it extends into the liquid.

9. Push the syringe plunger down all the way to displace the 1 ml of water into the bottle. *Keep the needle inserted*.

10. *Slowly* pull back on the plunger to fill the syringe and then *rapidly* push the plunger back down to produce mixing of the bottle contents. *Repeat* this step at least 5 times.

11. Draw about 2 ml of the mixed liquid from the first bottle and withdraw the needle. Repeat Steps 7, 8 and 9 and transfer 1 ml to the second bottle labeled 10^{-1}.[3] Repeat Step 10 and proceed to successively inoculate the remaining 4 bottles with 1 ml of liquid each. Be sure the liquid in each bottle is thoroughly mixed according to Step 10 before proceeding to the next bottle.

12. Discard the used syringe after inoculating the first 6 bottles. Use a new syringe for each group of 6 bottles and repeat Steps 1 through 11 for each group.

13. Store the inoculated bottles in a convenient place at a temperature of 35–45° C. After 5 days observe the phenol red broth bottles for color change from red to yellow. The bottles that have turned yellow will usually also look *turbid* when contents are *thoroughly shaken*. The color change *and* turbidity indicate positive growth of aerobic bacteria.

    By this time, some of the sulfate reducer bottles may have turned dark. However, these should be stored and observed after 21 days to be sure of a positive test. Anaerobic (sulfate reducer) bottles are considered positive when even slightly dark. Turbidity while remaining yellow is disregarded.

[1]Nutrient broth bottles are normally available from companies supplying biocides for oil field use.

Each bottle contains 9 ml of nutrient broth. One additional ml is added to each bottle during testing (10 ml total).

[2]Sterile disposable syringes may be obtained from medical or hospital equipment suppliers.

[3]1 ml of mixed solution from Bottle 1 contains 0.1 or 10^{-1} ml of original water. Thus, Bottle 2 is labeled 10^{-1}, Bottle 3 is labeled 10^{-2} and so on.

14. Estimate the most probable bacterial populations as follows:

Aerobic Bacteria Test		*Sulfate Reducing Bacteria Test*	
Bottle No. Positive After 5 Days	*Most Probable Population Cells/ml*	*Bottle No. Positive After 21 Days*	*Most Probable Population Cells/ml*
1	1–10	1	1–10
10^{-1}	$10–10^{2}$	10^{-1}	$10–10^{2}$
10^{-2}	$10^{2}–10^{3}$	10^{-2}	$10^{2}–10^{3}$
10^{-3}	$10^{3}–10^{4}$	10^{-3}	$10^{3}–10^{4}$
10^{-4}	$10^{4}–10^{5}$	10^{-4}	$10^{4}–10^{5}$
10^{-5}	$10^{5}–10^{6}$	10^{-5}	$10^{5}–10^{6}$

15. For subsequent tests of water from a single point source, the number of dilutions used may be increased or decreased as needed to bracket the usual count range.

APPENDIX 5A

NACE Standard TM-01-73
(Reprint)

NACE Standard TM-01-73
(1984 Editorial Revision)
Item No. 53016

technical practices committee

NATIONAL ASSOCIATION OF CORROSION ENGINEERS

NACE

Test Method

Methods for Determining Water Quality for Subsurface Injection Using Membrane Filters

Approved February 1973
Revision approved June 1976
Reaffirmed February 1983
Revision approved February 1984

National Association of Corrosion Engineers
P.O. Box 218340
Houston, Texas 77218
(713) 492-0535

Foreword

Corrosion engineers in the oil and gas producing industry are often charged with the responsibility of evaluating and controlling the quality of injection waters. Unfortunately, much of the data available is inadequate, misleading, or difficult to interpret. This Standard has been prepared by NACE Task Group T-1C-12, a component of Unit Committee T-1C on Detection of Corrosion in Oilfield Equipment, and is issued by NACE under the auspices of Group Committee T-1 on Corrosion Control in Petroleum Production to provide standard test methods for use in evaluating water quality for injection waters.

Section 1: General

1.1 This Standard describes two test methods for evaluating water quality for subsurface injection: Procedure A—Rate vs Cumulative Volume (for water quality monitoring) and Procedure B—Suspended Solids Test (for diagnosis or monitoring). The methods are intended to provide standardized water quality in the petroleum production industry. The test methods describe the apparatus required, test conditions, test procedures, reporting procedures, and supplementary tests. Interpretation of the results is beyond the scope of this Standard. The bibliography will supply a source of interpretation methods.

1.2 The injection behavior of subsurface formations varies widely, and interpretation of the results of water quality tests is empirical. Application of the results obtained by these tests, therefore, will be strongly influenced by the requirements of each subsurface injection project. This Standard provides procedures for monitoring water quality routinely to ensure that it meets previously determined empirical water quality requirements and supplies standard quantitative methods for use in diagnosing water quality problems and determining water quality for subsurface injection projects.

1.3 This Standard is applicable only when precautions are observed to ensure that the sample is representative of the water in the system of interest. It is not the purpose of this Standard to imply that the results or their interpretation may be arbitrarily applied to other water injection projects.

Section 2: Definitions

2.1 *Suspended solids,* as used in this Standard, are defined as the non-water, non-dissolved substances which exist in the water. These may typically include, but are not limited to, materials such as iron sulfides and oxides, precipitated carbonates and sulfates, sands and silts, oils and paraffins, and materials of biological origin. The suspended solids may also be considered as the materials in the water which may cause plugging and loss of injectivity in injection wells.

2.1.1 *Primary suspended solids* or *inline suspended solids* are those substances that exist in the water in the system at the time of sampling.

2.1.2 *Secondary suspended solids* include the primary suspended solids plus those that may be precipitated due to aeration, loss of dissolved gases, or aging of a water sample removed from the system.

2.2 *Membrane filter,* as used in this Standard, refers to porous disks composed of pure and biologically inert cellulose esters. Unless specified to the contrary, the membrane filters have a mean pore size of 0.45 μm ($\pm$0.02 μm), a diameter of 47 mm, a thickness of 150 μm ($\pm$10 μm), and an average total pore volume of approximately 80% of the total filter volume.

2.2.1 *Preweighed membrane filters* are those which have been weighed prior to the test.

2.2.2 *Matched weight membranes* are those obtained from the supplier in pairs with identical weights ($\pm$0.02 mg).

Section 3: Description of Test Methods

3.1 Rate vs Cumulative Volume Test (Procedure A)

3.1.1 This test method consists of passing a fixed volume of injection water through a membrane filter under constant pressure and measuring the flow rate and cumulative volume of water at intervals.

3.1.2 This test is primarily designed for monitoring injection water quality. A plot of the flow rate vs cumulative volume of water gives a general indication of the quality of the injected water.

3.2 Suspended Solids Test (Procedure B)

3.2.1 This test method consists of collecting samples of primary solids as they exist in a water system. The suspended solids from several liters of water are collected on a membrane filter in a manner that permits larger, more representative samples than those obtained from bottle samples.

3.3.2 This test method provides a simple procedure for collecting samples useful in the diagnosis of problems encountered in handling water for subsurface injection. The quantity of inline solids is determined. The composition of these solids also may be determined.

3.3 The suspended solids test and the rate vs volume test can be run concurrently.

Section 4: Test Apparatus

4.1 The test apparatus consists of the membrane filters, membrane filter holders, sample receiver, sample connections and fittings, and pressure regulation devices. As an alternative when the test cannot be run on line as in Figure 1, a sample reservoir may be used. Except as noted, the same apparatus is used for Procedure A and Procedure B.

4.2 Membrane Filter

4.2.1 The 47 mm diameter, 0.45 μm membrane filter as defined in Paragraph 2.2 should be used for the rate vs cumulative volume test (Procedure A). The diameter of the membrane filter is less critical when only suspended solids (Procedure B) are to be determined, but a fixed diameter for all comparative tests is desirable. The membrane should have a smooth surface to permit easy removal of filtered solids.

4.2.2 Matched weight membrane filters may be used to facilitate distinguishing of true suspended solids from dissolved solids remaining after evaporation of water held up on the filter. Since some waters are supersaturated in soluble salts such as $CaSO_4$, the matched weight membranes may give a more accurate representation of true

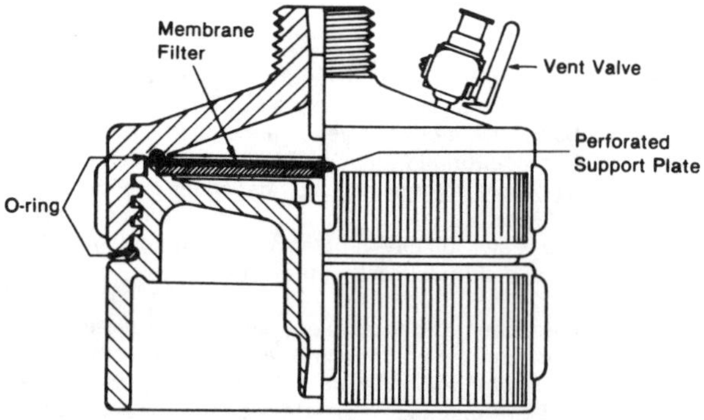

FIGURE 1 — Example of a filter for 47 mm membrane filter disk. Construction is of corrosion resistant materials capable of withstanding upstream pressure of 2000 psig. The holder is designed for quick connection to a water handling system to prevent detrimental aeration effects, and is also designed for rapid assembly or dismantling.

suspended solids in systems where these factors must be considered. Matched weight membrane filters (Paragraph 2.2.2) are placed in a filter holder, superimposing one of a pair on the other. The filter cake is deposited on the upper filter, and filtrate passes through both, leaving behind an equal amount of dissolved solids on both after evaporation of water held up by the filters.

4.3 Membrane Filter Holder

4.3.1 The membrane filter holder should be constructed of corrosion resistant materials. For safety, it should be capable of withstanding the pressure at the sampling point unless upstream pressure relief is provided.

4.3.2 The design should permit rapid assembling and dismantling. The holder should be capable of ready connection to the water system to prevent the effects of aeration. It may be constructed of clear material to permit observation of the membrane filter surface during the test. Gas is often trapped in the space over the filter during hookup. Provision should be made to remove this gas pocket before flow through the membrane is started.

4.3.3 The membrane filter holder should be constructed to prevent water from leaking around the edges of the membrane, and the filter support should be free of sharp edges which could perforate the membrane.

4.3.4 Figure 1 presents an example of a holder which embodies the basic requirements. It is not necessary to duplicate the illustrated device if the specifications in Section 4.3 are met. If filter diameters other than 47 mm are used for Procedure B, the holder must be modified accordingly.

4.4 Sample Receiver or Reservoir

4.4.1 A graduated cylinder may be used as a sample receiver when the equipment is connected as in Figure 2.

4.4.2 If the test is run as is illustrated in Figure 3, the sample reservoir should be constructed to withstand pressures above 20 psig (138 kPa). The reservoir should have a capacity of 3 to 4 liters (L) and be calibrated in 20 mL increments. It should be constructed of clear material to permit observation of the water level and measurements of the filtered volume. Reservoir components should be made of corrosion resistant materials.

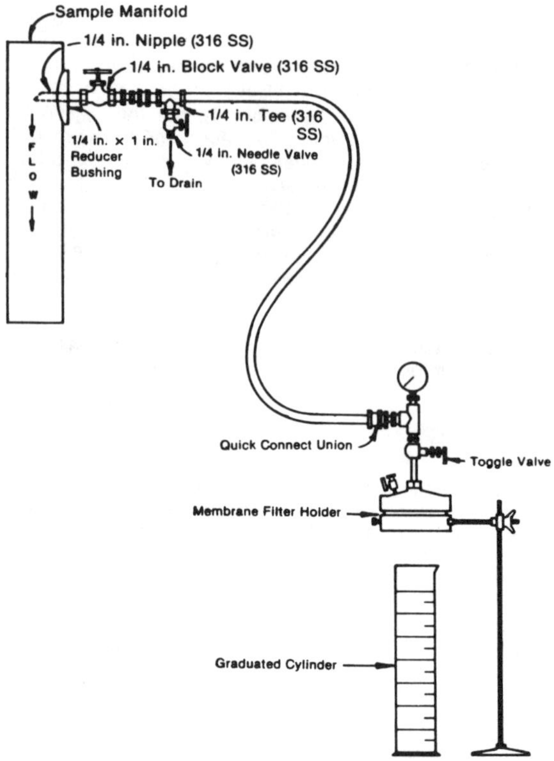

FIGURE 2 — Membrane filter test apparatus showing filter holder from Figure 1 connected to water handling system. The 1/4 in. block and needle valves near the top keep system pressure within set limits. Above filter holder is a quick-opening toggle valve to allow immediate full-stream flow vital to timing accuracy.

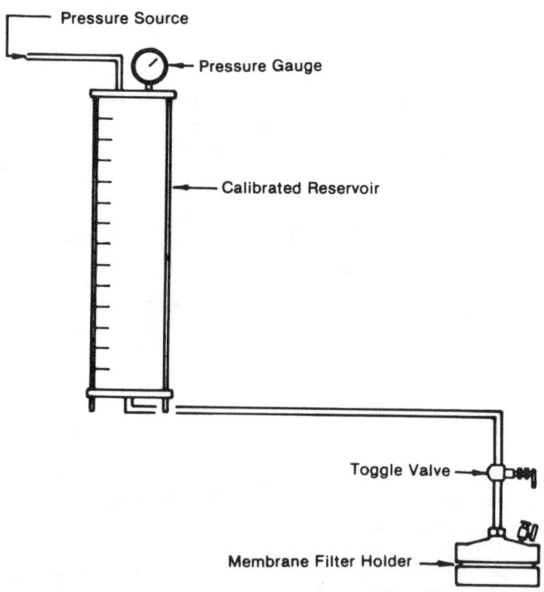

FIGURE 3 — Two stage apparatus with pressure gage and regulator for repressurizing and testing sample from reservoir rather than water handling system (Figure 2), used mainly when sample point cannot be easily adapted to on-stream application. Test conditions match Figure 2's when N_2 is used to raise pressure to 20 psig (see subsection 4.5.2).

4.5 Sample Connections, Fittings, and Pressure Regulation

4.5.1 Corrosion resistant fittings, valves, and tubing should be used to connect the apparatus. It is desirable to run the tests with the membrane filter holder connected directly to the system, using suitable pressure regulators and valves to set the test pressure (Figure 2). If the entire apparatus can withstand system pressure, no pressure relief is required. If, however, the system pressure is greater than the recommended operating pressure of any item of the apparatus, pressure relief should be provided with a pop valve, rupture disk, or other device.

4.5.2 If the reservoir (Figure 3) is utilized, a suitable two-stage pressure regulator will be required for repressurization with nitrogen.

Section 5: Preparation for Testing

5.1 Weighing (Taring) of Membranes for the Suspended Solids Tests (Procedure B)

5.1.1 The membrane filter should be dried and weighed prior to its use in Procedure B unless matched weight membranes are used.

5.1.2 The tare weights of membranes should be determined to the nearest 0.1 mg (constant weight).

5.1.3 These weights should be recorded on the container (petri dish, plastic case, etc.) for the individual membrane.

Section 6: Test Conditions

6.1 Cleanliness

6.1.1 Prior to each test run, all parts of the test apparatus must be clean.

6.1.2 The test points must be thoroughly flushed to ensure a representative sample.

6.2 Temperature

6.2.1 If the membrane holder is attached directly to the system, the temperature at the sample point shall be measured and recorded.

6.2.2 For a repressurized test, the temperature of the water in the sample reservoir shall be measured and recorded.

6.3 Pressure

6.3.1 The test pressure for rate vs cumulative volume should be 20 psig (138 kPa) ±10% at the membrane. Because permeability variations can occur with pressure-sensitive filter cakes, it is advisable to determine if a test conducted at higher constant pressure is more definitive of the effective water quality in the system.

6.3.2 Test pressure is obtained by suitable mechanical devices for on-stream sampling and by a pressure regulator between the nitrogen source and the sample reservoir for repressurized tests.

6.4 Sample Volume

6.4.1 For routine rate vs cumulative volume tests, the sample size shall be 2.5 L. Where water quality permits rapid filtration, samples up to 10 L or more may be required for meaningful interpretation. Sample volumes of 3 or 4 L are considered adequate to determine representative conditions in most systems using the suspended solids test.

6.4.2 Some waters are of such poor quality that almost complete plugging occurs before the recommended volume can be filtered. There is no worthwhile information to be obtained from continuing to measure effluent from the near-plugged membrane. In such instances the rate vs cumulative volume test should be terminated short of "test volume." The suspended solids test should be conducted on several membranes to provide an aggregate of 3 to 4 L of filtered solution. The results of the latter tests should be averaged.

6.5 Sample Points

6.5.1 Sampling must be done in a manner to provide a representative selection of the water to be tested throughout the complete system.

6.5.2 Although adequate and appropriate sampling points are seldom found in existing systems, planning for new systems should incorporate test sites as illustrated in Figure 2.

6.5.3 In existing systems, samples shall not be taken at high points which tend to collect entrained oil and drop suspended solids or at low spots which tend to lose oil and gain solids as compared to the mainstream.

Section 7: Test Procedures

7.1 Preparation

7.1.1 Carefully insert the membrane filter in the holder with tweezers to avoid damaging the membrane and to avoid bypassing water at an edge.

7.1.2 Moisten the membrane with filtered distilled water.

7.1.3 Purge the sample point and connecting tubing thoroughly to remove any sludge or deposits which may have accumulated in the valve or other fittings.

7.2 Repressurization

7.2.1 If it is not feasible to conduct an on-stream test, collect a water sample in the reservoir, after purging free of air, under an atmosphere of nitrogen. In some cases the pressure differential between the line and sample reservoir allows precipitation of solids that will invalidate conclusions concerning the quality of the water in the system if based on further data from a continuation of this test. In most cases this precipitation is visibly detected by rapid black or red coloration of the water.

7.3 Air Venting

7.3.1 Displace any air from the membrane holder with sample water by using the vent valve or by inverting the holder momentarily.

7.4 Procedures for Rate vs Cumulative Volume Test

7.4.1 Adjust the pressure to provide for a constant 20 psig (138 kPa) ± 10% at the membrane throughout the test. A higher constant pressure may be selected for pressure-sensitive filter cakes, or a series of tests at various constant pressures may be used (see Paragraph 6.3.1).

7.4.2 Initiate flow through the filter and start timing to coincide with passage of the first drop through the membrane.

7.4.3 Time and volume may be recorded in two ways: record the time for each 100 mL increment or record the volume at selected time increments. Continue the test until a specified volume is obtained or until 600 seconds have elapsed, whichever is first (see Section 6.4.).

7.4.4 Conduct two or more tests at each sample point.

7.4.5 Completion

7.4.5.1 Upon obtaining the required volume of filtrate, isolate the membrane holder from the pressure source.

7.4.5.2 Disconnect and dismantle the membrane holder and carefully remove the membrane with tweezers. If it is to be used in the suspended solids test, store the membrane in a container which will protect the filter cake from contamination.

7.4.5.3 The membrane filter should be kept moist with a wet blotter pad if it is to be sent to a distant laboratory for analysis. The moisture keeps the filter from adhering to the storage container as it dries out.

7.4.6 Calculations

7.4.6.1 Plot results from this test on semilogarithmic graph paper as illustrated in Figure 4.

7.4.6.2 Cumulative volumes and cumulative times recorded during the test are used to obtain the change in time (△t), the change in volume (△v), and the average rate (△v/△t) for each measured increment. The data are tabulated as in Table 1 and are graphically represented as in Figure 4. The cumulative volume in mL is plotted along the abscissa, and the rate in milliliters per second (mL/sec) is plotted on the ordinate.

7.5 Procedure for Suspended Solids Test

7.5.1 Timing is not required for this test.

7.5.2 Record the cumulative volume through the membrane.

7.5.3 Conduct two or more tests at each sample point.

7.5.4 Using tweezers, remove the membrane and filter cake from the container and prepare them for testing.

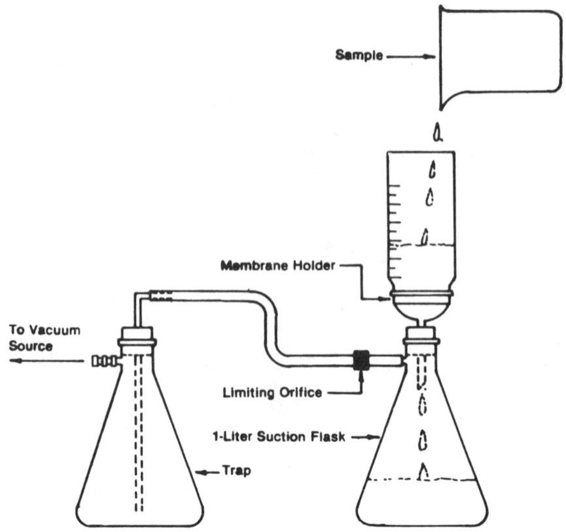

FIGURE 4 — Example of a graphic representation of water quality. Cumulative volume valves are plotted in milliliters (mL) on the abscissa, and the flow rate in milliliters per second (mL/sec) is plotted on the ordinate.

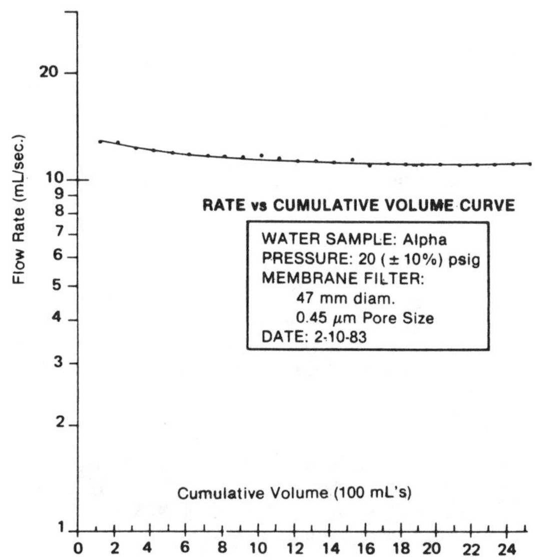

FIGURE 5 — Apparatus for testing aged samples (secondary suspended solids) by vacuum filtration. The sample is poured into the cylinder over the holder, which contains a preweighed membrane filter. The vacuum source may be either a vacuum pump or water aspirator.

7.5.5 If a single membrane is used, treat it to remove water soluble salts. (The matched weight membranes circumvent this step because the same quantity of water soluble salts should be trapped in each layer of the matched weight pair.) This step may be eliminated where supersaturated salt solutions are being examined.

7.5.5.1 Wash the single membrane in an apparatus as shown in Figure 5.

TABLE 1 — Example of Membrane Filter Test Data

Sample: Alpha — Membrane Filter: 47 mm Dia., 0.45μ Pore Size

Test Pressure: 20 (± 10%) psig — Time for 2500 ml: 3 min. 36 secs.

Date: February 10, 1964

Volume v (ml)	Time t (secs)	Δv (ml)	Δt (secs)	Rate Δv/Δt (ml/sec)	Comments
100	7.5	100	7.5	13.3	
200	15.1	100	7.6	13.2	
300	23.0	100	7.9	12.7	
400	31.0	100	8.0	12.5	
500	39.1	100	8.1	12.3	
600	47.3	100	8.2	12.2	
700	55.6	100	8.3	12.0	
800	64.0	100	8.4	11.9	
900	72.5	100	8.5	11.8	
1000	81.0	100	8.5	11.8	
1100	89.6	100	8.6	11.6	
1200	98.4	100	8.8	11.4	
1300	107.2	100	8.8	11.4	
1400	116.1	100	8.9	11.2	
1500	124.9	100	8.8	11.4	
1600	134.0	100	9.1	11.0	
1700	143.0	100	9.0	11.1	
1800	152.1	100	9.1	11.0	
1900	161.2	100	9.1	11.0	
2000	170.0	100	9.1	11.0	
2100	179.5	100	9.1	10.9	
2200	188.7	100	9.2	10.9	
2300	197.9	100	9.2	10.9	
2400	207.1	100	9.2	10.9	
2500	216.3	100	9.2	10.9	

7.5.5.2 Rinse the membrane until the filtrate is chloride-free as evidenced by no precipitate when 5% (0.25N) $AgNO_3$ is added to 10 mL of washings.

7.5.6 Dry the membranes in glass petri dishes (covers slightly ajar) in an oven at 90 C for 30 minutes. (Matched weight pairs must be carefully separated before drying.)

7.5.7 Allow the membrane to cool in its petri dish with cover slightly ajar. (For critical tests on high quality water, cooling in a desiccator may be required.)

7.5.8 Determine membrane weight to nearest 0.1 mg. If the matched weight pair is used, weigh each membrane separately or place each member of a pair on opposite sides of a pan balance.

7.5.9 Calculations

7.5.9.1 The weight of the filter cake in milligrams is divided by the volume filtered in liters. The results are expressed in milligrams per liter (mg/L).

7.6 Aged Sample Test (Secondary Suspended Solids Test)

7.6.1 This test method determines the total of suspended solids plus those solids precipitated by aeration, loss of dissolved gases, and temperature change. This information plus that of the "primary" suspended solids can provide means of further diagnosing problems in a water system.

7.6.2 Collecting Samples

7.6.2.1 Handling of samples being returned to a laboratory frequently imposes size limitations of 1 to 2 L for this test.

7.6.2.2 The container should be clean and capable of being capped for transport to the laboratory.

7.6.2.3 At the laboratory, the containers should be opened to permit air entry and circulation, but protected from contamination by dust particles. Because it is necessary to know the volume of the original sample and because some liquid may be lost by evaporation, it may be helpful to mark the liquid level on the sample container following collection or before opening the container (to permit air circulation).

7.6.2.4 The time lapse between collection and filtration of samples for this test should be at least 24 hours in order to allow solids to form and settle.

7.6.2.5 In special cases where it is desired to examine several various blends of separate waters, the time after mixing in order to conform to field conditions should be stated with results.

7.6.3.5 Filtered distilled water may be used to wash solids adhered to the sample container into the receiver, or a portion of the filtrate may be used.

7.6.4 At the completion of the test, wash the membrane and filter cake with distilled water until the filtrate is chloride-free. See Paragraph 7.5.5.2.

7.6.5 Dry and weigh samples following the same procedures described in Section 7.5 for field-filtered samples.

7.6.6 Make calculations as described in Paragraph 7.5.9.1.

7.7 The Appendix to this Standard gives further tests which may be performed to determine the composition and characteristics of the suspended solids.

7.6.3 Filtering the Sample

7.6.3.1 Place a tared membrane in a holder, as shown in Figure 5.

7.6.3.2 Agitate the sample bottle to resuspend the solids and then pour a small volume into the receiver above the membrane.

7.6.3.3 Start the vacuum pump or aspirator suction.

7.6.3.4 Continue filtration until the entire sample is consumed.

Section 8: Reporting Test Data

8.1 A report should be written after each series of tests and should contain the following details:

8.1.1 A complete description of the water handling system, including sources, mixtures, dispositions, and daily volumes.

8.1.2 A flow diagram with line sizes and types (steel, plastic, etc.).

8.1.3 Descriptions of the sampling points.

8.1.4 A brief discussion of the test methods used, noting any variations from the standard procedures.

8.1.5 Test Results: Tabulation and graphic representation of rate vs cumulative volume test (Table 1 and Figure 4) as described in subsection 7.4.6; tabulation of suspended solids data for each sample point in mg/L, ppm, or lb/1000 bbl (0.35 × mg/L.)

8.1.6 A description of the filtered solids as determined by one or more methods in the Appendix.

8.1.7 A discussion of the results and their implications, including comparisons to any prior test results.

8.1.8 Recommendations for improvements, based on observed results, and recommendations for future test methods and frequency.

8.2 Interpretation and Application

8.2.1 This Standard presents test procedures which will give reproducible results for comparative evaluations. As discussed in Paragraph 1.1, the use of the results is beyond the scope of this Standard.

Bibliography

American Conference of Government Industrial Hygienists. "Threshold Limit Values of Airborne Contaminants and Intended Changes," 1970.

American Society for Testing and Materials (ASTM). 19.32.05 "Proposed Standard Test Method for Determination of Suspended Solids in Waters for Subsurface Injections."

Barkman, J. H. and D. H. Davidson. "Measuring Water Quality and Predicting Well Impairment," Journal of Petroleum Technology, Vol. XXIV, pp. 865-873, July 1972.

Bell, W. E. and J. K. Shaw. "Evaluation of Iron Sequestering Agents in Water Flooding," Producers Monthly, Vol. 22, No. 3, March 1958.

Dosher, T. M. and L. Webber. "The Use of the Membrane Filter in Determining Quality of Water for Subsurface Injection," Drilling and Production Practice, American Petroleum Institute, pp. 169-179, 1957.

Farley, J. T. and D. G. Redline. "Evaluation of Flood Water Quality in the West Montalvo Field," Journal of Petroleum Technology, July 1968.

Appendix

Supplementary Suspended Solids Test Methods

The suspended solids test described in Section 3.2 of this Standard may be supplemented with additional tests to further identify the plugging material. When the effects of air on the sample are considered important in these determinations, the collection, transfer, and filtration of these samples should be accomplished under air-free conditions. Examples of procedures which may be used to determine the composition and characteristics of the suspended solids are listed below. (Obviously, many laboratories use other standard procedures which would be satisfactory for these determinations.)

I. Hydrocarbon-Soluble Suspended Solids

Materials: Trichlorotrifluorethane [Toluene (USP) may be used in lieu of trichlorotrifluoroethane taking care to use adequate ventilation.]
Petroleum ether (30 to 60 C).

Procedure: 1. Place the previously weighed membrane filter from the suspended solids test (Section 7.5) in the apparatus shown in Figure 5. If matched weight pairs are used, place the two membranes in the apparatus with the filter cake topmost.

2. Wash the membrane with 5 mL increments of trichlorotrifluoroethane until the filtrate is colorless or practically colorless. A water aspirator pump or portable vacuum pump with a suitable cold trap is recommended to pull the solvent through the membrane filter. The use of adequate ventilation and safety precautions is mandatory due to the nature of the solvents used.

3. Wash the membrane with 5 mL increments of petroleum ether until the last traces of solvent are removed and the filtrate is colorless. Use the same apparatus and safety precautions observed in the previous solvent wash.

4. Air dry solvent-extracted filter membranes overnight or until a constant weight is obtained.

5. Weigh membrane filter to nearest 0.1 mg. Record hydrocarbon-free membrane weight.

6. Calculate suspended solids:

$$\text{Hydrocarbon-Soluble Suspended Solids (mg/L)} = \frac{\text{Wt. after Water Wash (mg)} - \text{Wt. after Solvent Wash (mg)}}{\text{Volume of Sample through Filter (L)}}$$

NOTE: Oil carryover in the water cannot always be measured by this technique since oil can pass through the membrane filter. This method can only give a qualitative indication of the oil carryover. Other measurement methods are suggested if quantitative data are desired.

II. Acid Soluble and Acid Insoluble Suspended Solids Determination

Material: 6N hydrochloric acid

Procedure: 1. Place membrane filter from hydrocarbon-soluble suspended solids determination in apparatus shown in Figure 5.

2. Wash membrane filter with several 10 mL portions of hot 6N HCl until filtrate is clear.

3. Rinse membrane filter with distilled water until all traces of HCl are removed. (Filtrate pH should be >5 as determined by a suitable test paper.)

4. Dry membrane filter in drying oven at 90 C for 30 minutes.

5. Cool membrane at room temperature for at least 15 minutes.

6. Weigh cooled membrane filter to nearest 0.1 mg.

7. Calculate suspended solids:

$$\text{Acid Soluble Suspended Solids (mg/L)} = \frac{\text{Wt. after Solvent Wash (mg)} - \text{Wt. after Acid Wash (mg)}}{\text{Volume of Sample through Filter (L)}}$$

$$\text{Acid Insoluble Suspended Solids (mg/L)} = \frac{\text{Wt. after Acid Wash (mg)} - \text{Initial Wt. of Clean Membrane (mg)}}{\text{Volume of Sample through Filter (L)}}$$

For matched weight pairs, the initial weight is not needed since the lower membrane has been subjected to all liquids and its weight should be the same as that of the one supporting the filter cake (Paragraphs 2.2.2, 4.2.1, and 7.5.8).

Microscopic Examination

Although microscopic examination is not an absolute means of identification, it is a useful tool for determining the nature of the plugging material in injection systems. Examination by viewing the membrane filter with oblique incident light or transmitted light may be used.

Where large particulate material (>10 μm) is present, it can be viewed best by oblique incident light. Very small particulate matter is best viewed using transmitted light.

I. Microscopic Examination Using Oblique Incident Visible Light

Apparatus: The microscope used in this analysis typically has an inclined binocular body, a mechanical stage (2 axes), a multiple nosepiece to accommodate two or more objective lenses, a substage condenser, and wide field eyepieces (usually 10X). Total magnification capabilities of from 30 or 40X to 400 or 500X would be optimum for most any identification application using oblique incident visible light.

Any of the larger incandescent lamps which work directly on 110 volts provide a suitable light source for this application. These microscope lamps can be obtained from a number of manufacturers. They should be of the type which is mounted on an inclining stand and has a focusing condensing lens. An adjustable iris diaphragm and ground glass and daylight glass filters are desirable.

Materials: 2 × 3 in. (5 × 8 cm) glass slides
Petroleum jelly or other suitable grease
Stainless steel forceps

Procedure: 1. Use the dried membrane filter from either the rate vs cumulative volume test (Section 3.1) or the suspended solids test (Section 3.2).

2. Prepare a clean 2 × 3 in. (5 × 8 cm) glass slide by coating it lightly and evenly with a thin film of grease. Place the membrane filter on the greased slide. The grease will cause the membrane filter to lie flat and entirely in one focal plane.

3. Do not use a cover slide in this application as any contamination on the cover slide may be mistaken for material retained by the filter. If the filter is to be preserved for further examination, the top of a plastic petri dish placed over the filter will protect it from atmospheric contamination.

4. Adjust the incident light source to about 40 degrees from horizontal. Examine the filter at low magnification and then at suitable higher magnification. Manipulate the microscope stage making two mutually perpendicular traverses across diameters of the filters.

5. The oblique incident light will strike the contaminants on the membrane filter from the side, making differentiation between fibers, glass, and metals possible.

II. Microscopic Examination Using Visible Transmitted Light

Apparatus: Microscope and light source are the same as those used in the microscopic examination with oblique incident light except a 97X oil-immersion objective lens is useful for examining and identifying very tiny particulate matter.

Materials: 2 × 3 in. (5 × 8 cm) glass slides
45 × 50 mm No. 1 cover slides
Immersion oil—Index of refraction—1.51
Stainless steel forceps
Plastic petri dishes

Procedure: 1. Use the dried membrane filter from either the rate vs cumulative volume test (Section 3.1) or the suspended solids test (Section 3.2). If the membrane filter is not available, a suitable membrane filter may be prepared by passing 1 L of fresh sample through a membrane filter using the apparatus shown in Figure 5. This membrane filter should be rinsed and dried by the procedure described in the suspended solids test (Paragraph 7.5.5).

2. With the forceps, gently float the filter, sample side up, on a film of immersion oil in the cover of a plastic petri dish. Draw the filter over the rim of the cover to remove any excess immersion oil from the bottom of the filter. Roll the filter onto a glass slide.

3. Place a clean No. 1 cover slide over the surface of the membrane filter being careful not to trap any air bubbles under the slide.

4. Adjust the light path from the light source so it is reflected by the flat substage mirror up through the substage condenser, through the oil wet membrane filter, and into the objective lens. The immersion oil will render the membrane filter transparent and allow the particles on the surface of the membrane filter to be observed.

5. Examine the filter at low magnification and then under high powered oil immersion lens. When using the oil immersion lens, place a drop of immersion oil on the surface of the slide cover, carefully lower the oil-immersion objective, using the coarse adjustment, until the top of objective just makes contact with the oil drop. Do this carefully so as not to damage the objective lens by grinding it into the glass slide. Use the fine adjustment to bring the surface of the membrane filter into final focus.

6. Manipulate the microscope stage making two mutually perpendicular traverses across diameters of the filter.

7. The use of transmitted light with the high powered oil immersion lens (total magnification—970X) makes it possible to identify tiny objects such as diatomaceous earth, sand grains, slime masses, iron bacteria such as Sphaerotilus and Gallionella, fresh water diatoms, and rust fragments.

Microchemical Spot Test Methods

The spot testing technique lends itself to quick chemical analysis of matter collected on an inert surface (in this case, a membrane filter). Spot testing requires that the ion to be identified be dissolved or partially dissolved before reagents may be applied. After the sample on the filter has been solubilized and dried, the filter is cut into a number of wedge shaped segments. These segments are then individually tested for the various ions. The cutting blade may leave enough contamination at the edge of the segment to give a positive test, but the contamination will not extend into the center of the segment. The spot tests given in this section of the Appendix are for sulfate, silica, iron, and protein.

I. Sulfates

Materials: Lead nitrate reagent, saturated aqueous solution containing 1 mL of concentrated nitric acid per 25 mL of solution
6N nitric acid
6N hydrochloric acid
2 × 3 in. (5 × 8 cm) glass slides
47 mm plastic petri dish and cover
Fluorinated ethylene propylene (FEP) coated or lacquered stainless steel forceps
47 mm microfiber glass prefilter pad
Black membrane filter (if available)

Procedure: 1. Filter 1 L of fresh sample through a black membrane filter using apparatus described in Figure 5.

2. Place a 47 mm microfiber glass prefilter in a plastic petri dish and add 1.5 cc each of 6N hydrochloric and nitric acids.

3. Roll the filter containing the sample onto the acid-wet prefilter, sample side up, and cover the petri dish.

4. Allow the petri dish to stand for 10 minutes. To aid the dissolution of the sample, place the petri dish in a 60 C oven over this period.

5. Remove the filter and place it on a 2 × 3 in. (5 × 8 cm) glass slide. Allow the filter to dry.

6. Wet a microfiber glass prefilter pad with approximately 3 mL of lead nitrate. Roll the sample filter, sample side up, onto the pad and let it stand for 10 minutes. It is preferable to use a black filter for the collection of the sample.

7. Mount the filter on a glass slide and view it under low power (50X) microscopic examination. Use incident light. Sulfate will be indicated by white, grainy reaction spots. If it is necessary to analyze the sample collected on a white filter, view filter by transmitted light microscopy.

II. Silica

Materials: Benzidine,[1] 1% ethanol
Ammonium hydroxide, concentrated

Procedure: 1. Filter 1 L of sample using apparatus described in Figure 5.

2. Place the filter on wire gauze and heat gently over a hot plate.

3. Add one drop benzidine and develop over ammonia fumes.

Results: A blue color indicates the presence of silica.

[1]NOTE: Benzidine has been listed by the American Conference of Government Industrial Hygienists (ACGIH) as a cancer causing material.

III. Iron

Materials: 47 mm microfiber glass prefilter
47 mm plastic petri dishes with covers
6N hydrochloric acid
6N nitric acid
Potassium ferrocyanide solution—saturated aqueous

Procedure: 1. Filter 1 L of fresh sample through a membrane filter using apparatus illustrated in Figure 5.

2. Solubilize the iron on the membrane filter using the technique described in the procedure for detecting sulfates (See Appendix, Microchemical Spot Test Methods, I. Sulfates, Procedures 3-5). It is not necessary to dry the membrane filter.

3. While the filter is still slightly damp from the solubilizing procedure, add one drop of the potassium ferrocyanide solution. A vivid blue color indicates the presence of iron. This is a very sensitive test for iron; the results can be affected by the presence of very large quantities of copper or molybdenum, but this is seldom a problem.

IV. Protein

Since all bacteria are composed in part of protein, a spot test for the presence of protein suggests the presence of bacteria. If the test is positive, microbiological test procedures not included in this Standard may be used to further elucidate the nature of these organisms.

Materials: Stain solution (percent by weight in distilled water) composed of:
Ponceau-S dye 0.2%
Trichloroacetic acid 3.0%
Sulfosalicylic acid 3.0%
Acetic acid 5%

Procedure: 1. Filter 1 L of water through a membrane filter mounted in the apparatus shown in Figure 5.

2. Immerse the filter in staining solution for two minutes, agitate gently, remove, and blot between absorbent paper.

3. Place the filter in an acetic acid rinse and agitate gently. Rinse the filter in a second acetic acid bath, remove, and blot dry.

Results: The bottom side of the filter edge outside the filtration area will rinse free of the red stain; proteinaceous material will absorb the stain and remain a vivid red color.

APPENDIX 6A

Procedures for Determination of Oil in Water

API RP 45, *Recommended Practice for Analysis of Oilfield Waters*,[1] describes a procedure for measuring the concentration of oil in water. The oil is first extracted into clear chloroform solvent and then the degree of darkening (reduction in light transmittance) of the solvent is measured with a photometer instrument. The instrument reading is compared to a calibration curve (pre-prepared from known concentrations of the specific oil in the solvent) to obtain the parts per million by volume of oil in the original water sample.

Oil concentration in the extraction solvent can be measured directly regardless of color by infrared absorbance at a specific wavelength using a spectrophotometer.

With either analytical method (color comparison or spectrophotometric infrared absorbance measurement), one of the greatest detriments to accurate analysis of oil in water concentration is errors inherent to the water sampling process. When a water sample is collected and taken to a laboratory for analysis, the dispersed oil tends to float to the top and also stick to the walls of the water container (especially plastic containers). To overcome this problem, the following sampling and extraction procedure is recommended:

1. Collect water samples to be tested for oil content separately from any other samples being taken.

[1]Available from American Petroleum Institute, 211 N. Ervay, Dallas, Texas 75201-3688, U.S.A.

2. Obtain or prepare a volume-calibrated glass bottle with a screw-on cap for each sample to be taken. Calibrated prescription bottles obtained from a drug store or pharmacy are suitable.
3. Allow water from the source to flow for a few minutes and then fill the sample bottle *one-half* full of the water to be tested. For example, collect 500 milliliters (ml) of water (exactly) into a 1 liter bottle.
4. Add a measured amount of clear chloroform or carbon tetrachloride solvent to the half-filled water sample bottle, close securely and mix by inverting at least 10 times. The amount of solvent to be added is dictated by the amount of oil present. For example, if 500 ml of solvent is added to 500 ml of water, the final oil concentration in the solvent will be equal to the concentration originally present in the water. If only 50 ml of solvent is used, the final oil-in-solvent concentration will be 10 times greater and the solvent color will be proportionately darker.
5. In the laboratory, separate the water and the solvent. Measure the oil content of the solvent (by color comparison or spectrophotometrically) and calculate oil content of original water using the extraction volume correction factor from Step 4.

   Caution. Do not inhale vapors of the chloroform or carbon tetrachloride solvent. Use with adequate ventilation.

Alphabetic Index—Subject Locator

Acids, defined, 106, 107
corrosion by, 25, 67, 96, 108
examples, 4, 106
ionization (dissociation) of, 4, 14, 15
neutralizers, 20, 47, 83
stimulation, inhibitors, 12, 25, 67, 68
well cleaning, 126, 152–153
Acid gas,
sweetening, 75
Acidity, defined, 107
H_2S, CO_2, 14, 106
relation to pH, 107–108
Alkalinity, defined, 106, 107
calculation, 111, 112
components in water, 107, 108, 112
effect on corrosion, 14, 62, 83, 108
example calculation, 112
forms, 112
relation to pH, 107–108
Alloys,
aluminum, 11, 83
corrosion resistant, 11, 20, 55, 56–58, 62, 63, 73, 80–83, 84
for CO_2/H_2S environments, 83
plating, 42, 59
Anodic protection, 11, 20, 164–165
Anode, anodic, defined, 5–6, 160–161
half cell reactions, 5, 6, 13, 23, 29
materials, c.p., 9, 29, 30, 31, 64, 76–77

Bacteria, see Microorganisms
Bases, defined, 106, 107
examples, 106
reaction with acids, 106
Biochemical oxygen demand, BOD, defined, 109
Buffer capacity, 108

Carbon dioxide, CO_2, 4, 13, 14
acidity, 13, 108
corrosiveness and mechanisms, 14, 56, 67–69, 70, 73, 74, 79–82, 83
EOR systems, 79–83, 84
inhibition, 22, 23, 47, 67–72, 79–83, 169–180
partial pressure, 67
production wells, 67–72, 82
removal, 19, 47–48
Casing, extenal, 16–17, 34, 63–67
electrical resistance, 67
internal, 23, 67, 69, 70, 71, 80, 81
leak history, economics, 65
potential profiles, 35, 65–66
surface E log I tests, 65–66, 163–164
Cathode, cathodic, defined, 5–6, 160–161
c.p., 27–28
half cell reactions, 5–7, 13, 14, 27, 46, 161, 162
Cathodic protection, c.p., 9, 20, 27–38, 42
casing protection, 64–65
current density requirements, 27–28, 31, 36, 65–67, 77–78
effect on coatings, 46
interference, 16, 33–34, 64, 65
offshore structures, 29, 31, 32, 34, 37, 75–78
protection criteria, 27–28, 31–32, 35, 36, 37–38, 77–78
rectifiers, 28, 29–31, 32, 33, 34, 37, 64–65, 77
solar, thermoelectric generators, 30–31
tanks and vessels, 37–38, 72
Cavitation, 59–60
Cement, linings, concrete, 19, 40, 42–43, 64, 73, 80, 81, 82
joints, grout, 81
Chemical oxygen demand, COD, defined, 109
Cladding, internal, 58, 80
Clay blocking, dispersion, 127, 139
Coatings, 39–46
application/inspection, 43–45
conversion, 43
drill pipe, 41, 62
flowlines, 39, 40, 41, 72, 73, 76, 81, 82
for specific field uses, 46
marine, atmospheric, 39–43, 46, 76
marine, splash zone, 76
metallic, 42, 43, 80–82
pipe, external, 40–43, 46, 82
surface preparation, 43–44
tanks, internal, 40, 41, 46, 72, 73, 81
tapes, 40, 46
tubing, 44, 46, 62, 73, 80, 82
zinc rich, 39, 42, 43, 46, 76
Completion, packer, fluids, 63, 64, 67, 69, 80, 81
Concentration units, water analyses, 109–112
calculation of $CaCO_3$ equivalent, 110–111
calculation of meq/L, 110–111
Conversion factors, 88, 112, 155
Corrosion, defined, 1
cell, 5, 6, 7, 8, 9, 12, 16, 160, 161, 162

current, I_{corr}, 159, 162, 163, 164
potential, E_{corr}, 159, 162, 163, 164
prediction, 8, 9–10, 65, 67–68, 87–94
prevention, fundamentals, 19–20, 61–78
products, 4, 8, 11, 12, 14, 87, 97–98, 117, 136
rate expressions, 88
sensors, 87–93, 94–96
Corrosion fatigue, defined, 54
drill pipe, 61
Corrosion monitoring, detection, 87–98
drilling, 63–98
water injection systems, 73–74
Corrosiveness, defined, 14
Cracking processes
blistering, 55–56
chloride stress (CSC), 56, 80
corrosion fatigue, 54, 61
drill pipe, 54, 61
stress corrosion, (SCC), 56, 77
sulfide stress, (SSC), 54–55, 61, 67, 83, 96
Crevice corrosion, 58

Dalton's Law, 48
Deep hot wells,
inhibition, 70–71
resistant metals, 56, 57
Dehydration, 20, 74, 80
Density/Specific Gravity, (sp.gr.), defined, 109
calculated from TD5, 109
Dissimilar metal couples, 9–10
area effect, 9, 59
coating, 10
designing with, 59
Dissolved salts,
calculated from sp.gr., 109
effects, corr. rate, 5, 15, 35–36, 63, 67
in water, 104–106, 107
precipitation, 105–106, 117–126
Drilling equipment, 48, 61–63
corrosion test methods, 63
drill collars, 57, 62
drill pipe, 54, 55, 61–63
O_2 and H_2S removal, 47, 62
washout location, 61–62
Ductility, 53, 54

Elastic limit, 53
Elastomers, 82
Electric submersible pumps, 82
Electrical potential, polarity, 5, 7, 8–9, 28, 30, 31–32, 35, 63, 65, 160, 161
Electricity, defined, 159
current flow, 7, 16–17, 28–29, 35, 63–64, 65, 66, 160–161
Electrolyte, (water, soil), defined, 5
conductivity effect, 15
c.p. requiement, 28
electrical current flow, 7
in corrosion cells, 6
ion flow, 5, 7
resistance, resistivity, 7, 15, 35–36
Electrolytic, stray current, corrosion, 16–17, 33–34, 63–64
cells, 161

Electron, defined, 4
current flow, 5, 6, 7, 16, 17, 27, 28, 64, 161
Emf-pH, Pourbaix, diagrams, 165–166
EOR systems, 79–85
microorganisms in, 84, 131
Equipment handling precautions, 11, 56–58, 59, 62
Erosion, cavitation, 22–23, 59, 61, 73
ERW pipe, 58, 72, 80

Fatigue strength, defined, 54

Galvanic corrosion, defined, 8, 9–10
(See also dissimilar metal couples.)
Gas processing, 74–75, 83
Grounding devices, 34, 37

Hardness, 53–54
Heat treatment, 8, 10, 53, 54, 55, 58, 62, 72
Henry's Law, 48
Hydrogen, hydrogen ion, 4, 6, 9, 103, 106, 107–108, 161, 162
cathodic polarization by 6, 7, 14, 162
cracking, blistering, 54, 55
electrode (SHE), 8, 161
overvoltage, 12
Hydrogen sulfide, H_2S, 4, 14
acidity, 14
corrosiveness, 14, 38, 67–68, 73, 83
inhibition, 22, 23, 24, 56, 62, 68–71, 73
partial pressure, 55, 68
production wells, 67–68
removal, 47, 62
scavengers (drilling), 62
solubility, 47
Hydrolysis, defined, 106

Inhibitors, inhibition, 19, 21–25, 169–171
application principles, 24–25
application procedures, wells, 68–71, 80, 82, 147
surface, 72, 73, 74, 80
classification, 21–22, 163, 169–171
emulsion stabilization, 22, 23, 147
film formation, 19, 21, 22–23, 24, 169
formulation, 23–24
how inhibitors work, 21, 169
ionic character effect, 22
organic, basic structure, 22
Injection well cleaning, 147, 151–152
Injection well plugging, defined, mechanism, 135–139
bacteria, 131, 136
by oil, oily solids, 145, 150–153
plugging solids, sources, properties, 136–137
Insulators, wellhead, 34, 64
testing, 64
Ionic strength, calculation, 119
Ions, ionic, defined, 4
common in water, 105
current flow, 5, 7, 15, 28, 160
scale forming, 105–106, 118, 120–121, 205
Iron counts, COPRA correlation, 93–94

Metals, 53–60
heat treatment, 11, 54, 55, 62
properties, physical, 53–54, 57–58

strength, 53–54
Microorganisms, microbes, 129–134
classifications, 129–130
control methods, 130, 134
corrosion by, 15–16, 73, 84, 129–131
counts, methods, interpretation, 132–133, 179, 192–193
field problems, 15, 73, 129–132
in reservoirs, prediction, 131–132
occurrence, 131
prediction, 131
visual evidence, 133

Offshore, marine, structures
c.p., 28, 31, 34, 37, 91, 75–78
coating systems, 41, 42, 43, 45, 73, 193–196
potential surveys, 34, 77–78
riser pipe, 46, 76
Ohms Law, defined, 7, 32
graphic, 158
salt water corrosivity, 15
Oil carryover, produced waters, 144–147
disposal, 151–152
emulsion treatment, 144–147
oil concentration analysis, 149–151, 197
oil dispersion, stabilization, 146–147
oil removal from water, 147–149
Oxidation reactions, defined, 5, 9, 160–161
chemical, rust, 5, 6, 11, 12, 13, 84, 169
electrochemical, 5, 6, 16, 28, 129, 160–162
Oxygen, (O_2), 5, 6, 11, 12–15, 22, 27, 47–51, 169–171
bacterial growth, 15, 73, 129–131, 133
corrosiveness, 12–15, 48, 61, 70, 72–74, 75, 84
c.p. current requirements, 27–28, 77
differential concentration cells, 12–13, 58, 63–64, 131
dissolved, analysis, 63, 73–74
dissolved, inhibition, 22, 62, 73, 75, 84, 169–171
drillpipe corrosion, 62
exclusion, 48, 73, 75, 81
removal, 47–51, 62, 73, 80, 81, 83, 84
solubility, 16, 48

Passive, passivity, 11, 15, 20, 164–165
pH, defined, 107–108
estimation, field conditions, 119–120
natural waters, 108
optimum for solids coagulation, 142
spent acid, 108
pHs, defined, 113, 119–120
Plastics, 20, 39–42
pipe, 41–42, 81, 82
sucker rods, 82
tanks, 20, 39–42
Polarization, defined, 7–8, 17, 162
anodic, 5, 7, 162, 165
bacteria effect, 130–131
cathodic, O_2 effect on, 6, 14, 32, 162–163
diagrams, curves, 20, 162–165
inhibitor effects, 21, 163
linear, 65–66, 164
resistance, 88, 89–92, 163

Radicles, defined, 105
Records, economic analysis, 65, 95
Reduction, electrochemical, 5–6, 27, 161
Reference cells, electrodes, 7, 8, 9, 28, 32
hydrogen (SHE), 161
P/S potentials, 32–35
Ringworm corrosion, 82

Salinity, defined, 109
Salts, common, properties, 105–107
Scales,
inhibition, 23, 126–127
mineral, 105–106, 113, 117–127
Barium sulfate ($BaSO_4$),
occurrence, 105, 117, 120–121, 123–125
precipitation, prediction, 118, 120–121, 123–125
solubility, 118, 120–121, 123–125
Calcium carbonate ($CaCO_3$),
effect on c.p., 27–28, 77
occurrence, 106, 117, 119
precipitation, mechanism, 106
precipitation, prediction, 113, 119–120, 175
solubility, 106, 118
Calcium sulfate ($CaSO_4$),
occurrence, 105, 118, 120–121
precipitation, mechanism, 105
precipitation, prediction, 119, 120–123, 175–176
solubility, 105, 118, 120–121, 123
Strontium sulfate ($SrSO_4$),
occurrence, 117, 123–124
precipitation, prediction, 120, 124–125
solubility, 118, 124,
prediction in producing wells, 125–126
removal, 126
Scaling tendency, index, SI, 117, 119, 123
calculation examples, 175–176
Soils, corrosivity, 13, 33, 36
resistance, resistivity, 35–36
Solubility, salts, 105, 106, 107, 118
factors affecting, 117–118
limit, salts, 105, 118
Solution potential, 8, 162
Steam generators, 83
Steels and alloys, grades, 8, 9–10, 55, 56–59, 62–63, 80–83, 84
casing, steam injection, combustion operations, 55, 56, 84
compressors, 80
drill pipe, 55, 62–63
flowlines, 58
heat exchangers, dehydrators, 42, 59, 80
pumps and valves, 42, 58, 59, 73, 80–83
sucker rods, 58, 82
tanks, vessels, 56, 58, 83
tubulars, 54–57, 80, 84
wellheads, 58, 80
Stokes Law, particle setting, 145–146
Stress-strain, 53
Sucker rods, 42, 54, 58, 59, 82, 96–97
fiberglass, 82
rod pumps, 82
Surfactant/polymer corrosion, 84
Suspended solids removal, 141–144
clarifiers, 142
filters, produced waters, 144

filters, surface waters, 142–143
produced waters, 143–144

Tensile strength, defined, 53
Test methods, procedures,
bacteria count, 132–133, 177
calipers, 94–95
casing potential, 65–66, 164
coatings, 45–46
corrosion products, identifying, 96–97
corrosion rate, coupon/meters, 63, 74, 87–93
corrosion rate, metal loss, 94–96
E log I, 65–66, 164
iron count, 74, 93–94
O_2/H_2S concentration, 63, 73–74
oil in water, 149–151, 197
radiography, 96
tubing/casing calipers, 94–95
ultrasonic thickness gauges, 95–96
visual inspection, 96–97
water quality, membrane (Millipore), 183–193
water quality, rating, 138–139
Thread lubricants, 63, 82
Total dissolved solids, TDS, defined, 109
calculation from sp.gr., 109
Total suspended solids, TSS, defined, 109
determination, 138–139, 183–193
Turbidity, defined, 109
turbidimeters, 109–110

Velocity effects, 20, 22–23, 27, 59
bacteria growth, 131
c.p., 27, 77–78
limiting (critical), erosional, 20, 22–23, 59
on inhibitors, 22–23
scale deposition, 118

Water, defined, 103–104
classifications, characteristics, 104
constituents by analysis, 105, 106–107
properties by analysis, 107–109
Water analyses, examples, 112
diagrams, patterns, 114–115
Water hardness, defined, 111–112
calculation of, 111–112
example calculation, 113
Water mixtures,
bacteria in, 131
compatibility, 118–119, 120–121, 125, 126
determining incompatibility, 125
graphical calculation, 114–115
ratio in produced water, calculation, 114, 125, 176 (example)
Water quality, defined, 135
improvement, 141–149
mixtures, 125
rating methods, 138–139
Water softening, 83, 113
Welds, 10, 58, 72, 81

Yield strength, defined, 53
stress, 54